Financial Management for Farmers and Rural Managers

LANDSMAN'S BOOK SHOP LTD.
Bromyard, Herefordshire

# Financial Management for Farmers and Rural Managers

**Martyn F. Warren, BSc, MSc, MIAgrM**

Head of Land Use and Rural Management
Seale-Hayne Faculty
University of Plymouth
Newton Abbot, Devon
UK

**Fourth Edition**

Blackwell
Science

© Martyn F. Warren 1998
Blackwell Science Ltd
Editorial Offices:
Osney Mead, Oxford OX2 0EL
25 John Street, London WC1N 2BL
23 Ainslie Place, Edinburgh EH3 6AJ
350 Main Street, Malden
MA 02148 5018, USA
54 University Street, Carlton
Victoria 3053, Australia

Other Editorial Offices:

Blackwell Wissenschafts-Verlag GmbH
Kurfürstendamm 57
10707 Berlin, Germany

Blackwell Science KK
MG Kodenmacho Building
7–10 Kodenmacho Nihombashi
Chuo-ku, Tokyo 104, Japan

All rights reserved. No part of this publication may be reproduced, stored in a retrieval system, or transmitted, in any form or by any means, electronic, mechanical, photocopying, recording or otherwise, except as permitted by the UK Copyright, Designs and Patents Act 1988, without the prior permission of the publisher.

First published by Hutchinson Educational 1982
Reprinted, revised 1986
Reprinted 1987, 1990
Third edition published by Stanley Thorne (Publishers) Ltd 1992
Fourth edition published by Blackwell Science Ltd 1997

Set in 10/13 pt Bembo
by DP Photosetting, Aylesbury, Bucks
Printed and bound in Great Britain by
MPG Books Limited, Bodmin, Cornwall

The Blackwell Science logo is a trade mark of Blackwell Science Ltd, registered at the United Kingdom Trade Marks Registry

DISTRIBUTORS

Marston Book Services Ltd
PO Box 269
Abingdon
Oxon OX14 4YN
(*Orders:* Tel: 01235 465500
Fax: 01235 465555)

USA
Blackwell Science, Inc.
Commerce Place
350 Main Street
Malden, MA 02148 5018
(*Orders:* Tel: 800 759 6102
617 388 8250
Fax: 617 388 8255)

Canada
Copp Clark Professional
200 Adelaide Street West, 3rd Floor
Toronto, Ontario M5H 1W7
(*Orders:* Tel: 416 597 1616
800 815 9417
Fax: 416 597 1617)

Australia
Blackwell Science Pty Ltd
54 University Street
Carlton, Victoria 3053
(*Orders:* Tel: 03 9347 0300
Fax: 03 9347 5001)

A catalogue record for this title is available from the British Library

ISBN 0-632-04871-9

Library of Congress
Cataloging-in-Publication Data

Warren, Martyn F.
Financial management for farmers and rural managers/
Martyn F. Warren. – 4th ed.
p. cm.
Rev. ed. of: Financial management for farmers.
Includes bibliographical references and index.
ISBN 0-632-04871-9
1. Agriculture–Accounting. 2. Farms–Finance.
3. Farm management. I. Warren, Martyn F.
Financial management for farmers. II. Title.
HF5686.A36W37 1997
630'.68'1–dc21 97-17525
CIP

# Contents

*Preface to the Fourth Edition* ix
*Acknowledgements* xi

*Part 1:* **Basic Issues** **1**

**1. Why Bother?** **3**
1.1 In the beginning... 3
1.2 Building a framework 6

**2. Cash Flow, Profit and Capital** **9**
2.1 Cash flow 9
2.2 Net profit 10
2.3 Profit statements 14
2.4 Capital 16
2.5 Capital statements 18
2.6 Depreciation 20

**3. Interpretation of Financial Accounts** **28**
3.1 Adjusting for realism 28
3.2 Interpreting the profit and loss account 29
3.3 Interpreting the balance sheet 33

**4. Management Accounts** **42**
4.1 Management versus financial accounts 42
4.2 Gross margin accounts 43
4.3 Full cost accounts 44
4.4 Allowing for cost behaviour 45
4.5 Pricing 50

**5. Variations from the Norm** **53**
5.1 Farm management accounting 53
5.2 Partnerships and companies 56
5.3 Voluntary/non-profit organizations 60

*Part 2:* **Basic Budgeting** **63**

**6. A Profit Budget** **65**
6.1 Introduction 65
6.2 Compiling a profit budget 66
6.3 Home Farm profit budget 76

**7. A Cash Flow Budget for the Whole Business** **89**
7.1 Basic principles 89
7.2 Value Added Tax and overdraft interest 90
7.3 Cash flow budgets in practice 96
7.4 Reconciling cash flow and profit 99

**8. A Budgeted Balance Sheet** **106**
8.1 Compiling the budget 106
8.2 A budgeted balance sheet for Home Farm 107

**9. Budgeting for Incremental Change** **111**
9.1 Relevant costs and benefits 111
9.2 Partial budgets 113

*Part 3:* **Financial History** **121**

**10. Recording Cash Flow** **123**
10.1 Designing a cash recording system 123
10.2 The cash analysis system 125
10.3 Petty cash 131
10.4 Statutory records 136

**11. Recording Profit and Capital** **144**
11.1 Preparing a profit and loss account 144
11.2 Valuing stocks 150
11.3 Recording capital 156

**12. The 'Back-up' Records** **160**
12.1 Introduction 160
12.2 Transaction records 160
12.3 Physical records 162
12.4 Stock control 167
12.5 Background records 168
12.6 Organizing paperwork 168
12.7 Post-script 170

*Part 4:* **Controlling the Business** **173**

**13. Monitoring Cash Flow** **175**
13.1 Why monitor cash flow? 175
13.2 Comparison of actual and budgeted results 176
13.3 Interpreting the results 178
13.4 Annual cash flow monitoring 180

**14. Monitoring Profit and Capital** **183**
14.1 The importance of monitoring profit 183
14.2 Comparison with previous years 185
14.3 Interfarm comparisons (1): conventional profit and loss account 186

14.4 Interfarm comparisons (2): profit and loss accounts in enterprise account form 188
14.5 Budgetary comparisons 190
14.6 Variance analysis 191
14.7 Profit monitoring for Home Farm 194
14.8 Monitoring capital 197

*Part 5:* **More on Planning and Control** **201**

**15. The Planning Process – A Wider View** **203**
15.1 Tactics and strategy 203
15.2 The mission 204
15.3 Setting aims and objectives 204
15.4 Assessing 'internal' characteristics and external environment 207
15.5 Generating alternative plans 207
15.6 Selecting the optimal plan 208
15.7 Implementing and monitoring the selected plan 209
15.8 Involving other members of the organization 210

**16. Planning for Livestock Enterprises** **211**
16.1 Introduction 211
16.2 Feeding livestock 211
16.3 Allocation of forage costs 213
16.4 Estimating potential production 216
16.5 Replacement of breeding livestock 218
16.6 Livestock with long production cycles 220

**17. Labour and Machinery Planning in Farming** **224**
17.1 Introduction 224
17.2 Estimating level of use 224
17.3 Estimating labour and machinery costs 232
17.4 Investigating alternatives 234

**18. Capital Planning** **239**
18.1 Introduction 239
18.2 Sources of capital 239
18.3 Estimating the cost of capital 242
18.4 Comparing alternative uses of capital (1): simple measures 246
18.5 Comparing alternative uses of capital (2): discounted cash flow techniques 250

**19. Allowing for Risk and Uncertainty** **257**
19.1 Introduction 257
19.2 Allowing for risk and uncertainty in planning 258
19.3 Risk and investment appraisal 262

**20. Allowing for Inflation** **264**
20.1 Introduction 264

20.2 Inflation and financial accounts 264
20.3 Inflation and management accounts 265
20.4 Inflation and planning 267
20.5 Inflation and investment appraisal 268

**Appendix A. Personal Computers in Management** **271**
A.1 Introduction 271
A.2 Computers and farming 271
A.3 General business applications 273

*Index* *288*

# Preface to the Fourth Edition

The aim of this book is that you should learn how to use the basic skills of managing money, including budgeting, budgetary control, physical and financial recording. By the time you have read it, and completed the exercises, you should not only *understand* the ideas underlying the various techniques, but also be able to *do* the work involved. You should also be able to judge when and when not to use a particular technique.

Although the prime focus of the book is on owners and managers of farm businesses and farm diversifications, it is equally valuable for other types of rural business and organisation. It has proved popular on degree and diploma courses in Rural Resource Management, Countryside Management and Rural Estate Management, as well as those concerned with Agriculture, and the fourth edition is designed to be even more approachable for those who have a wider interest than farming alone. Although inevitably geared to the UK, most of the principles and techniques described are universally applicable, and the previous editions have proved popular in a wide variety of countries.

Financial management can easily become regarded, by practising managers and students alike, as a tedious, uninteresting pastime, obsessed with past history and epitomized by the chore of 'doing the books' on a wet weekend. This is little short of a tragedy, since financial management is not only vital to the well-being of farm businesses and their owners, but can be a fascinating process in itself.

In this book I have tried to transmit my own enthusiasm for the topic, and to make it attractive. One way in which I have attempted to stimulate interest is by starting, after discussion of basic concepts, with decisions about the future, rather than with records of the past. The reader learns first about potential problems and opportunities facing farm businesses, and realizes the need for careful planning. When studying the planning of a business, he comes to realize the need for control of the business's finances, and thus to appreciate that records and accounts play a vital role in the management process.

I have confined myself to the practical problems of using basic financial management techniques, and have had to resist the temptation to write at length on topics which are peripheral to (or too advanced for) this main theme. Where this is the case, I have suggested further reading for those who wish to widen the area of their study.

An essential component of this book is the *Home Farm Case Study*. This follows the progress of a young businessman through the planning and control of the various elements of a rural business, and is divided between the relevant chapters.

The business is designed to have a little of everything – crops, grazing and intensive livestock, service and manufacturing industry – and while the result may be a little fanciful, I hope that it will help the reader identify with the scenario.

For this fourth edition I have made substantial changes throughout, and especially to Part 1, as well as reworking most of the examples and tables. Increased emphasis has been given to general accounting principles and techniques, especially those of management accounting, cost-volume-profit and breakeven analysis, and pricing. The new edition also better accommodates the needs of non-agricultural enterprises, including service and manufacturing businesses, and non-profit organizations. In the process, it may do a little to help bring farm financial management techniques more into the mainstream, and out of the idiosyncratic backwater in which they have languished for too long.

Although the emphasis remains firmly on manual calculation (not everyone has a computer, by any means), the value of the personal computer is recognized throughout the book. In particular, Appendix A gives layouts and formulae for a number of financial spreadsheets, allowing most of the example calculations in the text to be readily reproduced as working computer-based management tools.

In the more agricultural parts of the book I assume that the reader has access to a good farm management handbook. The most comprehensive is *The Farm Management Pocketbook* by J.S. Nix, published annually by Wye College, University of London, and referred to hereafter simply as 'Nix'. Alternatives are the *Agricultural Budgeting and Costing Book*, published half-yearly by Agro Business Consultants Ltd, and the *Farm Management Handbook* published annually by the Scottish Agricultural Colleges. Many of the references given in this book are from the journal *Farm Management*, published by the Institute of Agricultural Management, and catering for a wide range of rural (not just farm) management interests. For anyone interested in agricultural management, a subscription to the Institute (including membership of a local branch and four copies of the journal per year) is an excellent investment. Details are available from IAgrM, Farm Management Unit, University of Reading, Earley Gate, Whiteknights, Reading RG6 2AT.

The use of masculine pronouns throughout the book is purely to avoid the clutter of he/she, his/hers, and so on. Inflation is mostly ignored, to allow concentration on the basics without unnecessary confusion. Some of the questions raised by the presence of inflation are discussed in Chapter 20.

A final point: before employing a management technique of any sort, however simple or complex, you should always be sure that the potential costs (in terms of time, effort and money) are outweighed by the potential benefits (long-term and short-term financial, physical and personal). The fact that a technique is described in this book does not mean that it should be used at every possible opportunity.

Keep it simple.

Martyn Warren
1997

# Acknowledgements

Fifteen years have passed since the first edition of 1982, during which I have become a manager myself (of a large academic department), and my two infants (now at university) have been joined by a third, now 12 years old. I have lost the support and guidance of Brian Camm, then my boss, who died in 1993. His imaginative, often anarchic approach to issues of all kinds was a major influence. Like all my colleagues, I miss his encyclopaedic memory, his experience (he was one of the pioneers of gross margin planning and linear programming in farming in the UK, as well as having been a farm business consultant), and most of all his friendship.

Nothing has changed enough, however, for me to want to change my original acknowledgements. I remain ever grateful to the colleagues, students and farmers who have provided inspiration and dissuaded me from my more wayward flights of fancy.

My family deserve particular gratitude for having suffered, usually with good grace, my absences of body and mind during the writing of the original and subsequent editions. It is to them – Jenny, Cate, Tim and Abigail – that this book is dedicated.

# Part 1

# Basic Issues

This part of the book deals with the basic principles of financial management. After discussing the need for financial skills in the running of a farm or other rural business, it examines the three indicators of the financial health of a business – cash flow, profit and capital – and the financial statements used to measure those indicators. It concludes with chapters on the principles of management accounting, and on variations to meet specific needs, including those of the agricultural industry.

Since each of the other parts draws on this material, Part 1 should be read in its entirety before moving on to the rest of the book.

# Chapter 1
# Why Bother?

## 1.1 In the beginning...

Let's start with a story. For the last 30 years, Mr John Haynes has owned and run a small rural business based on Home Farm, providing him with a modest profit to add to his other sources of income. It is a mixed farm of 58 hectares, including 4 hectares of unmanaged woodland, on reasonable soils, somewhere in England. There are three main farming activities: fattening bought-in weaner pigs; a dairy herd of about 80 cows with replacements bought in; and spring barley with grain and straw used in the dairy herd. In addition, Mr Haynes and his wife have run a successful farmhouse bed and breakfast enterprise using three bedrooms in the capacious farmhouse. There is also a small cheese-making venture, producing a speciality soft cheese for a strong but local market.

The dairy herd comprises about 80 Friesian dairy cows on average, with replacements bought in. Buildings include a herringbone parlour with 8 stalls and 4 units, a cubicle house with accommodation for 70 cows, and dry cow housing for 10 cows. There is a milk quota of 450 000 litres. Average milk yield in the previous year was 5300 litres per cow, but has been showing signs of increasing. About 1000 bought-in weaner pigs are fattened each year to cutter weight and sold on contract.

The labour force consists of a foreman, due to retire this Michaelmas, plus two full-time workers and a part-time cheese-maker. Casual labour is hired at busy times.

With his sixtieth birthday approaching, Mr Haynes has decided that it is time to break away from his business interests and see the world. He has no desire to sell his business though, and does not wish to let it to a stranger. He has found the solution to his problem in the shape of Adam Haynes, his nephew. Adam has been working for five years on a neighbouring estate as an assistant manager. By allowing Adam to rent the farm, Mr Haynes would be giving him a step on the business ladder, while at the same time solving his own dilemma.

Adam has proved to be amenable to the idea, though understandably apprehensive, and has agreed in principle to take the business over at the beginning of October 199X, at a rent of £9500. This will coincide with the retirement of the foreman, and Adam hopes that a combination of improved organization and hard work will avoid the need to replace him.

With his savings, an overdraft facility and a low-interest family loan, Adam should have sufficient funds to purchase the stocks and machinery already in the business (see Figure 1.1). Adam's wife, Evelyn, has a part-time job in a local country park, bringing home approximately £6000 per year net. Her hours at the country park can be flexible in order to accommodate the needs of the bed and breakfast enterprise, although she will need some paid domestic help.

| *Cropping* | | *Hectares* |
|---|---|---|
| Spring barley | | 12 |
| Grass leys | | 42 |
| Woodland | | 4 |
| Total | | 58 |
| *Tenant's stocks on the farm at the end of September 199X* | | £ |
| Livestock: | 77 dairy cows @ £650; 10 calves @ £80 | 50 850 |
| Barley, grain: | 42 tonnes | 3 780 |
| Barley, straw: | 26 tonnes | 390 |
| Pigs: | 550 | 24 750 |
| Cheeses: | 55 ready for sale | 1 370 |
| | 248 maturing | 3 720 |
| Dairy feed: | 20 tonnes | 2 800 |
| Silage | | 10 000 |
| Miscellaneous stores, cultivations, etc.: | | 5 684 |
| Total | | 103 344 |
| *Machinery and vehicles: written-down values* | | |
| Farm machinery and plant | | 39 000 |
| Vehicles | | 6 000 |
| Bed and breakfast equipment | | 2 000 |
| Cheese-making equipment | | 4 000 |
| Total | | 51 000 |
| (Note that the land, buildings and milk quota remain the property of the landlord, and thus do not appear in the valuation of the business.) | | |
| *Liabilities* | | |
| Overdraft (after purchasing stocks, machinery and vehicles) | | 21 000 |
| Family loan (parents cashing up life insurance) | | 34 000 |
| Creditors (taken on with the business) | | 6 800 |
| *Debtors* (taken on with the business) | | 10 000 |

**Fig. 1.1** Details of Home Farm.

Adam has never before been solely responsible for the running of a whole business. In his previous work he had experienced the day-to-day problems of managing, but had little to do with longer-term issues. Since he is intending to invest all his life savings in the business, he will be seeking to reassure himself that he will have a good chance of success. Others are also affected. As the potential landlord and guarantor of Adam's overdraft, Mr Haynes will be anxious to ensure that Adam will generate enough cash to be able to pay the rent. The bank manager, before lending Adam money, will expect to be reasonably satisfied that he will not only be able to repay the money in due course, but also to pay the interest charges incurred. Adam's parents, who have trustingly cashed in some life assurance in order to supply the loan capital, are likely to want confirmation that their money will be safe.

In order to obtain both the tenancy and the necessary finance, therefore, and to assess the potential of the business from his own point of view, Adam will need to prepare plans in some detail. But *how* is he to prepare those plans, and where should he start? How is he to use them in controlling the business? How is he to gather the information needed to use them effectively? How can he build into those plans an allowance for the considerable risks he is taking? It is questions such as these that this book is designed to answer.

It is easy, in examples such as that above, to see the logic in detailed financial planning and control of the business. In most businesses, though, problems of this scale arise infrequently. The manager's job is usually one of fine tuning: taking an established, well-tried system, and modifying it little by little in response to changes in the physical and/or economic climates. Why should he bother with the extra paperwork, when he could be out in the fields doing 'real' work, or putting his feet up by the fire? The answer is that he should not – provided, that is, he can afford the luxury, and does not resent paying for it. But he must be honest with himself, and recognize that it is a luxury and one which can cost dear. There are large numbers of business owners in this enviable position, often with no mortgage, or paying low rents, lending money to the bank rather than vice versa, and content to make just enough income to provide a comfortable life.

At the same time, a great many business people are under financial pressure, struggling to make a living in the face of inflation in costs, near-static product prices, and high interest rates. In order to get into business, or to pay death duties, or to undertake much-needed improvements, they have often borrowed heavily. This in turn has reduced the margin for error, making their businesses more vulnerable to the considerable risks in business. These people may not be facing problems as dramatic as those of Adam Haynes. Careful money management is, nonetheless, as important to them as it is to Adam. In the 'fine tuning' of their businesses, they must be able to judge the relative effects of various possible changes on the business finances. They must make reasoned forecasts of profits for the coming year, so that they can identify areas of inefficiency. They must be able to estimate the likely levels of borrowing in the future, so that they can persuade their bank managers, if necessary, to extend their overdraft facilities. They must be able to monitor closely the capital structure of their businesses so that, in the event of unexpected problems, changes can be made before too much damage is done.

In other words, the finances of a business must be managed as assiduously as any crop. A good farmer is thorough in his preparation for the crop, will monitor its growth and be ready to respond quickly to unexpected events such as epidemics of pests or diseases. Good money management is similar; indeed, it could be thought of as 'money farming'.

This book deals with the basic techniques of 'money farming'. A knowledge of financial techniques alone will not ensure good money management, any more than a knowledge of the principles of crop production will make a good arable farmer, or an understanding of the science of widget production make a good widget manufacturer. What it can do, however, is to provide a sound foundation on which the individual may build, using his or her intelligence, ingenuity, and experience. This book, then, is one of the building blocks – the rest is up to you.

## 1.2 Building a framework

Before embarking on a discussion of financial management, it is important to have some sort of framework for that discussion; not so much a definition as a description in terms simple enough to remember throughout the book.

Anyone pressed to suggest one distinguishing feature of the job of management is quite likely to include the word 'decision'. All tasks, from digging a trench to the administration of a multi-national company, involve decisions of one sort or another, but there is little doubt that the former is physical, and the latter managerial work. In the administration of a company, the greater part of the work is concerned with taking decisions, while decision-making is a very minor aspect of digging holes in the ground.

The process by which a manager makes a decision might be as follows.

(1) Determines what he or she wants to achieve, i.e. sets targets or 'objectives'.
(2) Makes as informed a guess as possible about future events, i.e. makes forecasts.
(3) Examines all possible ways of achieving the objectives.
(4) Selects, in the light of the forecasts, the action that seems most likely to lead to the objectives being achieved.
(5) Takes that action.

This process could be described as *planning*. When an immediate decision is called for, it will be made in the manager's mind, and may take only a second or two. When the decision concerns a major change in the running of a business, or in the personal life of the business owner, it may take many months before action is taken.

But, surely, management means more than just planning? A manager is not likely to achieve objectives merely by formulating plans at the beginning of a period of time and sticking to them rigidly, irrespective of events during that period. His planning includes guesses about the future: it is important that he recognizes that these guesses may well turn out to be wrong, and be ready to act accordingly. If his plans are based on the assumption of a warm, dry spring, and it turns out to be cold and wet, he must be prepared to revise his plans to make the best of the new situation.

Our description of management must include, therefore, the regular monitoring of the progress of the plans: comparison of *actual* results against the *expected* results built into the plans, followed by action to alleviate any problems (or to capitalize on any opportunities) shown up by the comparison. This process of *control* of a business can be compared to the control of a car: it is not enough to point the vehicle towards your destination; you must be continually checking for unforeseen hazards and be ready to modify your route accordingly.

Finally, it would be wise to acknowledge the dependence of the business on accurate information at all stages of the planning process. The less use the manager makes of information available from *outside* the business (from weather forecasts to reports of price trends and new technology), the less chance he has of his own forecasts being accurate. Without information on what is happening *within* the business, he is in no position to monitor its progress.

We can thus describe management as:

> planning and control of all or part of a business in an effort to meet the objectives of that business; a process which is heavily dependent on reliable information from both within and outside the business.

This is illustrated in Figure 1.2.

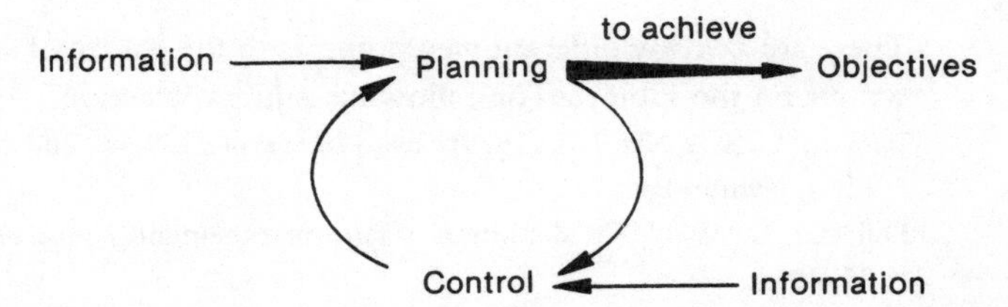


**Fig. 1.2** Illustration of the management process.

This description provides a useful framework for the study of any aspect of management. Since this book is concerned essentially with the management of money, Chapters 2–14 use the framework solely in the context of the finances of the business. This involves an assumption that the only objectives of the business are financial ones, and that a manager's work is primarily the planning and control of money. This is, of course, a narrow and unrealistic view, but it does help us to take a clear initial look at the techniques of financial management without too many complications.

By the time you reach Part 5 of the book you should have a clear enough understanding of the basic ideas and techniques to be able to take a wider look at management, and to cope with some of the more common complications that arise in the financial management of a business.

Before going on to discuss the techniques of financial planning and control, it would be sensible to decide just what we mean by 'financial'. The first thing that comes to mind is cash; it is obvious that a healthy *cash flow*, ensuring a steady bank balance, is vital to businesses and individuals alike. Cash flow is not a reliable guide to the operational performance of a business, however, and a special measure, *profit*, has to be used for this purpose. Neither cash flow nor profit give information about forms of wealth other than cash and trading stocks, so this is provided by a third measure, net *capital*.

For the purposes of study, all aspects of financial management can be classified in terms of cash flow, profit, and capital. Using an example relating to Home Farm, Adam Haynes' financial objectives for the first year might be expressed thus:

(1) Sufficient cash flow to avoid an overdraft of more than £30 000 while providing at least £5000 for Adam's private expenses (in addition to Evelyn's salary).

(2) A net profit of at least £10 000.
(3) A stable or growing net capital.

The remaining chapters in Part 1 take a closer look at these three measures which are so central to the whole financial management process.

## References

There are as many different viewpoints as to the nature of management as there are writers on the subject. The following is just a selection:

Barnard, C. S. & Nix, J. S. (1979) *Farm Planning and Control*, 2nd edn. Cambridge University Press, Cambridge.

Dillon, J. L. (1980) The definition of farm management. *Journal of Agricultural Economics*, **31**, 257–258.

Giles, A. K. & Stansfield, J. M. (1990) *The Farmer as Manager*, 2nd edn. CAB International, Oxford.

Harwood Long, W. (1981) The definition of farm management. *Journal of Agricultural Economics*, **32**, 225–227.

Nix, J. S. (1979) Farm management: the state of the art (or science). *Journal of Agricultural Economics*, **30**, 227–291.

# Chapter 2

# Cash Flow, Profit and Capital

## 2.1 Cash flow

### 2.1.1 *Basic calculations*

The term 'cash flow' means exactly what it says: the movement of cash through the business during a given period. It is rather akin to housekeeping; just as a family must keep a regular check on cash receipts and payments and their effect on the bank balance, so must the manager ensure that cash is available when required for the purchase of essential requisites and services. If the business runs so short of cash that it cannot pay the wages, buy the cattle feed, or pay the rent, it is doomed to failure.

The *annual net cash flow* of a business is calculated by taking the total of cash payments during the year from the total cash receipts. 'Cash' refers to money either in hand or in a bank current account; hence transactions by cheque or standing order are counted as cash transactions, although they do not fit the everyday use of the term. *Anything* which affects the cash balance must be included, even such items as money taken out to pay for the owner's holiday, income tax payments, or the receipt of a loan from a kindly aunt. Conversely, no item should be counted as part of the net cash flow until the money concerned has actually entered or left the business.

Given the amount of cash available at the beginning of the year (the *opening cash balance*), the *closing cash balance* is found by adding the annual net cash flow to the opening balance. As an example, take a business which starts the year with £1000 in the bank. During the year, cash receipts total £10 000 and cash payments total £6000. The annual cash net flow is thus £4000 which, added to the opening balance, gives a closing cash balance of £5000. If the cash payment had been £16 000 instead of £6000, the result would have been an annual net cash flow of –£6000, and thus a negative closing balance, i.e. a bank overdraft, of £5000.

It is unlikely that any businessman will be content with an annual review of cash flow. He will probably wish to keep at least a monthly check; in some trades weekly or even daily. Whatever time period is concerned, the principles are the same: opening balance plus net cash flow for the period gives the closing balance at the end of the period.

### 2.1.2 *From cash flow to profit*

As a means of trouble-spotting, the measurement of cash flow is invaluable. Forecasts of cash flow can identify periods where cash is likely to be short, and give clues to the alleviation of such shortages. Records of actual cash flow can be combined with forecasts to monitor progress throughout the year.

As a guide to the economic performance of the business, however, it has many limitations. Take, for instance, two businesses, A and B, each with an annual net cash flow of £5000 (£55 000 cash income and £50 000 cash expenditure) in 199X. On this basis their performances appear to be identical. Now consider various possibilities:

(1) A has sold all the goods it has produced during the year, while B still has goods worth £1000 in store.
(2) A has been paid cash for all the goods it has sold during the year; B, on the other hand, is owed £2000 for goods that it sold just before the end of the year.
(3) A made no investments in buildings or machinery during the year, while E spent £10 000 on equipment.
(4) The owner of A drew £3000 in cash for his living expenses for the year; the owner of B drew double that amount.
(5) The owner of A consumed none of the produce of the business, while the owner of B took home £500 worth of goods for the consumption of himself and his family.

So B finishes the year with £1000 worth of goods in store, £2000 due to be paid in the near future, and up-to-date equipment. Its owner has spent £6000 on himself and his family, and has consumed, without cash payment, produce of the business worth £500. The situation of A and its owner is but a shadow of this, and yet its annual net cash flow is identical to that of B.

What is needed is a measure of the operational performance of a business which takes these factors into account. Such a measure is net profit.

## 2.2 Net profit

### 2.2.1 Revenue and expense

Like cash flow, *net profit* is measured by the difference between the output and input of value during a given period. Whereas cash flow is concerned exclusively with *cash* values, however, profit is concerned with the revenues and expenses, cash or otherwise, arising from the productive activity of the business during that period.

*Revenue* is calculated by taking cash receipts arising during the period, excluding any capital receipts, e.g. grants, sale of machinery, injections of cash by the owner and including changes in valuation of stock of output items, benefits in kind, and adjustments for opening and closing debtors.

*Expense* is calculated by taking cash payments, excluding any capital payments, personal drawings, or tax payments, and including changes in the valuation of stocks of purchased inputs, benefits in kind, adjustments for opening and closing creditors, and depreciation of 'wasting' possessions such as buildings and machinery.

Thus in the example above, the net profits of business A and business B for the year are as shown in Figure 2.1. The following sections examine this process in more detail.

| | Business A £ | Business B £ |
|---|---|---|
| *Revenue* | | |
| Cash income | 55 000 | 55 000 |
| *plus* goods in store | 0 | 1 000 |
| debtors (money owed to business) | 0 | 2 000 |
| *less* benefits in kind (produce consumed) | 0 | 500 |
| | 55 000 | 58 500 |
| *Expense* | | |
| Cash expenditure | 50 000 | 50 000 |
| *less* capital purchase (machinery) | 0 | 10 000 |
| private drawings | 3 000 | 6 000 |
| | 53 000 | 34 000 |
| *Net profit before depreciation* | 2 000 | 24 500 |
| *less* depreciation @ 20% on machinery* | 0 | 2 000 |
| *Net profit after depreciation* | 2 000 | 22 500 |

* For simplicity this assumes that neither business owned any machinery at the start of the year.

**Fig. 2.1** Net profits of business A and business B for the year.

### 2.2.2 Change in valuation of stocks

*Stocks* can be defined loosely as possessions of the business (other than cash) that are likely to be consumed within one production cycle. They are conventionally classified as:

- Raw materials
- Work in progress
- Finished goods
- Goods purchased for resale

As an example, the stocks of a potter's shop may include clay, paint, and other materials; unglazed pots; completed pots; and various items bought in from other potters in an attempt to broaden the shop's range of goods.

The stocks of a farm typically include purchased stores: growing crops and tillages; trading and breeding livestock; and crops in store. Note also that although breeding livestock (e.g. dairy cows, breeding ewes and sows, etc.) do not strictly conform to the one-production-cycle definition, they are conventionally included in the valuation. (The 'herd basis' method of valuation is an exception. This is primarily an adjustment for tax purposes.)

If, inflation aside, a business ends the year with a greater value of stocks than it possessed at the beginning of the year, it has sold fewer goods than it has produced, and/or it has purchased more materials than it used in production. If the effects are ignored, profit will not be a fair reflection of the 'productive efficiency' of the business; the production of the business will be undervalued, and costs will be overstated. Any increase of valuations, whether of raw materials, work in progress, or of finished goods, must therefore be added to net cash flow in calculating profit.

The reverse is true if the valuation of stocks declines. Output will be overstated, since some of the sales will have arisen from depletion of stocks; costs will be understated due to the failure to replace stocks of materials used in production. A decline in stock valuation is thus *deducted* from net cash flow in calculating profit. Stock valuation is more extensively discussed in Chapter 11.

### 2.2.3 Debtors and creditors

A large proportion of the transactions of a typical business are credit transactions. The seller sends goods to the purchaser, together with an invoice, or bill. The purchaser is given a certain time to pay this invoice (say 28 days) and is not likely to do so until towards the end of this time.

It is very likely, then, that at any point in time a business will be owed a certain amount of money by its customers, and will owe money to its suppliers. These amounts are known as *trade debtors* and *trade creditors*, respectively. Those outstanding at the beginning of the year are opening debtors and creditors, and those at the end of the year are closing debtors and creditors.

In the example earlier in this chapter, business B was owed £2000 at the end of the year. This represents goods that, although produced and sold, have not been paid for (and are thus excluded from the net cash flow). Clearly the £2000 has arisen from the productive activity of the year; to exclude it from profit would be to understate the value of goods produced by the business during that time. This 'closing debtor' must therefore be added into the cash receipts of the year in calculating profit.

Similarly, closing creditors must be added to the cash payments of the year. If this is not done, the costs of the business will be understated. A consignment of dairy feed could, within a 28-day credit period, have been turned into milk. Although no cash has changed hands, the purchase has definitely contributed to the productive activity of the business. If it has not been consumed, the feed will appear in the valuation change and, if not balanced by the closing creditor, will be double-counted.

The closing debtors and creditors of one year are, inevitably, the opening debtors and creditors of the next. Shortly after the beginning of this next year these debtors and creditors will be paid. The cash receipts of the second year will thus include items arising from the productive activity of the first. To avoid double-counting, opening debtors of a given year are *deducted* from the cash receipts and payments of that year in the calculation of profit.

### 2.2.4 Prepayments and accruals

Certain services are paid for at infrequent intervals, either in advance or arrears. Insurance premiums, for instance, are typically payable yearly in advance; rural land rents half-yearly in arrears. At the end of the financial year, it is quite likely that the services represented by these payments will not have been fully consumed. The last half-yearly payment of rates, for instance, may have been made two months before the end of the financial year of the business. The local authority thus owes the business four months' worth of services; this is a special form of debtor called a *prepayment.*

Similarly, the business may have made its last half-yearly payment of rent, in arrears, three months before the end of the financial year. It has consumed a service (the use of the land) without payment for three months. This unpaid rent is a form of creditor known as an *accrual.*

The same arguments apply to receipts in advance (*deferred revenue* – a form of creditor) and in arrears (*accrued revenue* – a type of debtor). In calculating profit, these items are used to adjust payments and receipts in exactly the same way as 'ordinary' trade debtors and creditors.

### 2.2.5 Benefits in kind

The owner(s) and employees of a business often consume produce or goods and services bought by the business, without any cash payment. In a rural business these benefits may include produce such as food products (milk, meat, cheese, etc.) and firewood; and inputs such as housing, private use of car and telephone, electricity and heating fuel.

If profit is to be a reliable assessment of the performance of the business, the value of produce consumed as benefits in kind must be included in revenue. The fact that they have not been paid for in cash does not alter the fact that they arise from the productive activity of the business. Inputs consumed as benefits in kind must be deducted from payments: although they have been paid for by the business, they have not contributed to its productive activity. Since both of these adjustments increase profit, and involve relatively small amounts, they are frequently combined as a single item under 'revenue' (see, for example, Figure 2.2).

### 2.2.6 Depreciation

Included in the annual net cash flow of a business will often be items relating to the purchase or sale of assets such as machinery and equipment, buildings and improvements to property. Such assets can be expected to last longer than just one production cycle, and therefore relate to the productive activity of more than one financial year. It would be illogical to include the whole of the payment or receipt in the profit of the year of purchase or sale.

On the other hand, such items do represent a continuing cost to the business. They will not last for ever; they will wear out, or rot, or become obsolete, or all three. They are *wasting assets*. If their purchase or sale is removed from profit without making a corresponding allowance for this continuing cost, an exaggerated impression of profit will result. A new item is therefore included as an expense; that of *depreciation* of wasting assets. The calculation of depreciation is discussed in greater detail in Section 2.6.

### 2.2.7 *Exclusion of capital, personal and tax transactions*

Certain transactions appear in the cash flow which do not directly arise from or contribute to the productive activity of the business. One example has been considered in the previous section, the purchase and sale of wasting assets. These are converted into an annual depreciation charge, and are otherwise excluded from the profit calculation.

The business may make other capital payments (such as the purchase of land or investments and the repayment of borrowed money) and receive capital sums (such as the proceeds of the sale of land, capital grants, loans, gifts, and injections of cash by the owner). None of these are included in profit, since (for a farm business at least) they lie outside the productive activity of the business.

The same arguments apply to drawings of cash by the owner(s) of the business, and the payment of income and capital taxes. These items concern the *deployment* of income rather than its generation. They reflect the behaviour of the owner(s) rather than the productive activity of the business. They are therefore excluded from the profit calculation.

## 2.3 Profit statements

### 2.3.1 *Net, gross and operating profits*

The calculation of profit is shown in a profit statement, which should be clear, comprehensive, and consistent in form from year to year. The simplest form of profit statement is shown in Figure 2.2.

It is now more normal to encounter the two-stage *trading and profit and loss account (or budget)*. In the first section (the *trading account*), the sales (adjusted for opening and closing debtors) are balanced against opening and closing valuation of stocks, and

| | £ | £ |
|---|---|---|
| *Revenue* | | |
| Sales | | 81 000 |
| Change in valuation of stocks: | | |
| Closing stock | 51 000 | |
| Less opening stock | 50 000 | 1 000 |
| Other revenue | | 2 000 |
| *Expenses* | | |
| Purchases | 39 000 | |
| Labour | 9 500 | |
| Machinery and power | 12 000 | |
| Rent and rates | 9 000 | |
| Services | 1 500 | |
| Administration | 2 000 | |
| Finance charges | 2 000 | 75 000 |
| ***Net profit*** | | 9 000 |

**Fig. 2.2** Profit statement – simplified example.

the *cost of sales*. The latter is the cost of the inputs that are directly incurred in production of the sales: primarily materials and direct labour. The result is the *gross profit* of the business.

In the second section, the *profit and loss account*, the gross profit is balanced against all the other items of expense (the *indirect* or *overhead costs*). The latter are those expenses that are needed by the business to keep in operation, but which are not directly consumed in the process of production. Examples are managers' salaries, machinery costs, office expenses and property charges. The result of this calculation is the *net profit* of the business.

A simplified example of the two-stage trading and profit and loss account is shown in Figure 2.3.

| *Trading account* | £ | £ |
|---|---|---|
| Sales | | 81 000 |
| less cost of goods sold: | | |
| Opening stock | 50 000 | |
| Purchases | 39 000 | |
| | 89 000 | |
| less closing stock | –51 000 | 38 000 |
| Other revenue: | | |
| Training grant | 400 | |
| Rent received | 1 000 | |
| Interest on trade investment | 100 | |
| | 1 500 | |
| Benefits in kind | 500 | 2 000 |
| ***Gross profit*** carried down | | 45 000 |
| *Profit and loss account* | £ | £ |
| Gross profit brought down | | 45 000 |
| Other expenses: | | |
| Labour | 9 500 | |
| Machinery and power | 6 000 | |
| Rent and rates | 9 000 | |
| Services | 1 500 | |
| Administration | 2 000 | |
| Finance charges | 2 000 | |
| | 30 000 | |
| Depreciation | 6 000 | 36 000 |
| ***Net profit*** | | 9 000 |

**Fig. 2.3** Trading and profit and loss account – simplified example.

This two-stage calculation of net profit is referred to as 'the trading account', or 'the profit and loss account', but rarely by its full title. A third variant, which can be used in conjunction with either of the other two, involves the calculation of *operating profit* (or loss). This attempts to reflect more fairly the efficiency of the business by taking separate account of payments of finance charges on borrowed money (see Figure 2.4).

| *Profit and loss account* | £ | £ |
|---|---|---|
| Gross profit brought down | | 45 000 |
| Other expenses: | | |
| Labour | 9 500 | |
| Machinery and power | 6 000 | |
| Rent and rates | 9 000 | |
| Services | 1 500 | |
| Administration | 2 000 | |
| | 28 000 | |
| Depreciation | 6 000 | 34 000 |
| ***Operating profit*** | | 11 000 |
| Less finance charges | | 2 000 |
| ***Net profit*** | | 9 000 |

**Fig. 2.4** Operating profit – simplified example.

### 2.3.2 *The financial period*

Profit is concerned with revenues and expenses arising during a given period of time. The period used for the measurement of profit is usually a year, although it may be six months or even less. Large corporate businesses, where the owners (shareholders) of the firm have little say in the week-by-week decisions, may need to publish profit statements more frequently (e.g. quarterly) in order to keep shareholders informed.

Rural business is dominated by small, non-corporate businesses, with production cycles measured in months rather than days. The year is thus adequate as a financial period for most purposes. The financial year used by a business may start on any day of the calendar year. It is usually determined by the date at which the business was taken over by the present owners. Thus it is common in rural land-using businesses to find financial years starting at Michaelmas (29 September) or Lady Day (25 March), since these are traditional dates for the commencement of farming tenancies or the sale of farm land. Under UK income tax rules, the owner(s) of a business may choose (only once) to change the financial year, for the sake of convenience or tax saving.

Since both financial and management accounts are based on the same raw information, it is convenient to be able to use the same period for both, if possible. Where the financial year is very different from the production year, the management accounts and budgets should be based on the production year (known in farming as the 'crop year') (see Chapter 11).

## 2.4 Capital

### 2.4.1 *Introduction*

Of the three vital financial indicators – cash flow, profit and capital – only capital remains to be discussed. Unlike the other two, it is not concerned with flows of money during a financial period, but with the wealth owned by a business *at one particular point in time*. A hackneyed but useful analogy is that of a photograph; the picture shown is a representation of the state of affairs at the moment the shutter clicked, and at that moment only.

You will know from your own experience that wealth imparts financial security. The more possessions a person owns, the less likely he or she is to be bankrupted by some unexpected event such as an unwise gamble or the loss of his job. Money in the bank can be used as a substitute for income, and non-cash possessions can be sold or used as collateral for borrowing money. The same applies to a business.

Wealth also imparts freedom of action. The businessman with little wealth is able to take few risks where his income is concerned and has little chance of getting out of this position by expanding or diversifying his business. The wealthy person, on the other hand, is able to take more risks (since he or she has wealth to fall back on even if his income fails) and can more easily raise money to finance new ventures.

To summarize, then, capital provides stability in times of change, whether this change is a response to problems, or a response to opportunities. The business with a sure capital foundation can afford to adapt to change in its environment without risking its own downfall.

### 2.4.2 What is capital?

Capital is not exclusively concerned with possessions. Take the examples of businesses A and B. A has possessions worth £50 000 on 31 December; B owns £100 000. Looking at possessions alone, it appears that B has more capital. But suppose we find that A has debts of £1000, whereas B owes £60 000 to various people. Commonsense tells us that in fact A is the more wealthy, since it could pay off its debts and still retain £49 000, while B would have only £40 000 after settling its debts.

Wealth must therefore be considered in terms of *net capital*, which is calculated by deducting debts, or *liabilities*, from possessions, or *assets* (see Figure 2.5). Note that net capital is also known as *net worth* or *owner equity*. If the business is a partnership or a limited liability company, it will be shared between two or more people. The effects on the balance sheet and other accounts are considered in Section 5.2. Here and throughout most of this book a sole proprietorship is assumed.

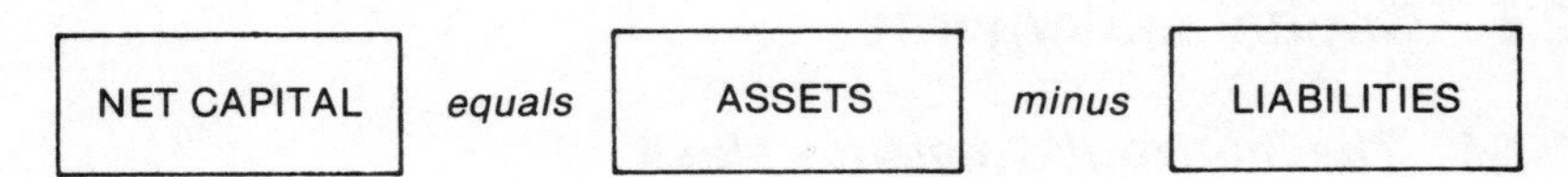


**Fig. 2.5** Accounting equation.

Typical business assets include:

- Cash in hand
- Cash in bank
- Trade debtors and prepayments
- Stocks

- Machinery and equipment
- Buildings and improvements
- Purchased production quotas
- Land
- Other investments
- Goodwill

Typical liabilities include:

- Bank overdraft
- Trade creditors and accruals
- Bank and other medium-term loans
- Mortgages and other long-term loans

### 2.4.3 Classification of assets and liabilities

Not all wealth is equally useful in a given situation. The man losing his job, for instance, will be in a better position if most of his wealth of (say) £5000 is in the bank, or is due to arrive any day in the form of a redundancy cheque, than if it is in the form of long-term investments which are difficult to cash in a hurry. On the other hand, in more happy circumstances, the long-term investment would be more beneficial to him (since it generates income) than money lying idle in the bank current account.

The same arguments apply to businesses, and it is useful to make some distinction between long-term and short-term assets and liabilities. Cash, debtors, and all stocks are normally regarded as short-term, or *current assets*; the remainder as long-term or *fixed assets*. It is often useful to take the distinction further and examine the very short-term assets (cash and debtors) separately; these are termed *liquid assets*. In the special case of a farm business, breeding livestock are generally regarded as fixed assets, while the remainder are included in the current assets as 'stocks'.

Liabilities are divided typically into short-term, or *current liabilities* (due for payment within one year), including bank overdraft and creditors; and long-term or *deferred liabilities*, which include all medium- and long-term loans.

## 2.5 Capital statements

### 2.5.1 The 'horizontal' balance sheet

The derivation of the net capital of a business is shown in a statement known as a *balance sheet*. A balance sheet may be drawn up to show the capital position at any time, but the most useful balance sheets are those drawn up at the beginning and end of the financial year (the *opening* and *closing* balance sheets respectively). Changes in capital between the beginning and end of the year may then be related to profit and cash flow generated during the same period. The closing balance sheet for one year is, of course, the opening balance sheet for the next.

The traditional (but now archaic) form of balance sheet is the *horizontal balance sheet*, where assets and liabilities are set out on either side of the page, net capital being the balance between the two sides. An example of this type of format is shown in Figure 2.6.

**Balance sheet at 31 December 199X**

| 1/1/9X | Liabilities | | 1/1/9X | Assets | | |
|---|---|---|---|---|---|---|
| £ | | £ | £ | | £ | £ |
| 25 000 | Bank overdraft | 40 000 | 30 | Cash in hand | | 50 |
| 7 000 | Trade creditors | 10 000 | | Cash in bank | | 0 |
| 65 000 | Mortgage | 60 000 | 1 970 | Trade debtors | | 7 950 |
| | | | | Valuation of stocks: | | |
| 97 000 | Total liabilities | 110 000 | 8 000 | Raw materials | 7 000 | |
| | | | 14 000 | Work in progress | 15 000 | |
| 141 000 | ***Net capital*** | 133 000 | 28 000 | Finished goods | 29 000 | 51 000 |
| | | | 26 000 | Machinery | 3 000 | 25 500 |
| | | | 10 000 | Buildings | | 8 500 |
| | | | 150 000 | Land and property (at cost 1985) | | 150 000 |
| 238 000 | | 243 000 | 238 000 | | | 243 000 |

**Fig. 2.6** Horizontal balance sheet – simplified example (related to the earlier profit statement examples).

Note that opening and closing balance sheets may be combined by use of extra columns; this saves space and eases comparisons between opening and closing figures. Note also that the assets and liabilities are shown in order of liquidity, i.e. shortest-term progressing to longest-term, or vice versa. A great many variations of this format are used. Some put the assets on the left, some on the right. Some start with current liabilities and assets at the top (as in Figure 2.6), others with long-term assets and liabilities. The most important thing is to ensure that the format adopted is clear, and consistent from year to year.

### 2.5.2 The capital account

Associated with the balance sheet is the *capital account*, a calculation which shows how the net capital has been increased or depleted over the year. There are only three ways by which net capital can be increased over the year: by injections of capital from the owner(s); by grants or gifts from organizations or individuals outside the business; or by reinvesting profits made by the business. Other sources of capital involve entries on both sides of the balance sheet, and thus leave net capital unaltered. (A loan of £1000, for instance, will increase the cash in the bank by £1000, but will also incur a liability of the same amount.)

The only ways in which capital can be depleted are by the withdrawal of cash for the personal use of the owner(s) or to pay tax due; by the business making a loss; or by the consumption of benefits in kind by the owner(s) and/or employees.

Therefore the capital account takes the net capital at the beginning of the year (from the opening balance sheet) and adjusts for the various items mentioned above, in order to arrive at the closing net capital (see Figure 2.7). Not only does this show how the owner's capital has been deployed, but it acts as a valuable check on the accuracy of the profit and loss account, cash flow statement and balance sheet. If any of these has been miscalculated, the capital account will not give the same result for net capital as that given by the closing balance sheet.

| | £ |
|---|---|
| Opening net capital | 141 000 |
| *plus* net profit | 9 000 |
| | 150 000 |
| *less* private drawings | 10 500 |
| tax paid | 6 000 |
| benefits in kind | 500 |
| Closing net capital | 133 000 |

**Fig. 2.7** Capital account – example.

### 2.5.3 The vertical balance sheet

The *vertical* or *narrative* form of balance sheet has rapidly gained popularity owing to its economy of space and ease of setting-out, and is now the standard format. As with the horizontal balance sheet, there are many variations on the theme. One version is shown in Figure 2.8, incorporating a classification of assets and liabilities. Note that in this form (standard in the UK) the balance sheet is balanced on total assets less current liabilities (also known as *net assets* or *capital employed*).

## 2.6 Depreciation

### 2.6.1 The concept of depreciation

The expense of depreciation[1] arises from the tendency of certain types of fixed assets, e.g. machinery and buildings, to 'waste' due to wear and tear, decomposition, and/or obsolescence. It may be thought of in two ways. First, it may be considered as a way of 'saving' money within the business (though not necessarily in cash form) to provide for the eventual replacement of the assets. Second, it may be considered as a means of spreading the initial cost of the assets 'fairly' between the years of their possession by the business. This is a slightly easier concept to understand, and at the time of writing is the convention used in the preparation of normal financial accounts. It is therefore used as the basis for the following discussion.

In a time of stable currency, the two conventions will give identical results. In times of inflation, however, the results will diverge, since the former is aimed at the *cost of replacing* assets at their future (inflated) price, while the latter is concerned with allocating the *historic* cost of assets between years (see Chapter 20). In order to allocate the initial cost of an asset over the years of its life it is necessary to:

| 1/1/9X | Balance sheet at 31 December 199X | | |
|---|---|---|---|
| £ | *Fixed assets* | £ | £ |
| 150 000 | Land and property (at cost 1985) | 150 000 | |
| 10 000 | Buildings | 8 500 | |
| 19 000 | Machinery and plant | 20 000 | |
| 7 000 | Vehicles | 5 500 | |
| 186 000 | Fixed assets | | 184 000 |
| | *Current assets* | | |
| 50 000 | Stocks | 51 000 | |
| 1 970 | Trade debtors | 7 950 | |
| 30 | Cash in hand | 50 | |
| 52 000 | Current assets | 59 000 | |
| | *Current liabilities (due within one year)* | | |
| 7 000 | Trade creditors | 10 000 | |
| 25 000 | Bank overdraft | 40 000 | |
| 32 000 | Current liabilities | 50 000 | |
| 20 000 | Net current assets | | 9 000 |
| 206 000 | Total assets less current liabilities | | 193 000 |
| | *Creditors due after one year* | | |
| 65 000 | Mortgage | | 60 000 |
| | *Capital account* | | |
| | Opening net capital | 141 000 | |
| | + net profit | 9 000 | |
| | + capital injections | | |
| | – private drawings | 10 500 | |
| | – income and capital tax | 6 000 | |
| | – benefits in kind | 500 | |
| 141 000 | Closing net capital | 133 000 | 133 000 |
| 206 000 | Capital employed | | 193 000 |

**Fig. 2.8** Vertical balance sheet.

- Calculate the net initial cost of the asset
- Estimate its likely useful life in the business
- Estimate its likely resale value at the end of that life
- Calculate the depreciation charge for each year of the asset's life

The net initial cost of an asset is the actual cash paid for it, after deducting grants and discounts received, but ignoring money allowed on part-exchange (this is part of the depreciation of the asset being sold, and not of that being bought).

Estimating the likely life of the asset is largely a matter of experience and judgement (if short of either, use someone else's – take advice). In agriculture and some other industries it is possible to obtain guides to the likely useful lives of particular types of machine (see, for instance, J. S. Nix, *Farm Management Pocketbook*, Section III). Such guides can only indicate the effects of wear and tear and corrosion, however, leaving the reader to judge the likely impact of obsolescence. Estimating eventual resale values is again a matter of judgement. An indication can often be obtained from current second-hand prices of similar assets.

### 2.6.2 Choice of calculation method

There are various ways of calculating the depreciation of wasting assets, each with its own advantages and disadvantages. No method can provide an exactly accurate measurement of depreciation, since the calculation depends on forecasts of asset life and future resale value. The aim must therefore be to find a method which most nearly represents the likely pattern of depreciation. Two methods are in common use in the UK: the *straight-line* method; and the *diminishing-balance* (or *reducing-balance*) method. It is essential to be consistent in the methods used: once a method of calculation has been chosen for a type of asset, it should be used thereafter.

### 2.6.3 The straight-line method

By this method, the difference between initial cost and eventual resale value is divided by the estimated life in the business (in years). The result is the annual depreciation charge. As an example (see Figure 2.9), an asset costs £10 000 to buy, net of discounts, grants, etc. Its estimated useful life in the business is four years, after which its resale value is likely to be £2000. The annual charge will then be (£10 000 – £2000)/4 = £2000. This charge may be used to find the *written-down value* (WDV) or *book value* at the beginning of the year by deducting the charge from the WDV at the beginning of the year. This is the value that is used in the closing balance sheet for the year concerned. Note that since the depreciation charge is constant throughout the life of the asset, the WDV declines at a constant or 'straight-line' rate (see Figure 2.10).

| *Year* | *WDV at beginning of year* | *Annual depreciation* | *WDV at end of year* |
|---|---|---|---|
| | £ | £ | £ |
| 1 | 10 000 | 2 000 | 8 000 |
| 2 | 8 000 | 2 000 | 6 000 |
| 3 | 6 000 | 2 000 | 4 000 |
| 4 | 4 000 | 2 000 | 2 000 |
| Accumulated depreciation | | 8 000 | |
| *plus* WDV at end of year 4 | | 2 000 | |
| Initial cost of asset | | 10 000 | |

**Fig. 2.9** Straight-line depreciation – example.

This method is easy to use, and is especially popular for budgets, where one calculation provides the depreciation charges for all the years of an asset's life. It is appropriate particularly for buildings and improvements, where the decline in value is likely to be relatively constant over the asset's life. Where machinery and equipment are concerned, however, it is reasonable to suppose that the asset will decline in value more in the early years of its life than later. This is partly a matter of wear and tear, and partly a matter of the likelihood of the asset being superseded by newer, better machines.

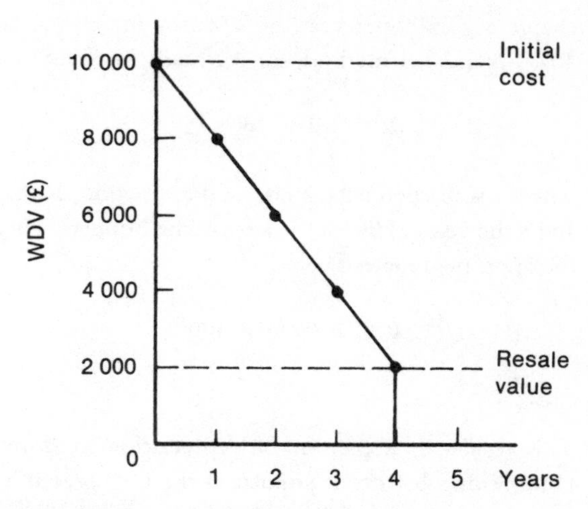


**Fig. 2.10** Straight-line depreciation – graphical representation.

An alternative depreciation method makes allowance for this by allocating a greater proportion of the total depreciation to the early years than to the later years of the asset's life.

## 2.6.4 The diminishing-balance or reducing-balance method

In this method, a percentage depreciation rate is applied to the WDV of the asset as at the end of the previous year. Using the same example as above, and depreciating at the rate of 20% per year, the annual charge would be as shown in Figure 2.11. A major problem with this method is the selection of a depreciation rate which will result in a WDV at the end of the asset's life in the business which is identical to the anticipated sale value of the asset; in the example above the rate used has resulted in a WDV in year 4 which is £2096 higher than the resale value assumed.

One way of dealing with this problem is to try a variety of rates until a suitable resale value is found. A useful starting point is to take a rate which is double the appropriate straight-line rate. (In the example above, the straight-line depreciation

| *Year* | *WDV at beginning of year* | *Depreciation at 20% of WDV* | *WDV at end of year* |
|---|---|---|---|
| | £ | £ | £ |
| 1 | 10 000 | 2 000 | 8 000 |
| 2 | 8 000 | 1 600 | 6 400 |
| 3 | 6 400 | 1 280 | 5 120 |
| 4 | 5 120 | 1 024 | 4 096 |
| Accumulated depreciation | | 5 904 | |
| *plus* WDV at end of year 4 | | 4 096 | |
| Initial cost of asset | | 10 000 | |

**Fig. 2.11** Diminishing-balance depreciation – example.

charge is £2000 per year, or 20%; the initial trial diminishing-balance rate is thus 40%.) A quicker method, for the mathematically minded, is to use the formula:

$$r = [1 - \sqrt[n]{(R \div C)}] \times 100$$

where $r$ is the percentage rate of depreciation, $R$ the resale value, $C$ the initial cost and $n$ the years of life of the asset in the business. Then, using the previous example, the appropriate rate is:

$$[1 - \sqrt[4]{(2000 \div 10\,000)}] \times 100$$
$$r = 33\%$$

This results in a pattern of depreciation as shown in Figures 2.12 and 2.13. Diminishing-balance is popular in the UK, probably because it is the method used for calculating capital allowances for machinery for income tax assessment purposes. Its biggest drawback is the necessity of fixing a percentage rate of depreciation. The impact of this drawback is reduced by the widespread use of 'pooling' in diminishing-balance calculations.

| *Year* | *WDV at beginning of year* | *Depreciation at 33% of WDV* | *WDV at end of year* |
|---|---|---|---|
| | £ | £ | £ |
| 1 | 10 000 | 3 300 | 6 700 |
| 2 | 6 700 | 2 211 | 4 489 |
| 3 | 4 489 | 1 481 | 3 008 |
| 4 | 3 008 | 1 008 | 2 000 |
| Accumulated depreciation | | 8 000 | |
| *plus* WDV at end of year 4 | | 2 000 | |
| Initial cost of asset | | 10 000 | |

*Note:* a small adjustment (£11) has to be made to the final year's depreciation to allow for rounding errors.

**Fig. 2.12** Diminishing-balance depreciation – example of use of formula.

### 2.6.5 Pool calculations

In most businesses, depreciation calculations of the form described earlier would have to be repeated many times in order to find the appropriate annual charge for each asset. This implies, in turn, making assumptions for each asset concerning its useful life and resale value. *Pooling* assets for diminishing-balance depreciation entails grouping similar types of assets together (e.g. for a farm business, tractors, cultivation equipment and motor vehicles) and treating each group as a block with a single depreciation rate. Additions and sales of assets can be incorporated in the calculations without separate treatment. The rate at which the group is depreciated is found in the first place by taking an average of the appropriate rates for the individual assets in the group. These rates can be determined either by using the formula

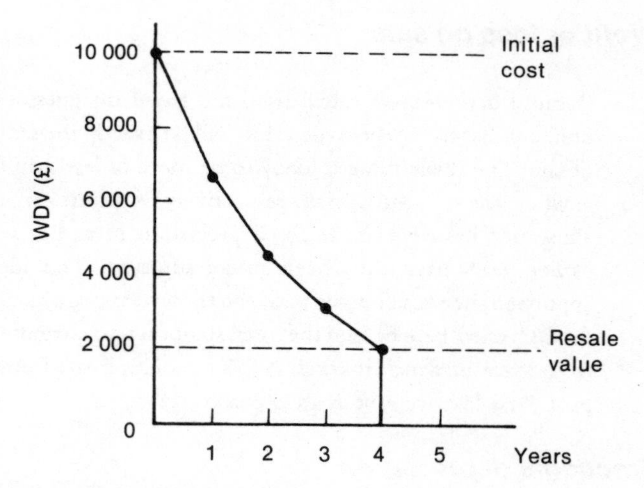


**Fig. 2.13** Diminishing-balance depreciation.

or by trial and error. Once the average rate has been calculated, it can be used thereafter; an adjustment of the rate will be needed only if the make-up of the group changes dramatically.

The mechanism of the pool calculation is shown in the following example. Suppose that the owner of a retail business starts the financial year with his delivery van 'pool' having a WDV of £30 000. During the year he buys a new van for £15 000, selling an old one in part-exchange for £2000. Using an annual depreciation rate of 20%, the annual charge for this year is calculated as shown in Figure 2.14.

| | £ |
|---|---|
| *Delivery vans* | |
| Opening value | 30 000 |
| *plus* purchases | 15 000 |
| *less* sales | 2 000 |
| | 43 000 |
| *less* depreciation charge @ 20% | 8 600 |
| Closing value | 34 400 |

**Fig. 2.14** Pool calculations – example.

It may seem that the use of pooling detracts from the accuracy of the depreciation calculation. It must be remembered, however, that in most farm businesses the expense of depreciation is a relatively small component of the total expenses. It needs a large error in the depreciation calculation to affect total expense by a significant proportion. The use of pooling thus costs little in accuracy while gaining much in convenience.

### 2.6.6 Profit or loss on sale

Because depreciation calculations are based on guesses about future values, it is unlikely that the *actual* resale value will be exactly the same as the WDV at the time of sale. The resale value is likely to be more or less than the WDV, giving rise to a *profit on sale* or a *loss on sale*, respectively. A profit on sale implies that too much allowance has been made for depreciation in earlier years. The profits of those earlier years have thus been underestimated. The ideal remedy would be to apportion the profit on sale between those years. In practice, this creates much effort for little extra benefit, and the usual solution is to attribute profit on sale as a revenue to the year in which it arises. A loss on sale is treated similarly, being shown in the profit and loss account as an expense.

### 2.6.7 Fractional depreciation

The calculations shown above assume implicitly that assets are bought at the beginning of the financial year. In practice this is not often the case; indeed a large proportion of purchases are made towards the end of the financial year, to take full advantage of capital allowances against income tax. It is possible to make allowances to remove this distortion. Depreciation can be calculated on a monthly or quarterly basis and allocated between financial years accordingly. This is known as *fractional depreciation*. It can be a complex and tedious exercise, especially as similar adjustments must be made when the asset is eventually sold.

An alternative is to ignore fractional depreciation and hope that the effect of acquisitions made part-way through the year is balanced by similarly timed disposals. This is particularly appropriate when dealing with groups of assets, and is the approach used in this book.

### 2.6.8 Appreciation of assets

It is possible for an asset to appreciate in value, as a consequence of inflation and/or relative scarcity. This increase in value does not arise from the productive activity of the business, and as such should not be included in the calculation of profit. If the asset is revalued in the balance sheet, the capital account should be credited with a 'holding gain' of an equal amount, to allow reconciliation (see Chapter 20).

## Notes

1. It is important to distinguish between depreciation and *capital allowances*. The latter, though calculated in a similar fashion to depreciation, are used solely in the assessment of UK Income and Corporation Tax.

## References

Berry, A. & Jarvis, R. (1991) *Accounting in a Business Context*, Chapman & Hall, London.

Broadbent, M. & Cullen, J. (1993) *Managing Financial Resources*, Butterworth Heinemann, Oxford.

Bull, R.J. (1990) *Accounting in Business*, 6th edn. Butterworth Heinemann, London.

Millichamp, A. (1995) *Finance for Non-financial Managers*, 2nd edn. DPP Publications, London.

Nix, J.S. (published annually) *Farm Management Pocketbook*. Wye College, University of London, Wye, Kent.

## Chapter 3

# Interpretation of Financial Accounts

## 3.1 Adjusting for realism

A businessman has a legal obligation to keep accounts, of which, together with the balance sheet, the profit statement is a central feature. This obligation exists largely to enable the Inland Revenue to verify the income of the owner(s) of the business and thus assess income tax (or corporation tax) liability. It is usually advisable to have these *financial accounts* prepared by a chartered accountant in order that the Inland Revenue may have a reasonable degree of faith in the figures shown. This has the added advantage that an accountant can often use his expertise to help the farmer exploit legitimate opportunities for reducing his tax liability.

Problems can arise when trying to use financial accounts as a guide to decision-making, however. The accountant works within various assumptions and conventions which are appropriate for the task which he has been asked to do, but may not be appropriate for the task of management. One of the key conventions is that of *historic cost*: assets and expenses are shown at their actual cost to the business, irrespective of how long ago the relevant transaction took place. This ensures that the costs used are clearly verifiable (from the original invoices and other documents) and values are unambiguous. Moreover, items which had no original cost are ignored; and unrealized gains are not taken into account. Fixed assets are valued at cost (less depreciation where appropriate), although where the original acquisition was a long time ago and/or values have changed markedly, there may be an occasional revaluation to current market values.

Accounts intended as a basis for tax assessment must be used with great care if one is trying to learn lessons for management purposes. This applies particularly to the values of stocks (especially breeding livestock), wasting assets and land, which in the presence of inflation are likely to be understated – although this depends also on the state of the market for those commodities. Depreciation is consequently likely to be understated. In the balance sheet shown in Figure 2.8, for instance, the value used for the land is the price of purchase many years ago. Depending on the way land prices have moved in the interim, the current market value could be double – or half – the historic cost. There is no value for the milk quota, although the current sale price could run into six figures, since it was acquired without cost when quotas were introduced: it was 'given' to the business. The value of the dairy herd, based on the cost of producing home-reared replacement heifers, is a fraction of the actual value.

Thus although the use of historic cost is fundamental to the preparation of financial accounts, it limits the value of the 'financial' accounts as an aid to decision-making. One way of dealing with these problems is to make the necessary adjustments to the 'financial' profit in order to arrive at a figure that is more useful in decision-making – a 'management' profit – with corresponding adjustments to the balance sheet. It is generally more satisfactory to establish a separate system of *management accounts*, as discussed in Chapter 4. In the absence of such management accounts, however, we have little choice.

Caution is needed. The values used in adjustment should be 'forced sale' values, i.e. the price one would be likely to receive, net of transport, legal and other pre-sale costs, if one had to sell in a hurry. Be careful in revaluing production quotas: the value could disappear as quickly as it arrived. Particular care should be taken to check whether machinery is leased before revaluing. Finance leasing (see Chapter 18) is a popular way of borrowing money for machinery and other medium-term assets. However, since it is technically a lease, neither the outstanding lease payments nor the value of the leased asset need appear on a sole trader's balance sheet. There is a temptation, on seeing the resulting low value for machinery, to substitute a current value. This ignores the fact that the assets do not belong to the business, and that there is a contractual obligation to make payments for the rest of the duration of the lease.

Before making a major decision on the strength of your revised balance sheet it is wise to check with your accountant in case you have made an expensive mistake. He or she will be able to advise you on other hidden items which will only become due when you sell an asset, such as tax liabilities.

## 3.2 Interpreting the profit and loss account

### *3.2.1 Historical analysis*

Once necessary adjustments have been made, the simplest form of analysis is to compare the profit and its make-up with that of previous years. Usually last year's accounts are presented in a separate column against this year's figures, but it is advisable to compare with several years' accounts if at all possible, reducing the likelihood of your judgements being distorted by short-term fluctuations in values. Any adjustments you have made to this year's accounts will, of course, have to be applied to the comparison accounts as well.

For your particular business, identify both key indicators and trends over time. Start with the 'bottom line' – the net profit – and work back, trying to gain some understanding of how changes in profit were influenced by variations in individual elements of revenue and expense. Then follow this up by more detailed investigation of invoices and physical records to find out how much any variation was caused by movements in price, as opposed to changes in volume: how much you used or produced.

This type of analysis is relatively quick and easy, relying more on commonsense and an understanding of your industry than complex techniques. It allows you to identify trends within the business, and to attempt an explanation of the consequences of decisions made in the past. On the other hand, it is insular, taking no account of how others in the same type of business are progressing.

### 3.2.2 Interfirm comparison

One way of overcoming this problem is to attempt a comparison of your accounts with those of other similar businesses. Ideally this would be with aggregated accounts of many businesses in the same industry: the more businesses involved, the more confident one can be that the averaged results are representative of the whole industry. In those rural businesses whose production depends on climatic conditions, topography, soil types, etc., a regional or even local comparison may be necessary. Where business sizes vary considerably, it is usually better to use a *common-size comparison*, relating all key indicators to a specific factor. The most common such factor is probably sales revenue or *turnover* (so both business and comparison accounts are recalculated to show net profit, gross profit, overhead costs, direct costs, and so on, as a percentage of sales revenue) (see Figure 3.1). More common in rural land-using business is to express such indicators per hectare or acre of land.

| | *Own business* | | *Comparison accounts* | |
|---|---|---|---|---|
| | £ | *% of sales* | £ | *% of sales* |
| Sales | 81 000 | 100 | 165 000 | 100 |
| *less* cost of goods sold | 37 000 | 46 | 87 000 | 53 |
| Other revenue | 1 000 | 1 | 0 | 0 |
| ***Gross profit*** | 45 000 | 56 | 78 000 | 47 |
| Other expenses | 34 000 | 42 | 58 000 | 35 |
| ***Operating profit*** | 11 000 | 14 | 20 000 | 12 |
| Finance charges | 2 000 | 2 | 5 500 | 3 |
| ***Net profit*** | 9 000 | 11 | 14 500 | 9 |

**Fig. 3.1** 'Common-size' comparison, profit statement.

Availability of comparison data varies considerably. Farming and associated activities are particularly well-served in most countries, with government departments and other public agencies producing comparison data for different farming types and by region. For other industries it may be necessary to seek help from relevant professional bodies. An increasing trend is for larger firms of accountants and consultants to provide comparison data based on aggregating the results of their clients in particular industries: although such data is unlikely to be representative of all firms in the industry, it can nevertheless prove valuable. Further discussion of interfirm comparison can be found in Chapter 14.

### 3.2.3 Ratio analysis

*Ratio analysis* is a device that can allow more useable information to be extracted from potentially confusing accounts, when combined with historical analysis and/or interfirm comparison. Whole books have been written about ratios (see, for instance, Tyran, 1986), but a relatively small number is necessary to allow reasonable insight. For the profit and loss account, they can be grouped into those concerned with profitability, and those concerned with efficiency or activity. The example figures used in this section and Section 3.3 relate to the accounts shown in Figures 2.3 and 2.4. For simplicity, it is assumed here that the valuations of all assets are accurate, although the first task would normally be to adjust the values for realism.

**Profitability ratios** include various measures of *percentage return on capital* and of *percentage profit margins*. The general form of the return on capital equation is:

$$\frac{\text{profit}}{\text{capital}} \times 100 = x$$

Assessing whether net profit is good, bad or indifferent is difficult in isolation. The return on capital allows the owners of a business to look at profit in relation to the money they have invested, in the same way (and using the same measure) as one would judge the rate of return on money invested in a building society account.

Perhaps its most useful expression is *return on owner equity*, calculated by:

$$\frac{\text{net profit}}{\text{owner equity}} \times 100$$

Example:

$$\frac{9000}{133\,000} \times 100 = 6.8\%$$

Equity is the owners' stake in the business, so this is what they will be especially interested in. The resulting percentage can be compared with the interest rate the business is paying on borrowed capital, if any, and the return which could be obtained if the capital were realized and invested elsewhere (remembering however that interest on capital invested in government stocks or a building society is virtually risk-free – as well as work-free).

At the time of writing the rate of return shown here is not much higher than the return available from safe and effort-free investments outside the business. Judged purely on financial grounds, the owner might be better off selling the business, but it would be unwise to make a judgement on the results of just one year. Moreover, the long-term gains from capital growth may outweigh the lack of profits in the short term, and in any case the business owner(s) may have other, non-profit objectives that outweigh the relative lack of profitability.

Another popular measure is *return on capital employed*:

$$\frac{\text{operating profit}}{\text{capital employed}} \times 100$$

Example:

$$\frac{11\,000}{193\,000} \times 100 = 5.7\%$$

This measures the return on all the long-term capital employed in the business, whether owned or borrowed, and is thus a broader measure of performance of the business. Operating profit is used to exclude interest charges which themselves represent part of the return on capital.

Percentage profit margins allow comparison of profit figures on a common basis. One key to business profit is keeping the margin of profit to sales revenue as high as possible (the other is ensuring high volume of sales, of course). The *gross profit margin* is calculated by:

$$\frac{\text{gross profit}}{\text{sales}^{(2)}\ \text{(turnover)}} \times 100$$

Example:

$$\frac{45\,000}{81\,000} \times 100 = 55.5\%$$

This can be tracked over time through past accounts, and compared to performance of similar businesses where possible. It gives a measure of the sales and production performance which is vital in itself: a business which is failing on gross profit margin has fundamental problems. The *net profit margin* allows the effect of overhead costs to be examined: if gross profit margin is higher than average but the net profit margin is lower, careful examination of the overhead cost structure is called for.

$$\frac{\text{net profit}}{\text{sales}^{(2)}\ \text{(turnover)}} \times 100$$

Example:

$$\frac{9000}{81\,000} \times 100 = 11.1\%$$

**Efficiency, or activity, ratios** allow a closer look at the control of potentially expensive activities: stock-holding and credit control. The *stock turnover period* compares the value of stocks to the value of sales, and expresses the result as a number of days-worth of stocks on hand:

$$\frac{\text{value of stock}}{\text{sales}} \times 365\ \text{days}^{(3)}$$

Example:

$$\frac{52\,000}{81\,000} \times 365 \text{ days} = 234 \text{ days}$$

Since stocks tie up valuable working capital, the lower the figure the better, in general (although too low a figure could risk very expensive 'stockouts', costing production (input stocks) or sales (output stocks)). Ideally the average of opening and closing stocks should be used: further precision is given by splitting into two ratios, i.e. finished goods to sales, and materials/work in progress to cost of sales.

*Debtor turnover* uses a similar calculation to examine the bills owed to the business in the course of trade:

$$\frac{\text{debtors}}{\text{sales}} \times 365 \text{ days}$$

Example:

$$\frac{7950}{81\,000} \times 365 \text{ days} = 36 \text{ days}$$

Commonsense suggests that the less owed to the business the better, so good credit control is reflected in a low debtor turnover period. A high period indicates a need to ensure swifter payment, by means of a few well-placed phone calls and/or letters, for instance. A debtor period of more than a month would be regarded by most as too high: the business is financing these outstanding bills through its overdraft.

*Creditor turnover* takes a similar approach to money owed *by* the business to others, and is perhaps less important:

$$\frac{\text{creditors}}{\text{sales}} \times 365 \text{ days}$$

Example:

$$\frac{10\,000}{81\,000} \times 365 \text{ days} = 45 \text{ days}$$

Here one might aim for as high a figure as possible, while bearing in mind the loss of goodwill, or even withdrawal of supply, that can result from an over-zealous application.

## 3.3 Interpreting the balance sheet

### 3.3.1 Historical analysis

Providing that adjusted balance sheets are available for at least the beginning and the end of a particular year (and preferably for previous years), it is possible to examine the way in which the capital structure has changed over time. This is just as important as the historical analysis of the profit and loss account, and as with the latter, starts with a 'commonsense' look at movements in certain key indicators. Foremost among these is net capital; others include the bank balance, levels of debtors and creditors, and loans. It can also be useful to examine changes in stock values, especially in times of inflation. Any significant increase in fixed assets should be accompanied by an increase in liquid assets, in order to avoid 'overtrading'; the use of all available funds to finance expansion and running short of working capital as a result.

Historical analysis can be assisted by *flow of funds analysis* or *source and allocation of funds analysis*. This technique works by tracing the flow of cash through the business during the year, and in the process monitoring changes in the various assets and liabilities. It comprises two lists: one of sources of cash, and the other of uses of cash. Sources of cash can be grouped into four categories: the trading activities of the business; external sources; increase in liabilities; and reduction in assets. Similarly the uses of cash may be classified as losses; external uses; reduction in liabilities; and increase in assets.

These main categories may be further subdivided to give a list such as that in Figure 3.2. (There are various ways of presenting a flow of funds statement, but they all include the same components.) Here cash from trading is separated from other sources: note that because we are concerned with the flow of cash, it is necessary to adjust the profit or loss in order to remove the effects of depreciation, changes in debtors, creditors and stock valuations, and benefits in kind. Similarly, in accumulating the various uses of cash, commitments such as private drawings, tax payments and loan repayments are identified separately. This latter distinction is more useful in using the flow of funds as a budgeting aid (where it is important to know what is left to spend after basic financial obligations have been met) than in analysis of a historic balance sheet. The total uses of cash should exactly match the total cash available. If it does not, the analysis should be checked for mistakes.

The aim of the technique is to show the effect of decisions on long-term uses of cash, and to ensure that there is a sufficient supply of liquid funds available when required. It is useful both as a historical analysis, and as a budgeting device to test the likely effect of an expansion (or a foreseen difficulty) on the overdraft. It can act only as a summary budget, however, and a cash flow budget should be used for detailed analysis (see Chapter 7).

A particular benefit of funds flow analysis is its ability to link the sources of cash with the manner of its use. Among the points to watch for are significant increases in liabilities (especially in conjunction with slow-growing profits), growth in fixed assets at the expense of current assets, growth in fixed assets financed largely by short-term liabilities, and private drawings which seem excessive relative to the other calls on cash. Applying the above to the example business in Figures 2.3 and 2.8, it can be seen that net capital is down by £8000 (5.7%), with the overdraft rising by £15 000 (60%) and creditors by £3000 (43%). The mortgage has declined, presumably due to normal repayment of the principal. Debtors are up by £6000 (305%) but there is little change in the other assets. The capital account (Figure 2.7)

| | **FLOW OF FUNDS ANALYSIS** | | **£** |
|---|---|---|---|
| **Sources** | *Net profit (loss)* | | .............. |
| | Depreciation (+) | | .............. |
| | Change in debtors (increase –, decrease +) | | .............. |
| | Change in creditors (increase +, decrease –) | | .............. |
| | Valuations (increase –, decrease +) | | .............. |
| | Benefits in kind (–) | | .............. |
| **Cash from trading** | | | .............. |
| **Add** | Capital introduced | | .............. |
| | Sale of land, property, investments | | .............. |
| | Increase in overdraft | | .............. |
| | Increase in loans | | .............. |
| | Reduction in cash balances | | .............. |
| **Cash from other sources** | | | .............. |
| **Total cash available** | | | .............. |
| **Uses** | *Commitments* | | |
| | Private drawings | | .............. |
| | Tax payments | | .............. |
| | Reduction in loans | | .............. |
| **Total commitments** | | | .............. |
| **Add** | Purchase of land, property, investments | | .............. |
| | Net purchase of buildings | | .............. |
| | Purchase of machinery | .............. | |
| | Less sale of machinery | .............. | .............. |
| | Repayment of overdraft | | .............. |
| | Increase in cash balances | | .............. |
| | | | .............. |
| **Total cash used** | | | .............. |

**Fig. 3.2** Flow of funds worksheet.

shows a high level of private drawings and tax payments: the implication is that 199X has been a less successful year than 199Y, and that part of the problem lies in poor profitability.

For a flow of funds analysis, we need to refer to the calculation of depreciation shown in Figure 3.3. The flow of funds analysis itself is shown in Figure 3.4. It shows that the major sources of funds were short-term borrowing together with trading. The funds were used partly to replace machinery and make loan repayments, but mainly to allow substantial private drawings and tax payments. The latter point is particularly worrying, as the money is leaving the business altogether.

| *Machinery* | £ | *Buildings* | £ |
|---|---|---|---|
| Opening value | 26 000 | Opening value | 10 000 |
| *plus* purchases | 6 000 | *plus* purchases | 0 |
| *less* sales | 2 000 | *less* sales | 0 |
| | 30 000 | | |
| *less* depreciation @ 15% | 4 500 | *less* depreciation @ 15% | 1 500 |
| Closing value | 25 500 | Closing value | 8 500 |

**Fig. 3.3** Depreciation calculations for 199X.

| | **FLOW OF FUNDS ANALYSIS** | | **£** |
|---|---|---|---|
| **Sources** | *Net profit (loss)* | | 9 000 |
| | Depreciation (+) | | 6 000 |
| | Change in debtors (increase –, decrease +) | | –5 980 |
| | Change in creditors (increase +, decrease –) | | 3 000 |
| | Valuations (increase –, decrease +) | | –1 000 |
| | Benefits in kind (–) | | –500 |
| **Cash from trading** | | | 10 520 |
| **Add** | Capital introduced | | 0 |
| | Sale of land, property, investments | | 0 |
| | Increase in overdraft | | 15 000 |
| | Increase in loans | | 0 |
| | Reduction in cash balances | | 0 |
| **Cash from other sources** | | | 15 000 |
| **Total cash available** | | | 25 520 |
| **Uses** | *Commitments* | | |
| | Private drawings | | 10 500 |
| | Tax payments | | 6 000 |
| | Reduction in loans | | 5 000 |
| **Total commitments** | | | 21 500 |
| **Add** | Purchase of land, property, investments | | 0 |
| | Net purchase of buildings | | 0 |
| | Purchase of machinery | 6 000 | |
| | Less sale of machinery | –2 000 | 4 000 |
| | Repayment of overdraft | | 0 |
| | Increase in cash balances | | 20 |
| | | | 4 020 |
| **Total cash used** | | | 25 520 |

**Fig. 3.4** Flow of funds analysis – worked example.

### 3.3.2 Interfirm comparison

Again the historical analysis on its own results in a rather insular view, and it can be advantageous to use interfirm comparison providing suitable comparisons are available (e.g. farm management handbooks for agricultural enterprises). This needs caution, however: the capital make-up can vary considerably between businesses that are very similar in terms of production and profitability, as a result of different ways of financing the means of production (e.g. renting as opposed to buying; using personal finance rather than borrowing, and so on).

### 3.3.3 Ratio analysis

Balance sheet ratios can help shed further light on the capital structure of the business, and in particular on overall stability (gearing[(4)]) and short-term stability (liquidity). Some indication of *overall stability* or *gearing* may be gained by examination of net capital, but only by relating it to other aspects of the balance sheet. Take two businesses, C and D. Their respective balance sheets on 31 December look like this:

| | C | D |
|---|---|---|
| Assets | 60 000 | 250 000 |
| Liabilities | 10 000 | 200 000 |
| Net capital | 50 000 | 50 000 |

Their net capital figures are identical, yet it is obvious that D is less stable than C. A proportionately small increase in liabilities would bring D to the point where the assets were outweighed by liabilities; in other words, where D would be 'insolvent'.

The quickest and simplest guide to the overall stability of a business is the *percentage owned* or *percentage equity*. This is a measure of the owner's stake in the business, expressed as a percentage of the total assets, i.e.:

$$\frac{\text{net capital}}{\text{total assets}} \times 100$$

Thus the percentage equity of C is 83%, and that of D is 20%. In the example balance sheet in Figure 2.8, the figure would be:

$$\frac{133\,000}{243\,000} \times 100 = 54.7\%$$

Suggesting an appropriate target percentage is difficult, as it will vary with the type of business, the level of profitability and the general risk level. In an industry with relatively low profitability and significant seasonal fluctuations, such as agriculture, a percentage owned of less than 66% should give rise to some concern.

Further investigation can make use of the *long-term debt to equity ratio*:

$$\frac{\text{long-term debt}}{\text{equity}} \times 100$$

Long-term debt is specified because it typically gives rise to a constant drain on the finances of the business in the form of interest charges and the repayment of principal, thus adding to the risk level. Short-term liabilities can also act as a drain on the business, but they are usually balanced to some extent by short-term assets. As an example, let us return to businesses C and D, and assume that C has long-term liabilities of £1000, and D has long-term liabilities of £40 000. Their long-term debt to equity ratios are, respectively, £1000/£50 000 or 20%, and £40 000/£50 000 or 80%.

Business D has increased its capital considerably by borrowing, while C employs little more than its owner's own capital. As long as D is obtaining a higher rate of return on capital than the interest rate payable on the borrowed money, its profits are likely to be higher than those of C. In a bad year, however, the return on capital could be below the rate of interest, and could indeed be negative. Since payments of interest and capital have to be made in good years and bad, the profit and loss position of business D in a poor year could easily be worse than that of business C (see Figure 3.5). Thus a high gearing implies a high level of risk. The equivalent ratio for the example balance sheet would be:

$$\frac{60\,000}{133\,000} \times 100 = 45.1\%$$

| | Business C | Business D |
|---|---|---|
| | £ | £ |
| Owner equity | 50 000 | 50 000 |
| Long-term loans | 1 000 | 40 000 |
| Gearing | 2% | 80% |
| *Good year* | | |
| Net profit before interest | 12 150 | 36 000 |
| Interest on long-term loans at 15% | –150 | –6 000 |
| Profit after interest | 12 000 | 30 000 |
| *Poor year* | | |
| Net profit before interest | 2 550 | 6 000 |
| Interest on long-term loans at 15% | –150 | –6 000 |
| Profit after interest | 2 400 | 0 |

*Note:* the example does not account for the repayment of loan capital, which would exacerbate the effect.

**Fig. 3.5** Effect of gearing on risk levels.

The maximum 'safe' long-term debt/equity ratio again depends on the type of business, and in particular on its profitability in relation to the capital employed. In most rural businesses, a ratio of 40% is probably the highest that can be tolerated, and then only if the business is specializing in high-margin products. Ratios of double that level can be sustained in some other industries, such as electronics, but these are characterized by high turnover and high return on capital.

Other measures of overall stability include *long-term debt to capital employed*, the *solvency ratio* (total assets as a percentage of total liabilities) and the *debt/equity ratio* (total liabilities as a percentage of net capital). A useful measure for linking finance charges with profitability is the *interest cover* (or *times covered*):

$$\frac{\text{operating profit}}{\text{finance charges (including leasing)}} \times 100$$

Example:

$$\frac{11\,000}{2000} \times 100 = 550\% \text{ (or 5.5 times)}$$

This shows to what extent the profit before interest can cover the finance charges. There is no rule of thumb (although one would hope that it was more than 100%), but the measure can be extremely useful when investigating trends or making interfirm comparisons.

It is not enough to look at the overall stability of the business: investigating the *short-term stability* or *liquidity* is vital. Even if the gearing is healthy, shortage of cash for purchase of essential materials and services can force a business out of operation. The best guide to the availability of this cash, or 'working capital', is a cash flow budget (see Chapter 7). If only a balance sheet is available, an alternative is to calculate a short-term safety margin, the *acid test ratio*(5).

$$\frac{\text{liquid assets}}{\text{current liabilities}} \times 100$$

Example:

$$\frac{8000}{50\,000} \times 100 = 16\%$$

This is a test of the ability of the business to survive should all the short-term (or current) liabilities be called in. The normal definition of *liquid assets* is current assets minus stocks (i.e. cash and debtors). The normal definition of *current liabilities* is overdraft and creditors. In the past, where the business owner has had a particularly good relationship with his bank manager, it has been possible to think of the overdraft as a medium-term liability, and thus exclude it from the liquidity calculation. This is increasingly unrealistic, however: remember that any overdraft can be called in almost immediately by the bank.

In theory the acid-test ratio should be at least 100%, indicating that the business could meet short-term demands on its funds without resorting to the sale of assets and thus possibly reducing potential profitability. In practice it is rare for this to be the case in a going concern.

A further short-term ratio may also be used; the *current ratio*, which is current assets (i.e. cash, debtors and stocks) as a percentage of current liabilities:

Example:

$$\frac{59\,000}{50\,000} \times 100 = 118\%$$

A target figure often quoted is a minimum of 200%, but this is highly dependent on the type of business, and particularly on the type and level of stocks carried, and in practice is unlikely. With both the current and acid-test ratios, historical and interfirm comparisons are more important than application of arbitrary 'targets': but even so, the example business shows signs of short-term instability. Figure 3.6 shows a summary of the ratio analysis.

| *Ratio analysis* | *Opening* | *Closing* |
|---|---|---|
| **Profitability ratios (%)** | | |
| Return on owner equity | | 6.8 |
| Return on capital employed | | 5.7 |
| Gross profit margin | | 55.6 |
| Net profit margin | | 11.1 |
| **Activity ratios (days)** | | |
| Stock turnover period | | 234.0 |
| Debtor turnover period | | 36 |
| Creditor turnover period | | 45 |
| **Gearing ratios (%)** | | |
| Percentage owned | 59.2 | 54.7 |
| Long-term debt to equity | 46.1 | 45.1 |
| Interest cover | | 550 |
| **Liquidity ratios (%)** | | |
| Acid-test ratio | 6.3 | 16.0 |
| Current ratio | 162.5 | 118.0 |
| **Growth (%)** | | |
| Growth in net capital | | –5.7 |
| Growth in assets | | 2.1 |
| Growth in liabilities | | 11.8 |

**Fig. 3.6** Ratio analysis of profit statement and balance sheet.

### 3.3.4 Postscript: the example

The example business appears to be in a weak, and declining, position. The first action must be to verify the valuations of assets and to check on the nature of the overdraft. Next the reason for the high level of private drawings should be ascertained, in order to check the likelihood of it recurring next year. An investigation of profitability should also be carried out, in an attempt to raise both profits and the level of cash generated by the trading activity of the business (see Chapter 14). The nature of the debtors should be determined, in case there is a possibility of reducing the burden of unpaid bills on the business. Very tight control should be exercised over cash flow (see Chapters 7, 11 and 13).

The above remarks assume that the balance sheet in question is a historical statement. If it were a budgeted balance sheet, there would be the further option of amending the current year's plans in order to alleviate the situation.

## Notes

2. Adjusted for debtors.
3. Assuming use of annual accounts: adjustment should be made for shorter periods.
4. Also known as 'leverage'. The term 'gearing' is often applied to specific ratios – but unfortunately to different ratios by different authorities, so it is used here as a generic term.
5. Also known as the liquidity ratio. As with gearing, this term is applied to various measures, and it is used here in its generic sense.

## References

Barnard, C.S. & Nix, J. S. (1979) *Farm Planning and Control*, 2nd edn. pp. 542–551. Cambridge University Press, London.

Berry, A. & Jarvis, R. (1991) *Accounting in a Business Context.* Chapman & Hall, London.

Broadbent, M. & Cullen, J. (1993) *Managing Financial Resources.* Butterworth Heinemann, Oxford.

Bull, R.J. (1990) *Accounting in Business*, 6th edn. Butterworth Heinemann, London.

Millichamp, A. (1995) *Finance for Non-financial Managers*, 2nd edn. DPP Publications, London.

Shepherd, C.G. (1991) *An analysis of performance by size, type and form of ownership of farms.* Unpublished Diploma in Farm Management project report, Seale-Hayne Faculty, Polytechnic South West.

Tyran, M.R. (1986) *Handbook of Business and Financial Ratios.* Prentice-Hall, Hemel Hempstead.

Warren, M.F. & Pierpoint, P.R. (1991) Vulnerability of farm businesses in the far South West of England. *Farm Management*, **7**, 12.

Warren, M.F. (ed.) (1993) *Using your Accounts.* Book 2 in series: Finance Matters for the Rural Business. ATB Landbase/Seale-Hayne Faculty of Agriculture, Food and Land Use, University of Plymouth.

# Chapter 4
# Management Accounts

## 4.1 Management versus financial accounts

The previous chapter was concerned with *financial accounts*: those accounts prepared, usually by an accountant, to give shareholders, lenders and other stakeholders a view of the business performance. It also described some ways of extracting information from them in an attempt to make them more useful for decision-making. Such analysis is better than nothing, but is a very blunt instrument, particularly with respect to the profit and loss account. Financial accounts are based on historic cost (and thus may give a very false impression of stock values, for instance), and are often heavily influenced by taxation considerations (giving incentive to show low profits).

The biggest limitation of financial accounts is that they treat the business as a single, homogeneous unit, whereas most businesses are more logically thought of as combinations of profit-making activities, referred to as *enterprises* in farm management, but *profit centres* in normal parlance. Large businesses may also have further divisions termed *cost centres,* allowing allocation of the costs of production and service functions (such as a personnel department, office administration, and so on) which have no specific income stream. In smaller businesses, profit centres are usually identified with specific products: thus Home Farm comprises six profit centres, i.e. dairy, spring barley, fattening pigs, bed and breakfast, cheese, and miscellaneous. If winter barley or breeding pigs were present, they would be treated as specific profit centres in their own right, since the production processes are different from spring barley and fattening pigs, respectively. Since the forage is consumed within the business, and then by the dairy herd only, it can be treated as part of the dairy enterprise in this case, although it is more normal to find it shared between several types of livestock.

In controlling the business, it is vital to be able to monitor each of these enterprises individually. If one of the enterprises does not appear to be pulling its weight, steps can be taken to improve its performance or to replace it in part or in whole by another enterprise. Using financial accounts in making such decisions risks confusion, at the least; the information is too highly summarized to show enterprise performance. It is difficult to check the account in Figure 2.3, for instance, since there is so little information presented. By grouping outputs together (e.g. 'livestock', 'cereals') and inputs together ('materials', representing fertilizers, seeds, feeds, etc.) it enables omissions and errors of judgement to be easily overlooked. Moreover, it is impossible to relate the output of a particular enterprise to the costs directly incurred in its production.

For his profit and loss account to be truly useful, Adam thus needs to formulate it in such a way that the performance of each enterprise is shown clearly. This implies the production of *management accounts* in either gross margin or full costing form.

## 4.2 Gross margin accounts

The simpler of the two systems is *gross margin accounting*[6]. Here the revenue from each product is allocated to a profit centre or enterprise (milk, calves and cull cows to dairy, accommodation receipts to bed and breakfast, for instance). The *direct costs* involved in producing the goods or service concerned are also allocated. These are costs that can easily be identified with the product, such as the materials used in production (seed, fertilizers, and sprays for the barley crop; rennet and milk for the cheese, for instance). Sometimes it is also possible to identify labour costs incurred specifically by that product, or *direct labour*. Subtracting direct costs (also known as *prime costs*) from revenue gives a *gross margin* for each enterprise (roughly equivalent to the gross profit in financial accounts). All the other costs, termed *overhead* or *indirect costs*, are kept together, with no attempt to allocate them. The result is a *net margin*, for the whole business, found by subtracting overhead costs from the gross margin (see Figure 4.1).

| | *Total* | *Enterprise 1* | *Enterprise 2* | *Enterprise 3* | *Enterprise 4* |
|---|---|---|---|---|---|
| Revenue | 145 000 | 20 000 | 50 000 | 60 000 | 15 000 |
| *less* direct costs | 70 000 | 10 000 | 30 000 | 25 000 | 5 000 |
| ***Gross margin*** | 75 000 | 10 000 | 20 000 | 35 000 | 10 000 |
| *less* overhead costs | 40 000 | | | | |
| ***Net margin (profit)*** | 35 000 | | | | |

**Fig. 4.1** Gross margin accounting.

Gross margin accounting gives a degree of comparison of profitability between enterprises, without going to the full extent of attempting to allocate all the overhead costs. It is especially useful in small firms, where time and facilities for detailed record-keeping may be limited, and in businesses where many resources are shared (in the Home Farm business, for instance, the rent, labour and machinery costs). It is open to criticism in not producing a net margin for each enterprise, and thus failing to make allowance for different levels of use of the shared resources: the Home Farm dairy will use more labour than the spring barley, for instance. On the other hand, in most small businesses the decision-maker is close enough to the production processes to be able to make subjective judgements about these differences, without the formality of a strict calculation.

There are other limitations. Gross margin accounting does not give much help in valuing stocks, where cost of production is a basis, since it accounts for only part of those costs. Nor is it helpful in pricing decisions in the long term: ideally one would know the cost of production in order to use it as a 'bottom line' guide to pricing (see Section 4.5).

## 4.3 Full cost accounts

An alternative approach is to complete a profit and loss account for each enterprise individually. The net profits, or *net margins*, of the enterprises are accumulated to give the net profit of the whole farm. The result is a *full costing* or *full cost absorption* account (see Figure 4.2). Despite its apparent simplicity, however, the full costing approach is fraught with difficulties. Awkward, and usually arbitrary, decisions have to be made concerning the allocation of overhead expenses between enterprises. Labour and machinery costs must be estimated in great detail and allocated carefully. In a typical mixed rural business, this implies the careful recording of the amount of labour and machinery used in each enterprise. Elaborate timesheets must be kept by all employees, since the person who fed the cattle and counted the sheep before coffee-break may well have top-dressed wheat, barley and grass before lunch, and spent the afternoon mowing the caravan site – and all with the same tractor.

| | *Total* | *Enterprise 1* | *Enterprise 2* | *Enterprise 3* | *Enterprise 4* |
|---|---|---|---|---|---|
| Revenue | 145 000 | 20 000 | 50 000 | 60 000 | 15 000 |
| *less* direct costs | 70 000 | 10 000 | 30 000 | 25 000 | 5 000 |
| ***Gross margin*** | 75 000 | 10 000 | 20 000 | 35 000 | 10 000 |
| *less* overhead costs | 40 000 | 5 000 | 10 000 | 10 000 | 15 000 |
| ***Net margin (profit)*** | 35 000 | 5 000 | 10 000 | 25 000 | –5 000 |

**Fig. 4.2** Full cost absorption accounting.

Leaving aside the arguments about the likely accuracy of timesheets, the collation of this information will add considerably to the manager's paperwork burden. In most rural businesses, the benefit gained does not justify the huge effort involved. The exceptions are those where either the task is relatively easy (such as self-contained enterprises in intensive livestock, horticulture, leisure or food manufacture), or where the potential benefits are particularly large (where, for instance, the information is to be used in education or research, or for determining the price to bid in tendering for a contract). An alternative, particularly for more general overhead costs, is to use a formula for allocation, such as dividing office costs between enterprises in proportion to their materials costs, but this can result in allocations that are so arbitrary as to be at best useless, and at worst dangerous.

A further, more serious criticism of the use of full costings in decision-making is that there is a danger of treating all costs as equally variable[7]. Suppose a farmer were to use full-cost budgets in order to test the pros and cons of scrapping a potato enterprise in favour of winter wheat (see Figure 4.3). If he is unwary, he may well deduce that the proposed change would increase the profits of his business by £50/ha (£70 – £20). But many of the costs associated with potato production will still be there, whether or not potatoes are grown on the farm.

| | *Main-crop potatoes (£/ha)* | *Winter wheat (£/ha)* |
|---|---|---|
| *Output* | 2000 | 600 |
| *Costs* | | |
| Fertilizer | 250 | 100 |
| Seed | 500 | 40 |
| Sprays | 200 | 70 |
| Harvesting, casual labour, transport, etc. | 300 | 5 |
| Regular labour | 170 | 20 |
| Machinery, running cost | 150 | 40 |
| Machinery, depreciation | 210 | 55 |
| Miscellaneous overhead costs | 200 | 200 |
| | 1980 | 530 |
| ***Net margin*** | 20 | 70 |

**Fig. 4.3** Budgets in full-cost form for potato and wheat enterprises.

To take a single example, the change may give rise to no savings in the cost of regular labour. Unless the farm is heavily overstaffed, the replacement of a small area of potatoes by wheat is not likely to enable the business to dispense with the labour of one person for the whole year. The wages will have to be paid, whether or not potatoes are produced. If this cost is eliminated from the calculation a rather different picture emerges (see Figure 4.4). Therefore, the effect of dropping potatoes in favour of wheat is to *reduce* profit by £100/ha (£190 – £90).

## 4.4 Allowing for cost behaviour

### *4.4.1 Marginal costing*

*Marginal costing* is a variant of gross margin accounting which takes account of the ways that different costs behave. It distinguishes between variable costs and fixed costs[8]. A *variable cost* is one that varies in direct proportion to the scale of production: thus in the example above, wheat seed cost would double if wheat area were doubled – it is a variable cost. The other costs are *fixed costs*, as in the case of labour in the example above: they will change eventually as the scale of the enterprise rises or falls, but not continuously. Deducting the variable costs of an enterprise from its

| | Main-crop potatoes (£/ha) | Winter wheat (£/ha) |
|---|---|---|
| *Output* | 2000 | 600 |
| *Costs* | | |
| Fertilizer | 250 | 100 |
| Seed | 500 | 40 |
| Sprays | 200 | 70 |
| Harvesting, etc. | 300 | 5 |
| Machinery, running costs | 150 | 40 |
| Machinery, depreciation | 210 | 55 |
| Miscellaneous overhead costs | 200 | 200 |
| | 1810 | 510 |
| ***Net margin*** | 190 | 90 |

**Fig. 4.4** Revised enterprise budgets, excluding regular labour.

revenue gives the *contribution* (i.e. the contribution of the enterprise towards profit). Fixed costs are not allocated, but are deducted from the total of the enterprise contributions to give a net margin for the whole business (see Figure 4.5).

| | Total | Enterprise 1 | Enterprise 2 | Enterprise 3 | Enterprise 4 |
|---|---|---|---|---|---|
| Revenue | 145 000 | 20 000 | 50 000 | 60 000 | 15 000 |
| *less* variable costs: | | | | | |
| materials | 70 000 | 10 000 | 30 000 | 25 000 | 5 000 |
| overtime and piecework | 3 500 | 2 000 | 1 000 | | 500 |
| variable power costs | 2 500 | 1 500 | | 1 000 | |
| | 76 000 | 13 500 | 31 000 | 26 000 | 5 500 |
| ***Contribution*** | 69 000 | 6 500 | 19 000 | 34 000 | 9 500 |
| *less* fixed costs: | | | | | |
| basic salaries | 10 000 | | | | |
| fixed power costs | 7 000 | | | | |
| administration | 8 000 | | | | |
| rent and rates | 7 500 | | | | |
| finance charges | 1 500 | | | | |
| | 34 000 | | | | |
| ***Net margin (profit)*** | 35 000 | | | | |

**Fig. 4.5** Marginal costing.

Marginal costing is valuable for decision-making, as one can now compare enterprise contributions in a 'what-if?' context: the effect of changing the size of an enterprise can be clearly seen by multiplying the contribution by an appropriate factor (e.g. to test the effect of increasing the scale of enterprise 2 by 10%, multiply its contribution by 10% to give £1900). It is particularly useful in short-term pricing decisions (see Section 4.5).

Caution is needed, however, since the variable/fixed distinction applies only over a certain range of production scale (the *relevant range*). Most so-called fixed costs are in fact 'stepped' or 'lumpy' costs. For instance, the cost of employing a full-time worker is fixed over a certain range, but there comes a point where overtime payments are incurred, and then another point where the only way of coping with increased workload may be to employ more labour, implying a sudden step up in labour costs[9] (see Figure 4.6). Because the fixed/variable distinction only holds good for the relevant range, marginal costing is not suitable for stock valuation and longer-term pricing decisions.

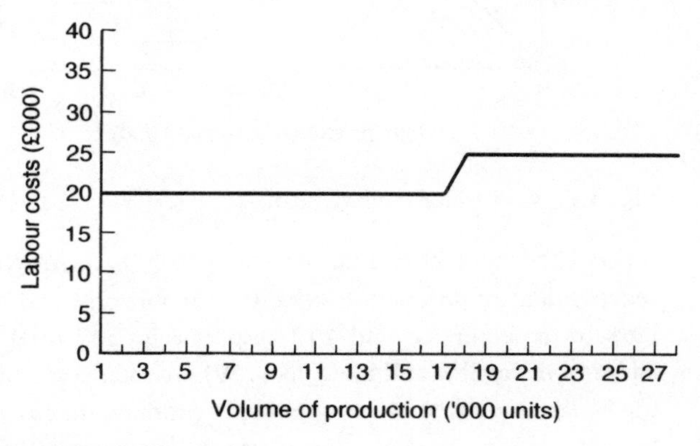


**Fig. 4.6** 'Stepped' cost – example.

## 4.4.2 Cost-volume-profit analysis

Cost-volume-profit (CVP) analysis is a simple way of illustrating, and to some extent analysing, the behaviour of costs. It uses simple assumptions about a specific production process: one product (or a constant sales mix), one variable cost (or group of costs), with the remaining costs fixed over the range of production relevant to the decision. All costs and revenues behave in a linear fashion, stock values do not change, and the only factor affecting revenues and costs is volume of production. If these assumptions are not too limiting, CVP charts can be useful as a decision guide, allowing 'what if? . . .' questions to be explored.

The graph in Figure 4.7 is based on the following information for product A:

| | |
|---|---|
| Expected sales (number of units) | 6000 |
| Sale price per unit | £10 |
| *Variable cost* | |
| Materials cost per unit | £5 |
| *Fixed cost* | |
| Administration costs | £25 000 |

Revenue, variable cost and fixed cost are simply plotted on a graph with value on the vertical axis, and volume of production on the horizontal axis. Fixed cost is

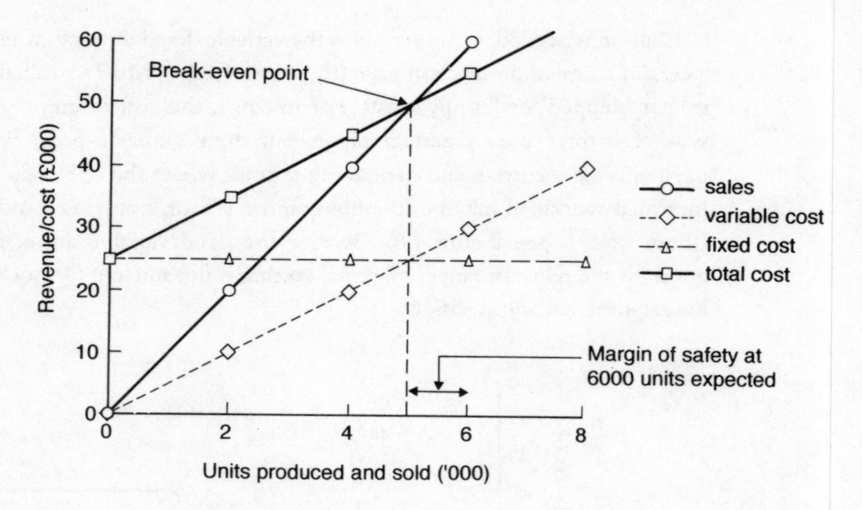


**Fig. 4.7** Cost–volume–profit chart.

represented by a horizontal line at £10 000, since over the relevant range it is independent of production volume. Variable cost and revenue are represented by lines of gradients 5:1 and 10:1 respectively, and total cost is simply the vertical addition of variable and fixed costs. Where total cost and total revenue lines cross is the *breakeven point*: at lower volumes of production, costs exceed revenue and thus a loss results; at higher volumes, an excess of revenue over costs gives rise to a profit. In this example, the breakeven volume is 5000 units.

In the first instance, the graph can be used to check the position of the break-even point with respect to normal or expected output. The difference between expected sales and break-even sales is know as the *margin of safety*: if, in this example, there was good reason to expect a production of 6000 units in the period concerned, then the margin of safety would be 6000 – 5000 = 1000 units. This can also be expressed as a percentage of the expected sales (in this instance 16.6%). More usefully, the effect of variations in cost and price can be shown by varying the gradient of the lines accordingly. Figure 4.8 shows the effect on break even of reducing the price by £1 per unit: an increase in break-even point to 6250 units, which is beyond the level of expected sales.

Graphs can be clumsy devices for analysis, and it is often better to reduce the break even to a formula. Profit is calculated by sales revenue less variable and fixed costs, or:

$$P = (S \times n) - (VC \times n) - FC \tag{4.1}$$

where $P$ = profit
$S$ = sales price per unit
$n$ = number in units
$VC$ = variable cost per unit
$FC$ = total fixed cost

At break even, profit is zero:

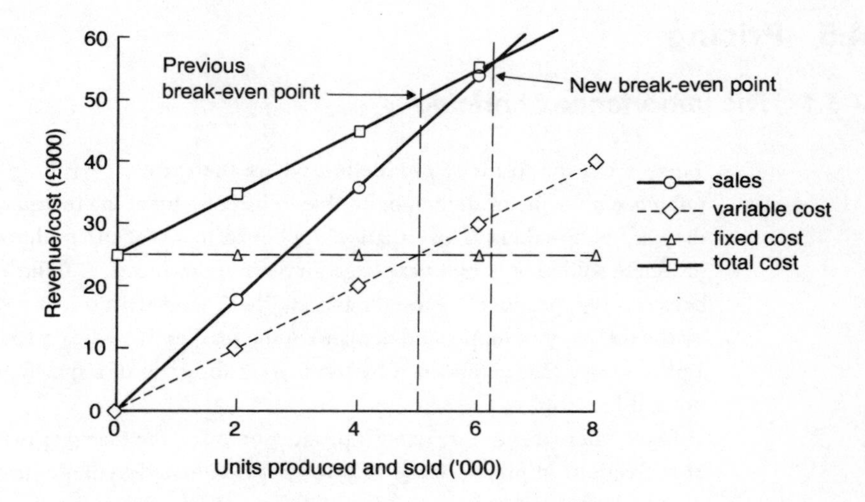


**Fig. 4.8** Effect of reducing sale price by £1.

$$(S \times n) - (VC \times n) - FC = 0 \tag{4.2}$$

Thus:

$$(S \times n) - (VC \times n) = FC \tag{4.3}$$

$$n(S - VC) = FC \tag{4.4}$$

$$\text{and } n = \frac{FC}{S - VC} \tag{4.5}$$

In other words, break-even volume $= \dfrac{\text{fixed cost}}{\text{contribution per unit}}$

For break-even price ($S$) we can go back to Equation (4.3) and divide both sides by $n$, giving:

$$S = \frac{VC \times n + FC}{n} \text{ or } \frac{\text{total costs}}{\text{units sold}}$$

Thus in the example above, break-even volume is 25 000 ÷ 5 = 5000 units, and break-even price at the expected sales volume of 6000 units is 55 000 ÷ 6000 = £9.17 per unit.

Given these formulae, it is a simple matter to investigate changes in fixed cost, variable cost and sales price. For instance, if fixed costs could be reduced to £20 000, break-even volume would fall to 20 000 ÷ 5 = 4000 units, and break-even price would fall to £50 000 ÷ 6000 = £8.33. A reduction in price per unit of £1 would increase break-even volume to 25 000 ÷ 4 = 6250 units: this would be the new minimum production level before beginning to make a profit (see Figure 4.8).

## 4.5 Pricing

### 4.5.1 The importance of pricing

There are some forms of production where the manager has little opportunity to influence price: he or she might be able to haggle a bit at the margins, but in the end they are price-takers. This situation applies to many of the traditional agricultural products, sold in bulk by a very large number of producers, with little differentiation between one producer's product and another's, and with prices determined solely by the interplay of supply and demand in the market. If a wheat producer tries to set a price above that determined by the market for grain of a specific quality, he will not sell his crop.

Even where it is a more specific product on offer, the manager may still have very little freedom in price-setting. In a market dominated by large firms, for instance, the smaller ones may have to adopt a 'follow-the-leader' approach. As an example, if Adam were considering branching out into agricultural contracting, and if several large professional contractors were already in operation in the area, the price he could quote for equivalent services would have to be closely linked to theirs.

However, in the case of products which have an element of uniqueness about them, the pricing decision is one of the most important to be taken. This uniqueness may derive from various factors, including a quality of the product itself (e.g. an unusual cheese), the way it is distributed (e.g. mail-order lamb), the location (e.g. a farm shop or a tourist visitor attraction), and so on. It implies that there will be no leader to follow in pricing. In these circumstances (which apply to a large proportion of farm diversifications, and the majority of non-farm businesses), pricing becomes a crucial part of the decision-making process. Price is a key factor in customers' decisions, and getting it wrong is often one of the main factors in business failure. But how does one decide on the prices?

### 4.5.2 Cost-based pricing methods

Perhaps the most obvious approach would be to find a price which will cover all the costs of production and make a profit. This implies using a full costing to determine the total cost of the product, with overhead costs allocated (see Section 4.3), and then adding a profit margin or mark up. If Adam were hoping to market a new cheese, for instance, and worked out that the full cost of production would be £4 per kg, then he would add a mark up of, say, 10%, to arrive at a selling price of £4.40.

Variants of this approach include *break-even pricing*, which involves working out the break-even price for the product (see Section 4.4.2), and then applying a mark up. This allows for variations in cost behaviour, rather than assuming the cost relationships are independent of the scale of production. *Profit-orientated* pricing starts from the profit required from the profit centre (either in absolute terms or as a rate of return (see Chapter 18)), and uses the full costing to work back to the price needed to achieve this.

All these methods are subject to the limitations of full cost absorption, and particularly the problem of arbitrary allocation of general overhead costs. One answer might be to use *direct cost pricing*, where the mark up is intended to cover both profit margin and overheads. Using the example above, Adam may not know precisely what share of the overheads should be allocated to his new line, but might be able to identify direct costs of production of £3.40 per kg. He might then apply a mark up of 30% to the direct costs (as opposed to the 10% used on full costs) to arrive at a price of £4.42 per kg.

Such pricing methods also leave open the question of how to decide the appropriate mark-up percentage in the first place. They also ignore the fact that there can be times when it is right to charge a price which does not cover all the costs, let alone make a net margin. In the short term, it might be better to charge a price which generates sales and covers fixed costs, than to stick to a price which covers all costs in theory, but in practice results in no sales. In the above example, Adam is committed in the short term to the costs of employing a cheese-maker, and financing specialist equipment: he is also producing a perishable product. Here the 'relevant' cost is more important than the full cost: a concept explored further in Chapter 9.

### 4.5.3 Market-orientated approaches

Another criticism of cost-plus methods is that they ignore the workings of the market. A manager setting a price for a product needs to be aware that this decision will affect consumer demand for that product – and will affect other firms' decisions as to the degree they try to compete with equivalent products. Factors affecting demand for a product will include the numbers of potential consumers of the product, the levels and distribution of income of those consumers, and the tastes of consumers (local customs, regional food consumption habits, and so on). The nature of the demand itself is also important, particularly its responsiveness (elasticity) to price. Some products are highly sensitive to price, for example some of the more 'luxury' products such as restaurant meals or exotic foods. Others, such as basic food commodities, are much less responsive (see Brassley, 1997 or any introductory economics text to learn more about demand and supply, and elasticities).

*Market-orientated pricing* attempts to take these considerations into account in trying to judge 'what the market will stand'. Its use implies considerable research of the market, which may include searches of statistics (for instance, population and employment census data, and household expenditure statistics), surveys of consumers, test marketing on a small scale, checking on competitors' advertisements, and even shadier practices such as posing as potential customers of competing firms in order to find out about *their* pricing practices.

### 4.5.4 The manager's judgement

There is, unfortunately, no formula for market-orientated pricing, and it cannot be a straight alternative to cost-based pricing since, in the long run at least, it is important to make a margin over costs. In practice, the manager is likely at first to have to make a considered judgement on price, taking into account both cost and market considerations, as well as other managerial factors such as the broader objectives of the business (for some business owners, for instance, the pleasure of offering a good, cheap service to local people may override the urge to charge the maximum price possible).

In setting initial prices, the manager may wish to employ specific pricing tactics, such as *penetration pricing* (charging a lower price than the competition in order to get a toe-hold in the market); *skimming pricing* (charging a particularly high price for a new product, while the novelty is attractive to customers, and before the competition has had time to react); and *loss-leader pricing* where one product is offered at a very low price in order to attract customers who might then be tempted to buy other products.

Thereafter will come a period of refinement in the light of experience, perhaps eventually reducing to a mark-up system for convenience, but always maintaining alertness to market conditions.

## Notes

6. 'Gross margin' here has a specific agricultural definition: see Section 5.1.1.
7. This also applies to some extent to the classic gross margin approach.
8. Both 'variable cost' and 'fixed cost' have specific agricultural definitions: see Section 5.1.1.
9. Unless, of course, hourly-paid casual labour is employed, which is a true variable cost.

## References

Attrill, P. & McLaney, E. (1994) *Management Accounting: An Active Learning Approach.* Blackwell Business, Oxford.

Brassley, P. (1997) *Agricultural Economics and the CAP.* Blackwell Science, Oxford.

Broadbent, M. & Cullen, J. (1993) Chapter 4 – Cost classification, and Chapter 9 – Pricing. In: *Managing Financial Resources*. Butterworth Heinemann, Oxford.

Giles, A.K. (1987) Net margins. *Farm Management*, **6**, 6.

Hill, B. & Ray, D. (1987) *Economics for Agriculture: Food, Farming and the Rural Economy.* MacMillan, Basingstoke.

Millichamp, A. (1995) Unit 5 – Costing and pricing a product, and Unit 7 – Marginal costing. In: *Finance for Non-financial Managers*, 2nd edn. DPP Publications, London.

# Chapter 5
# Variations from the Norm

## 5.1 Farm management accounting

Management accounting has specific variants in many industries: the tourism and hospitality industry has its own needs in terms of stock and cost control, for instance (see, for instance, Atkinson, Berry and Jarvis (1995)). In most cases these variants are merely small adaptations of the basic model. Farming, however, at least in the UK, went its own way for several decades, and although the standardization of accounting practice and the spread of computerisation have begun to bring it back to the fold, it retains some peculiarities in management accounting.

### 5.1.1 'Agricultural' gross margins

When management accounting was introduced in UK agriculture, economists sought a pragmatic solution which:

- Recognised the difficulty of allocating overhead costs in businesses with many shared costs, but few or no office staff
- Allowed fixed and variable costs to be treated differently
- Resulted in a margin for each enterprise showing its contribution to net profit of the business.

In other words they wanted a system which met the accounting need for ease of allocation, while satisfying the decision-maker's need to consider only those costs that would change as a result of a given decision. The resulting system has been used for decades in farming, and is easily adapted to the needs of other businesses. It is known as the *agricultural gross margin* system, and brings the Great British Compromise into play (see Figure 5.1).

A *variable cost*, for agricultural gross margin purposes, is redefined as a cost which:

- Is easy to allocate to a specific enterprise *and*
- Varies directly with small changes in the size of the enterprise.

In other words, the agricultural version of a 'variable cost' is one which is *both* direct and variable in the normal meaning of the words. Remember also that the word 'enterprise' is used here to mean a specific profit centre, rather than a whole business (as is commonly understood in the non-agricultural world).

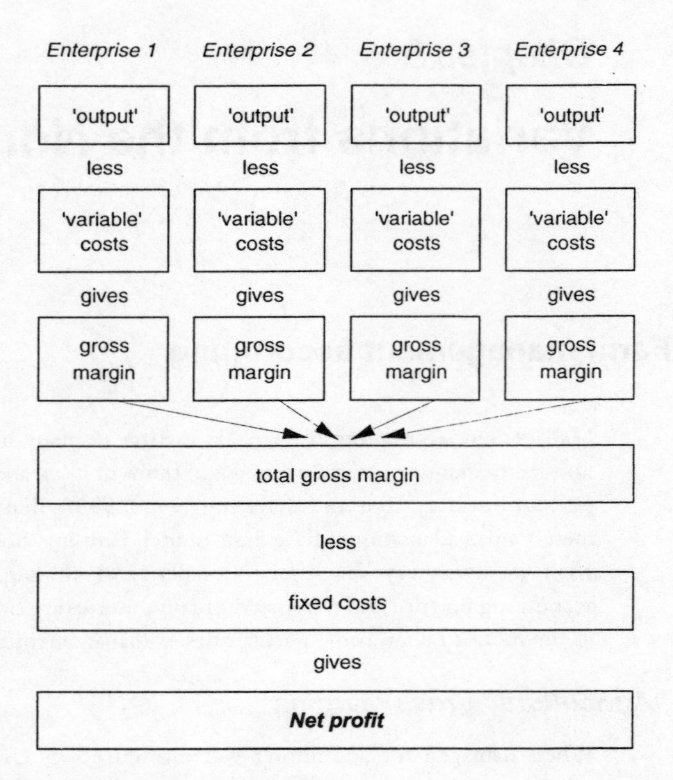


**Fig. 5.1** The agricultural gross margin system.

A *fixed cost*, for agricultural gross margin purposes, is any cost which does not meet *both* of the above criteria. The most common examples are regular labour, machinery depreciation and running costs, property rent and repairs, finance charges and general administration costs. Note that most variable costs are also direct costs, and most fixed costs are also overheads, but regular labour, machinery depreciation and running costs are never allocated. Even if a person's labour can be entirely allocated to an enterprise (a cheese-maker or a dairy herd manager, for instance), their employment as regular labour means that small changes in size of the enterprise are unlikely to have any effect on the labour bill, except marginally through overtime. Similar arguments apply to machinery.

As an example, the cost of wheat seed is easy to allocate to the wheat enterprise, and will vary with small changes in enterprise size; it is therefore a variable cost. The cost of diesel fuel used in drilling the wheat will vary with enterprise size, but will be difficult to allocate without very detailed timesheets. It is thus classified as a fixed cost. The cost of depreciation of the grain store is not likely to vary with small changes in the cereal hectarage, nor, if there is more than one cereal enterprise, will the cost be very easy to allocate. It is therefore a fixed cost. Some typical agricultural variable and fixed costs are shown in Table 5.1.

Another difference between the agricultural gross margin system and marginal accounting is that the revenue of the enterprise is now called the *output*. In most respects, output is calculated in the same way as the revenue in a marginal cost-

**Table 5.1** Examples of variable and fixed costs.

| *Variable costs* | *Fixed costs* |
|---|---|
| Feedstuffs (including forage) | Regular labour |
| Veterinary, medical and A1 fees | Power and machinery running costs (except contract hire) |
| Fertilizers | Machinery and building depreciation |
| Seeds | Rent and/or landowning expenses |
| Sprays | Rates |
| Casual labour | Interest charges |
| Contract hire of machinery | Miscellaneous overheads, e.g. office expenses, accountant's fees, fencing costs, property insurance |
| Miscellaneous, e.g. baler twine, recording services, transport | |

ing system, taking account of internal transfers and benefits in kind as well as sale proceeds. Valuation changes of products, livestock and crops in the ground are usually accounted for within the output (and, theoretically, input stock changes are dealt with in the variable cost calculation, though in practice this is often neglected).

One peculiar aspect of the output calculations used in most farm management handbooks, costing services and consultants' reports is that, for reasons that are no longer relevant, the cost of purchase or inward transfer of livestock is deducted in calculating the output, rather than being treated as a variable cost. This practice is deeply entrenched in agricultural accounting, although things are gradually changing: it is important to be aware, particularly in making comparisons between accounts, of the way in which livestock inputs are treated.

### 5.1.2 Livestock enterprise gross margins

Crop enterprise gross margins are usually quite straightforward. Livestock gross margins can be more difficult, though. As well as variations in the treatment of purchased and transferred livestock mentioned above, there is also an important distinction between *trading livestock* – those which are effectively 'work in progress' being grown on to sell – and *breeding livestock*, where the main revenue comes from byproducts of the breeding process: milk, eggs, youngstock, etc.

With breeding livestock it can be difficult to determine the size of the enterprise for gross margin calculations: if the size of a breeding herd or flock is likely to vary a little during the year, do your best to estimate the typical herd or flock size. Thus if the dairy herd fluctuates between 100 and 107 cows, but is usually nearer the latter, an appropriate average may be 105.

Where a breeding and a fattening enterprise are highly interdependent (for instance a pig-breeding herd producing weaners for fattening on the same farm), it can be appropriate to calculate a single gross margin for the two. In other enterprises, however, the progeny from breeding stock may be kept on to form quite separate enterprises from the breeding unit. Rearing dairy replacements is an example, or keeping back bull calves to fatten for beef. The breeding could exist without the rearing, or vice versa, so they should be treated as separate enterprises from the point of weaning.

It is best to think of a young stock enterprise in terms of the number produced each year. Thus if enough heifer calves are normally retained to produce 25 down-calving heifers a year, the size of the enterprise is expressed as 25 replacement units. If young stock are purchased to rear or fatten, exactly the same method should be used.

## 5.2 Partnerships and companies

So far in this book the sole proprietor form of business ownership has been assumed. Other forms of ownership are subject to the same general financial management principles, but differ in detail. The purpose of this section is to give a brief introduction to the two most common forms of ownership and their influence on financial accounts.

### *5.2.1 Partnerships*

In its most basic form, a partnership consists of two or more people working together in order to make a profit for themselves. Ideally, their association will be formally constituted by means of a legally binding partnership agreement which will set out the way in which profits will be shared, voting procedures, and so on. Each partner is jointly and severally liable for the debts of the partnership, which means that if the business gets into financial difficulties, each of the partners has to participate in paying off the debts to the full extent of their personal assets. The actions of one partner could bankrupt another. As long as there is at least one fully liable partner, one or more partners may elect to be 'sleeping' partners, with their liability limited to the amount they have invested in the partnership. In doing so, however, they sacrifice the formal right to vote on decisions.

There are two main motives for forming a partnership. The first is to enable individuals to pool their resources – usually capital (including machinery and property) but also physical labour and specific abilities. The second is to reduce the collective tax liability of the partners – usually important only in the context of a family business, since the tax savings arise through the principal of the business sharing income with other members of the partnership. Other motives for forming a partnership include spreading risk, sharing responsibility, gaining reflected 'respectability' by taking a well-known and credible partner and providing a means whereby younger members of a family may gradually build up investment in, and control of, the family business.

The accounts of a partnership are identical to those of an equivalent sole proprietorship, except that additional information has to be provided concerning the division of profits and capital between the partners. An extra account called the *profit and loss appropriation account* shows how profits are shared: the division will depend on what is written in the partnership agreement. This may allow for interest on the capital injected by the partners (allowing for different levels of investment by the partners); for a 'salary' to allow for the fact that the partners may put different amounts of work into the business, and for a split of the remaining profits in agreed proportions.

Suppose, for instance, that Adam Haynes' mother had formed a partnership with Adam, rather than lending him the money. If we jump ahead in time, referring to the 'actual' profit and capital statements for Home Farm at the end of the year (Figures 11.6 and 11.7), their profit and loss appropriation account might be as shown in Figure 5.2. This assumes that Mrs Haynes contributed £34 000 of the initial net capital, with Adam supplying the remainder, and also injecting a further £200 during the year. The partnership agreement for Adam and his mother specifies that the profit should first reward the capital contributions of the pair by an interest charge of 3.5%, then Adam's work with a nominal 'salary' of £8000, with the remainder divided fifty–fifty. Note that if Adam's 'salary' had been £18 000 rather than £8000, the final item would be sharing a residual deficit, rather than profit.

| | £ | | £ |
|---|---|---|---|
| Interest on initial capital | | | |
| Mrs Haynes (3.5% of £34 000) | 1 190.00 | Net profit brought down (no loan interest) | 20 749.00 |
| A. Haynes (3.5% of £58 744) | 2 056.04 | | |
| 'Salary': A. Haynes | 8 000.00 | | |
| Share of remaining profits (50% each) | | | |
| Mrs Haynes | 4 751.48 | | |
| A. Haynes | 4 751.48 | | |
| | 20 749.00 | | 20 749.00 |

**Fig. 5.2** Example of profit and loss appropriation account.

The other account to be affected is the capital account, which must show how the capital stakes of the partners are influenced by the profitability of the business, and the various external demands on it. This is often divided to show a joint capital account, showing initial and subsequent injections of capital, and a joint current account showing the accumulation of profits. Figure 5.3 shows joint capital and current accounts for the Home Farm situation described above. The current account indicates that both parties have increased their capital stakes.

The accounts show one of the possible disadvantages of 'equity' as opposed to loan capital: since the business is making good profits compared to the capital invested, Mrs Haynes is receiving more from the business than she would from loan interest. She also has a right to be fully involved in policy-making in the business which, depending on relations between Adam and his mother, could be a high price to pay. On the other hand, in a poor year she would have to share in the hardship,

| *Capital accounts* | | | | |
|---|---|---|---|---|
| Mrs Haynes | | 34 000 | Initial capital | 92 544 |
| A. Haynes: initial | 58 544 | | Capital injection | 200 |
| plus capital injection | 200 | 58 744 | | |
| | | 92 744 | | 92 744 |
| *Current accounts* | | | | |
| Mrs Haynes | | | | |
| Private drawings | | | Balance brought down | 0 |
| Tax | | | Interest on capital | 4 080 |
| Benefits in kind | | 2 000 | Share of profits | 810 |
| Closing balance | | 2 890 | | |
| | | 4 890 | | 4 890 |
| A. Haynes | | | | |
| Private drawings | | 3 520 | Balance brought down | 0 |
| Tax | | 856 | Interest on capital | 7 049 |
| Benefits in kind | | 2324 | Salary | 8 000 |
| Closing balance | | 9 159 | Share of profits | 810 |
| | | 15 859 | | 15 859 |

**Fig. 5.3** Example of joint capital and current accounts.

whereas a loan agreement would insist on regular payments of interest, whatever the financial circumstances of the business. Moreover, the cash flow of the business would be helped by not having to make regular repayments of loan capital (£3000 in the first year).

Note that in normal circumstances one would expect the partnership to be wider, including both Adam's wife and his father, but this would have made the example unnecessarily complex.

### 5.2.2 Limited liability companies

The fact that partners are jointly and severally liable for debts is often a significant deterrent to their formation, and an alternative is the limited liability company. Here the investors each make a contribution to the capital of the company at its inception, and profits are subsequently allocated in proportion to those shares. Profits that are not distributed to shareholders are retained as 'reserves', helping to finance the company in much the same fashion as does the current account in the partnership.

In the typical company, each shareholder's liability is limited to that money he or she has invested, i.e. original investment plus the appropriate share of accumulated reserves. This reduces the risk to the shareholder, but at the potential expense to creditors, who, if the business fails, will only be able to recover what is left in the company, irrespective of the wealth of the individual shareholders. For this reason small companies often find it difficult to obtain credit, or directors are asked to provide personal security such as the title deeds of a house, or a personal guarantee.

Companies can take many forms, but the most common is the private company, usually owned by the members of a family. The shareholders can be as few as two. All or some of the shareholders will be elected as directors, giving them the right to participate in decision-making between general meetings (otherwise the only way in which shareholders can formally influence matters). The formation of a company requires formal registration with Companies House, involving registration and legal fees, and some loss of privacy, in that accounts and other information must be filed annually with the Registrar of Companies and are open to public scrutiny.

Apart from limited liability, the attractions of forming a company are much the same as those identified above for a partnership. Income tax is levied on the company, rather than the individual shareholders, in the form of Corporation Tax, giving some potential tax savings for the more profitable farm businesses (Foreman, 1996, pp. 530–552). A valuable benefit is the aspect of continuity: while a partnership is automatically dissolved by the death of one of the partners, a company, providing that at least two shareholders are left, can live on. On the other hand, the company structure can reduce the freedom of action of the business and/or incur considerable costs when significant change is required.

In the accounts, the most noticeable difference is likely to be that a statement of share capital and reserves appears in the balance sheet, in place of 'net capital'. Usually the only 'reserve' is the *profit and loss account reserve*, indicating accumulated surpluses over the life of the business, but others such as a *general reserve* may also appear. Figure 5.4 shows an example. Share capital plus reserves are equivalent to the net capital of a sole trader or partnership and can be used as such in ratio analysis.

| *Financed by* | |
|---|---|
| ***Share capital*** | |
| (authorised and issued) | 1 000 |
| ***Reserves*** | |
| Provision for future taxation | 2 500 |
| General reserve | 9 034 |
| Profit and loss account | 80 210 |
| | 92 744 |

**Fig. 5.4** Capital account for company – example.

It is likely that the profit and loss account will include 'directors' salaries' as an expense. Since sole traders and partnerships are not allowed to include salaries of owners in their financial accounts, profit comparisons between farms of various forms of ownership (and between company farms and survey data) must be made with care. These salaries, together with 'dividends' or payouts to the shareholders (shown in an appropriation account) are roughly equivalent to the private drawings of a non-corporate business.

### 5.2.3 Co-operatives

The co-operative is essentially a variation on the limited liability company, whether or not it is constituted under Company law or under Friendly Society law. The essence of a co-operative is that it exists to serve its members, rather than make a return for shareholders. Although there are many examples of people forming a co-operative as their only business (e.g. the 'worker co-operative' that was much-vaunted in the 1970s), the majority tend to be formed to provide services to support members in their business or personal life. Examples of the latter include credit unions (co-operative banks), community enterprises, bulk-buying food groups, and even baby-sitting circles.

Examples of business support co-operatives include marketing groups (e.g. regional food producer groups, farmhouse accommodation groups, grain marketing groups), buying groups (e.g. providing supplies for farm, tourist, food producer businesses at lower prices than elsewhere) and service provision (e.g. machinery rings, manpower groups). In each case, the main purpose of the co-operative is not to make a profit to be distributed to shareholders, but to keep prices for services or commodities low.

Where financial and management accounting are concerned, there is little difference between the needs of a co-operative and those of a limited company.

## 5.3 Voluntary/non-profit organizations

Both in his social and professional life, Adam is likely to be involved with various forms of not-for-profit organizations: charities, clubs and other associations which exist for purposes other than making money for the participants. Examples might include the rugby club, the church, the school, the Chamber of Trade, Cancer Relief, Age Concern, Oxfam, and so on.

In terms of management accounting, there is little difference in the needs of not-for-profit organizations and businesses. Unless the organization is very small, there will still be a need to monitor and analyse costs and revenues in order to make sure that the resources of the organization are being well husbanded. The approach to pricing may vary, in that some organizations exist to keep prices low (for instance a charity established to provide low-cost rural housing) rather than to make maximum profit. This is little different to the business owner-manager who has objectives other than maximum profit, however, and is more a difference of attitude than of method.

### 5.3.1 Income and expenditure accounts

Some differences in method do arise in financial accounting. Not-for-profit organizations are usually subject to the same financial reporting requirements as businesses, and the accounts are prepared using the same conventions and principles.

The equivalent of the business profit and loss account is the *income and expenditure account*, of which an example is shown in Figure 5.5. Note that the emphasis is now not on profit, but on surplus of income over expenditure.

| **Barsetshire Rural Business Club**<br>**Accounts to year ended 31 March 199Y** | | | |
|---|---|---|---|
| 199X | | | 199Y |
| £ | *Income and Expenditure Account* | £ | £ |
| | *Income* | | |
| 900.00 | Sponsorship | 950.00 | |
| 470.00 | Subscriptions | 495.00 | |
| 85.58 | Interest received | 102.36 | 1547.36 |
| | *Expenditure* | | |
| 595.00 | Speakers' expenses | 598.00 | |
| 700.00 | Room hire and refreshments | 752.00 | |
| 58.00 | Annual outing | 90.00 | |
| 72.58 | Printing, postage and stationery | 75.35 | 1515.35 |
| 30.00 | ***Surplus*** | | 32.01 |
| £ | *Balance sheet* | £ | £ |
| | *Assets* | | |
| 1528.65 | National Savings Bank | 1756.45 | |
| 428.93 | Bank deposit account | 327.25 | |
| 0.00 | Sundry debtors | 125.00 | |
| 621.42 | Bank current account | 198.31 | |
| 2579.00 | | | 2407.01 |
| | *Liabilities* | | |
| 579.00 | Sundry creditors | 375.00 | |
| | *Reserves* | | |
| 500.00 | Scholarship reserve | 500.00 | |
| 1470.00 | Capital balance brought forward | 1500.00 | |
| 30.00 | Add surplus on Income and Expenditure A/c | 32.01 | |
| 2579.00 | | | 2407.01 |

**Fig. 5.5** Not-for-profit organization – example accounts.

## 5.3.2 Balance sheet

The balance sheet is prepared on the same principles as that of a business. Reserves will include the accumulated surplus from the current and previous years, but may also include reserves for specific purposes such as a building fund or a hardship fund.

# References

Anthony, R.N. & Young, D.W. (1984) *Management Control in Non-profit Organizations*, 3rd edn. Irwin, Illinois, United States.

Atkinson, H., Berry, A., Jarvis, R. (1995) *Business Accounting for Hospitality and Tourism.* Chapman and Hall, London.

Blume, H. & Norton, M. (1980) *Accounting and Financial Management for Charities.* Directory for Social Change, London. [Out of print, but still available in libraries.]

Bull, R.J. (1990) *Accounting in Business*, 6th edn. Butterworth, London.

Centre of Management in Agriculture (1986) *Practical Share Farming.* British Institute of Management, Corby.

Foreman, A. (1996) *Allied Dunbar Tax Handbook 1996–97.* Pitman, London. Published annually.

Furlong, L.A.C. (1987) Farming as a limited company. *Farm Management*, **6**, 5.

Millichamp, A. (1995) *Finance for Non-financial Managers: An Active Learning Approach*, 2nd edn. Accounting in the not-for-profit sector, pp. 220–227. DPP Publications Ltd, London.

Solomon, J.P. (1981) *Financial and Accounting Handbook for Service Industries.* CBI Publishing Company, Inc., Boston, Mass, United States.

# Part 2

# Basic Budgeting

We left Adam Haynes, at the end of Chapter 1, facing the task of making major decisions about the future of Home Farm. Ideally, he should undertake a systematic and comprehensive review and replanning of the whole business. This ideal is difficult to achieve in Adam's case, since he lacks experience of managing Home Farm, and has only a small knowledge of the options open to him and of the constraints on the business. Moreover, he has little time for consideration, as decisions must be made very soon concerning the coming year.

It would be quite sensible, therefore, for Adam to continue operation of the existing system, leaving major replanning until the end of his first year, when he will have got the 'feel' of the business. One advantage of this approach is that Adam would be looking at the financial consequences of just one course of action, rather than wrestling with several alternatives. It would enable him (and us) to become accustomed to the techniques of budgeting, before becoming enmeshed in the complexities of a major replanning.

For the moment, then, we follow this course, leaving discussion of a complete 're-think' to Part 5. The first requirement is the preparation of budgets to show the effect on profit, cash and capital of continuation of the existing system. The obvious starting point in this process is to build up a picture of the operational aspects of the business. This involves the compilation of a profit and loss budget, showing the anticipated profit over the first year of operation and, just as important, how that profit is to be achieved.

# Chapter 6
# A Profit Budget

## 6.1 Introduction

Adam is faced with the task of preparing a *whole-business profit and loss budget*[10], referred to hereafter as a *profit budget*. This will involve two processes: first, forecasting the physical and financial performance of the business during the coming year; and second, assembling this information in such a way as to give an estimate of net profit for that year.

### 6.1.1 Forecasting prices and quantities

Forecasting for a business can be extremely difficult. To the uncertainty of the physical environment (e.g. climate, disease, technological development) is added the considerable uncertainty created by political, economic, social and legislative factors. And yet, as indicated in Chapter 1, the greater the uncertainty, the greater the need to plan carefully. How, then, can Adam, a novice in business management, ensure that his guesses about the future are the best possible?

First, he could use history; look at past performance and use it as a base for decisions about the future. History can be particularly valuable for forecasting physical performance. The facts that a certain patch of land has consistently yielded only 4 tonnes per hectare of wheat in the past, or that typical occupancy rates for a tourist enterprise have never risen above 60% outside July and August, provide a useful guide for next year's budget.

Past events are not always a reliable guide to the future, however. In estimating *prices* of most agricultural inputs and outputs, for instance, it is unwise to forecast purely on the basis of historical trends. Prices can be unexpectedly and dramatically affected by natural, political and economic events on the other side of the world, particularly as protective trade barriers are progressively reduced or removed.

A second way of improving the quality of forecasting is to use the expertise of others. Academic, commercial and government researchers are employed to make forecasts in various areas, such as the rate of inflation over the next five years; the world production of wheat during the coming year; or the weather in the next four weeks. These forecasts are published in various media, and regular study of trade journals, radio programmes, etc. can pay dividends. At a more local level, Adam can learn a great deal from discussions with neighbours and local advisers, and from telephone calls to merchants, dealers and professional associations. Chambers of

Trade, Enterprise Agencies, Business Links and similar organizations can often provide useful information, or at least point Adam in the right direction.

He has still no guarantee that he will make the right decisions. On the other hand, by blending historical evidence with the expert predictions of others, and adding a measure of his own intelligent reasoning, he is likely to reduce the uncertainty surrounding his forecasts.

## 6.2 Compiling a profit budget

### 6.2.1 What form?

Let us assume that Adam has made his best possible forecasts of the inputs and outputs of Home Farm for the coming year. He has to arrange this information in such a way that:

- It results in the calculation of anticipated net profit
- It provides sufficient information for precise control throughout the year.

The simplest and quickest method of calculating profit would be to compile a budget on the lines of a conventional profit and loss account, as described in Chapter 3. This can be valuable as an initial test of business profitability, but it omits so much detail that it is virtually useless for precise control. Far better to refer to the management accounting models described in Chapter 4, and particularly the gross margin format. Since most of the activities of the business are agricultural, the 'agricultural gross margin' described in Chapter 5 will be used as the basis for this and later chapters, assuming the following equivalence for working purposes:

| **Agricultural use** | | **General use** |
|---|---|---|
| Enterprise | *equivalent to* | Profit centre |
| Output | *equivalent to* | Revenue |
| Variable cost | *equivalent to* | Direct cost |
| Fixed cost | *equivalent to* | Overhead or indirect cost |

### 6.2.2 How much detail?

The division of the budget into enterprise sections enables it to spread over several pages, and thus allows space to be used for extra detail. The degree of detail provided can make the difference between a useless budget and an invaluable one. At the same time, the provision of too *much* detail can confuse rather than enlighten, and incur unwanted extra work.

The first problem is how far to take the division into enterprises. Of the barley grown on a farm, for instance, some may be winter barley and some spring barley. Of the winter barley, some may be destined for seed, some for animal feed. The feed barley area may be drilled with three different varieties. Should all the barley be treated as one enterprise; should each separate variety be treated as an enterprise; or does the answer lie somewhere between the two extremes?

As a rule, a division should be made where there are likely to be significant differences in output and/or variable costs. In the above example, therefore, the barley should be treated as at least three enterprises: spring barley, winter barley (seed) and winter barley (feed). Each has a specific product and a specific management regime. Therefore, each needs to be budgeted separately if the farmer is to make sensible decisions about both the management of each crop and their relative proportions.

On some farms, it may even be worth budgeting the different varieties separately, especially where large areas are involved and there are distinct differences in performance between varieties. In most cases, however, the benefits gained would not justify the time and effort involved (remember that every such division increases not only the effort involved in budgeting, but also the burden of recording of control information).

The second problem is that of how much to present within each enterprise gross margin budget. Consider a possible budget for the output of a winter wheat enterprise shown in Figure 6.1.

| *Enterprise output* | £ |
|---|---|
| Grain sales | 18 000 |
| Straw sales | 1 600 |
| | 19 600 |

**Fig. 6.1** Aggregated output calculation.

This budget is adequate if the only purpose of it is to estimate the likely profit of the whole business. If, however, the intention is to *use* the budget constructively in financial control of the business, this format is virtually useless. When the time comes to compare actual and budgeted results, the manager will want to know what assumptions he made in the budget concerning yields and prices. Only then can he draw conclusions about his management of the enterprise. The budget shown in Figure 6.2 gives much more detail about the components of the output, at the expense of little extra effort.

The manager has estimated the likely yield of the crop at the next harvest, and has multiplied this by the price he hopes to receive. Variable costs would be treated in the same way, working from the rate of application and the anticipated prices per unit. This approach has the great virtue of simplicity, and is particularly useful for rapidly testing a number of alternative options.

### 6.2.3 A full budget

As an accurate reflection of the performance of the enterprise and its contribution to the business, however, this simple approach can leave much to be desired. It takes

| Enterprise output (40 hectares) | Yield per hectare (tonnes) | Total yield (tonnes) | Price per tonne (£) | Total output (£) | Output per hectare (£) |
|---|---|---|---|---|---|
| Grain sales | 4.5 | 180 | 100 | 18 000 | 450 |
| Straw sales | 2 | 80 | 20 | 1 600 | 40 |
| | | | | 19 600 | 490 |

**Fig. 6.2** More detailed output calculation.

no account, for instance, of the possibility of variation of stocks between the beginning and the end of the year. Figure 6.3 includes opening and closing valuations of grain, reflecting a situation where the manager's estimates of sales ignore the fact that he has an unusually high stock of grain left unsold from the previous harvest, which will be disposed of during the year. The use of some of the

| **Winter wheat** | | | **Gross margin** | | |
|---|---|---|---|---|---|
| *Hectares: 40* | *Total yield (tonnes)* | *Yield per hectare (tonnes)* | *Price per tonne (£)* | *£ total* | *£ per hectare* |
| *Grain* | | | | | |
| Sales | 100 | 2.5 | 104.4 | 10 435 | 261 |
| Transferred to livestock | 95 | 2.4 | 100.0 | 9 500 | 240 |
| Closing valuation | 45 | 1.1 | 95.0 | 4 275 | 105 |
| Less opening valuation | –60 | –1.5 | 101.0 | –6 060 | –152 |
| *Straw* | | | | | |
| Transferred to livestock | 85 | 2.1 | 25.0 | 2 125 | 53 |
| Closing valuation | 22 | 0.6 | 20.0 | 440 | 12 |
| Less opening valuation | –35 | –0.9 | 20.0 | –700 | –18 |
| ***Output*** | | | | 20 015 | 501 |
| *Variable costs* | *Total input (tonnes)* | *Input per hectare (tonnes)* | *Price per tonne (£)* | *£ total* | *£ per hectare* |
| Seedbed fertilizer | 12 | 0.3 | 180.00 | 2 160 | 54.0 |
| High-nitrogen fertilizer | 8 | 0.2 | 150.00 | 1 200 | 30.0 |
| Seed | 8 | 0.2 | 200.00 | 1 600 | 40.0 |
| Sprays | | | | 1 200 | 30.0 |
| Casual labour | | | | 400 | 10.0 |
| ***Variable cost*** | | | | 6 560 | 164 |
| ***Gross margin*** | | | | 13 455 | 337 |

**Fig. 6.3** Example gross margin calculation.

grain and straw for feeding livestock in other enterprises on the same farm is also acknowledged. The final gross margin is the same, but the budget is more explicit, and thus easier to relate to the realities of the business.

It is when trying to estimate the true profit of the farm that such adjustments become essential, rather than just making the budget easier to follow. Net profit, for reasons discussed in Chapter 3, must incorporate changes in valuation of stocks, changes in debtors and creditors, benefits in kind and depreciation. A profit budget in gross margin form, rather than a back-of-the envelope test of alternative possibilities, must therefore incorporate all these. Furthermore, it should incorporate any transfers between enterprises (such as barley used for feeding dairy cattle; milk used in the cheese-making and the bed and breakfast enterprise). To be 'fair' to each enterprise, these must be shown as non-cash 'sales' and 'purchases' at the market values prevailing at the time of transfer(11).

In a full gross margin, the components would be calculated as follows:

*Enterprise output*

Sales (adjusted for debtors) and subsidies
*plus* Transfers to other enterprises (at market value)
*plus* Increase (or less decrease) in valuation of stocks of products
*plus* Benefits in kind (products consumed by families of owner and employees)

*Variable costs*

Purchases (adjusted for creditors)
*plus* Transfers from other enterprises (at market value)
*less* Increase (or plus decrease) in valuation of stocks of inputs
*less* Benefits in kind (inputs consumed by families of owner and employees).

Note that an increase in valuation of stocks of produce is *added* to output, whereas an increase in the valuation of stocks of inputs is *deducted* from variable costs. In the former case, sales and transfers alone do not show the full value of all that has been produced, and any increase in the level of stocks must be added in order to give a true reflection of enterprise performance. In the latter, an increase in stock levels implies that (inflation apart) more inputs have been purchased than were actually used in producing the output. To obtain a fair picture of the relationship between inputs and outputs, therefore, the increase must be deducted from the purchases.

A similar argument applies to benefits in kind. These arise where the business has produced goods which do not appear in sales or transfers and/or (less likely) has purchased goods which, having been diverted outside the business for no cash payment, have made no contribution to the performance of the enterprise.

It is often the case that some minor items of output and variable costs do not fit exactly into the enterprise classification. An increase in the value of miscellaneous stores, for instance, may be so small that the effort of allocating it accurately between enterprises is not justified; the same may apply to casual labour hired for harvest work. Such items may be either divided arbitrarily between the relevant enterprises (on the grounds that they are too small for any consequent distortions to have a significant effect on the enterprise gross margins), or deducted from the *total* gross margin without any attempt to allocate them. The latter is the approach adopted here (see Figure 6.4).

| | £ |
|---|---|
| Sum of enterprise gross margins | 70 000 |
| *plus* unallocated output items | 2 000 |
| *less* unallocated variable cost items | –1 500 |
| ***Total gross margin*** | 70 500 |
| *less* total fixed costs | 55 500 |
| ***Net profit*** | 15 000 |

**Fig. 6.4** Financial summary.

The fixed costs of the enterprise must now be estimated. These will include all or some of the following:

- Regular labour costs
- Running costs and depreciation of machinery, buildings and improvements
- Rent and/or costs of maintaining property
- Miscellaneous overhead costs
- Interest and other finance charges

Any relevant benefits in kind (e.g. private use of car, telephone or fuel) should be deducted from fixed costs. Total fixed costs may now be deducted from total gross margin to give net profit.

### 6.2.4 A gross margin worksheet

To allow for all these adjustments and still be able to understand the results (and to have confidence in your calculation) takes a certain amount of discipline. It helps greatly to use a worksheet such as the one shown in Figure 6.5 (computer spreadsheet formulae for this worksheet are given in Figure A.1). Everyone has their own needs and preferences in setting out gross margin calculations, and you may prefer to design your own.

The worksheet is in two parts, the lower part showing the familiar gross margin calculation. The upper part, which is where the raw material of the gross margin is gathered, is in three sections: output calculation, variable cost calculation and miscellaneous valuation changes. In the output section, the various products of the enterprise are identified, and, working from the cash to be received, the adjustments are made to arrive at an output for each product. This output is calculated in both physical (total quantity) and financial terms, thus giving two figures that can be transferred down to the relevant columns (shaded) in the gross margin calculation below. From that point it is a simple matter to divide each by the number of head to arrive at quantity produced per head and financial output per head (third and sixth columns in the gross margin).

The variable cost section works in the same way, although in some categories it is not appropriate to include quantities (in this case 'forage', brought forward from another worksheet, and 'other'). The 'miscellaneous valuation changes' section allows the inclusion of valuation changes which are not easy to relate to specific products or cost categories; particularly breeding livestock number and value, and values of growing crops.

There are good arguments for alternative ways of setting out the worksheet. In budgeting, for instance, it is often more convenient to *start* with output, calculating either cash sales or closing stock valuations as a residual figure. The format shown in Figure 6.5 provides a good compromise between the needs of budgeting future performance and recording that of the past, however, and leaves plenty of scope for you to experiment with variations on the theme.

To make budget preparation even more organized, the use of a gross margin assumptions worksheet is recommended (see Figure 6.6). This allows you to set out clearly the basis of all your estimates for each product and variable cost. Not only does this make completing the gross margin worksheet much easier, but it gives you a clear record of what you assumed in your budget – and why. This could save you precious time and avoid needless frustration when, in the future, you come back to the gross margin to adjust it, or to use it in control (see Chapter 14).

If you are used to a more casual use of gross margin budgets, the above might seem to be bureaucratic overkill. It cannot be stressed too highly, however, that an organized approach at the outset is the key to efficient use of time and accurate calculation. Other benefits will become clear when budgeting for cash flow and capital: you will find that your gross margin worksheet is an excellent starting point for both, with many of the assumptions and calculations already made.

A number of difficulties may be encountered in applying the theory of profit budgeting to actual problems. The following points may be of some assistance.

(1) The easiest way to start a profit budget is to work out a plan in physical quantities. Forecast prices may then be applied to the physical data to enable calculation of the gross margins. This leaves only the fixed costs to estimate in order to arrive at a net profit. (Chapter 17 gives some guidance on the estimation of labour and machinery costs for agricultural enterprises.)

(2) It is not enough to budget for only one year at a time. This is especially true where the budget is for a plan which will take some time to become established, but ideally should apply whatever the circumstances. In most cases it will be sufficient to prepare plans for the second and subsequent years in outline only. These outline budgets may then be updated, and the detail filled in, before the end of the previous budgeting year. The result is a 'rolling plan': as the business moves through time, the budgets are extended progressively so that there is always a minimum period of time covered. Life does not stand still

| *Enterprise:* Dairy | *Ha or head?* 80 | | | | *Year* 199X/9Y<br>*Version* 1 | | | |
|---|---|---|---|---|---|---|---|---|
| *Products* | Milk | | Calves | | Culls | | | |
| | Qty | £ | Qty | £ | Qty | £ | Qty | £ |
| Cash received | 420 500 | 84 100 | 74 | 7 400 | 20 | 11 000 | | |
| Closing debtors | 12 000 | 2 400 | 7 | 700 | | | | |
| –Opening debtors | –12 000 | –2 600 | –5 | –500 | | | | |
| **Sales** | 420 500 | 83 900 | 76 | 7 600 | 20 | 11 000 | | |
| Transfers out | 20 000 | 4 000 | | | | | | |
| Benefits in kind | 700 | 147 | | | | | | |
| Closing valuation | | | 10 | 800 | | | | |
| –Opening valuation | | | –10 | –800 | | | | |
| ***Output*** | 441 200 | 88 047 | 76 | 7 600 | 20 | 11 000 | | |

| *Inputs* | Purch. livestock | | Conc. feed | | Straw | | Vet/med | | Forage | Other | | |
|---|---|---|---|---|---|---|---|---|---|---|---|---|
| | Qty | £ | Qty | £ | Qty | £ | Qty | £ | £ | £ | £ | £ |
| Cash paid | 25 | 21 250 | 86 | 12 900 | | | 1 | 3 500 | 5 755 | 4 000 | | |
| Closing creditors | | | 20 | 3 000 | | | | | | | | |
| –Opening creditors | | | –25 | –4 000 | | | | | | | | |
| **Purchases** | 25 | 21 250 | 81 | 11 900 | | | 1 | 3 500 | 5 755 | 4 000 | | |
| Transfers in | | | 42 | 3 780 | 26 | 390 | | | | | | |
| –Benefits in kind | | | | | | | | | | | | |
| Opening valuation | | | 20 | 2 800 | | | | | | | | |
| –Closing valuation | | | –20 | –2 800 | | | | | | | | |
| ***Variable cost*** | 25 | 21 250 | 123 | 15 680 | 26 | 390 | 1 | 3 500 | 5 755 | 4 000 | | |

**Fig. 6.5** Gross margin worksheet used for Home Farm dairy enterprise.

| ***Miscellaneous valuation changes*** | Breeding livestock | | Crops in gd. | Total |
|---|---|---|---|---|
| | No. | £ | £ | £ |
| Closing valuation | 82 | 53 300 | | 53 300 |
| –Opening valuation | –77 | –50 050 | | –50 050 |
| **Valuation change** | 5 | 3 250 | | 3 250 |

| *Dairy* | ***Gross margin*** | | | | |
|---|---|---|---|---|---|
| *Ha or head* 80 | *Total quantity* | *Qty/ha or /head* | *£ per unit quantity* | *£ Total* | *£/ha or /head* |
| Milk | 441 200 | 5 515 | 0.20 | 88 047 | 1 101 |
| Calves | 76 | 0.9 | 100 | 7 600 | 95 |
| Culls | 20 | 0.2 | 550 | 11 000 | 137 |
| Misc. valn. changes | xxxxxxxx | xxxxxxxxx | xxxxxxxxx | 3 250 | 41 |
| ***Output*** | | | | 109 897 | 1 374 |

| ***Variable costs*** | *Total quantity* | *Qty/ha or /head* | *£ per unit quantity* | *£ Total* | *£/ha or /head* |
|---|---|---|---|---|---|
| Purch. livestock | 25 | 0.3 | 850 | 21 250 | 266 |
| Conc. feed | 123 | 1.5 | 127 | 15 680 | 196 |
| Straw | 26 | 0.3 | 15 | 390 | 5 |
| Vet/med | | | | 3 500 | 44 |
| Forage | | | | 5 755 | 72 |
| Other | | | | 4 000 | 50 |
| ***Variable cost*** | | | | 50 575 | 632 |
| ***Gross margin*** | | | | 59322 | 742 |

**Fig. 6.5** (cont'd)

| ***Gross margin assumptions*** | | | Budget title: Home Farm Year 1<br>Year: 199X/9Y<br>Version No: 1 | | |
|---|---|---|---|---|---|
| Enterprise | Item | Amount bought/sold | Price per unit (£) | Total rec'd/paid | Notes |
| Example: dairy | Milk sales | 440 200 litres | 0.20/litre | £88 047 | Last year's yield + 2%. Price forecast by dairy company. |
| | Transfers | 20 000 litres | 0.20/litre | £4 000 | Used in cheese-making: assume same price as liquid sales. |

**Fig. 6.6** Gross margin assumptions worksheet.

at the end of one year while you sort yourself out ready for the next 12 months.

(3) When budgeting for the coming year, it is important to forecast prices and quantities as accurately as possible for that year, using the best outlook information available. When preparing outline budgets for subsequent years, use the first year's values, except where you have strong feelings about likely price movements in particular commodities. You will have the opportunity to revise these values as the first year passes and it becomes easier to make specific forecasts for later years. (Chapter 20 includes discussion of the problems of budgeting in times of inflation.)

(4) Be careful to retain a sense of proportion when preparing the budget. Avoid spending hours calculating items which will have an insignificant effect on profit. This commonly applies to benefits in kind and valuations of miscellaneous items.

(5) When more than one grazing livestock enterprise is present on the farm, it is necessary to make some attempt to allocate the variable costs of grass and other forage crops between them. In order to avoid undue complications at an early stage, this topic is treated in full in Chapter 16.

(6) Forecasts, and thus budgets, are based on guesses about the future. Techniques such as sensitivity analysis and break-even budgets are useful in showing the likely effects of those guesses being wrong (see Chapter 19).

(7) Interest charges on overdrafts and loans must be deducted in calculating net profit; they are part of the fixed costs of the business. It is advisable at this stage to first calculate profit before interest, leaving the calculation of net profit until a cash flow budget has been compiled. This will show the likely levels of overdraft and allow calculation of interest on that overdraft (see Chapters 7 and 18).

(8) You *will* make mistakes in calculations. Therefore always use pencil for initial drafts, and have an eraser ready. Remember that it is far easier to trace errors by systematic checking at each stage of compiling the budget, than to start looking for them when the budget is complete.

(9) As a final test, calculate profit by the conventional method and check against the result obtained by the gross margin method. If the two figures do not tally, check and check again until they do.

### 6.2.5 Normalized budgets

The most common use of the whole-farm profit budget is in examining the likely effects of a specific year's trading, e.g. 199X–199Y for Home Farm.

Where a farm plan incorporates changes which will take some time to establish, e.g. one that involves a programme of investment in stock, machinery and buildings over many years, a *normalized* budget may be used. This depicts the make-up of profit or loss of the business after all the changes have been completed and the new system is running smoothly. The assumption is that economic and physical conditions will be those of a 'typical' year. This allows the exclusion of valuation changes from the calculation, since fluctuations between years are averaged out. For the same reason, a normalized budget can be calculated on a 'crop year'[(12)] basis irrespective of the dates of the financial year. All depreciation must be calculated by the straight-line method, since the other methods are designed to provide specific depreciation charges for particular years.

Although the business system that actually develops is unlikely to resemble exactly that planned (since most businesses are continually evolving, rather than stabilizing at a particular point), the normalized budget is valuable in giving an impression of the eventual worthwhileness of a long-term reorganization. It may also be used in the 'short-list and budget' method of choosing between alternative farm systems (Chapter 16), where the problem is too complex for partial budgets (Chapter 9) and limited time precludes the calculation of several fully detailed whole-business budgets. The exclusion of valuation changes and crop/financial year adjustments is particularly helpful in this respect.

## 6.3 Home Farm profit budget

Adam Haynes has completed his gross margin budget for the first year, using the procedures described above. You have already seen one of his budgeted gross margins (Figure 6.5): the rest follow, together with calculations of forage variable cost and depreciation on a 'pool' basis (Figure 6.7: 20% depreciation rate has been used throughout for simplicity). The remaining fixed costs have been estimated on the basis of previous years' figures adjusted for expected increases (see Figures 6.8–6.13). Note that the opening values of stocks, machinery and bank balance have been brought forward from Figure 1.1. If Adam's assumptions are correct, he seems likely to achieve his profit objective of at least £10 000 per year: it remains to be seen whether his other goals are realistic.

| | *Farm machinery* | *B & B equipment* | *Cheese equipment* | *Vehicles* | *Total* |
|---|---|---|---|---|---|
| | £ | £ | £ | £ | £ |
| Opening value | 39 000 | 2 000 | 4 000 | 6 000 | 51 000 |
| *plus* purchases | 15 000 | | | 7 000 | 22 000 |
| *less* sales | 5 000 | | | 4 000 | 9 000 |
| Sub-total | 49 000 | 2 000 | 4 000 | 9 000 | 64 000 |
| *less* depreciation @ 20% | 9 800 | 400 | 800 | 1 800 | 12 800 |
| Closing ('written-down') value | 39 200 | 1 600 | 3 200 | 7 200 | 51 200 |

**Fig. 6.7** 'Pool' depreciation, diminishing balance method.

Once you are sure you understand how the budget works, try producing your own gross margin budget for a business you know or, failing that, a fictitious business (perhaps one that you would like to own in the future!). Don't be too adventurous at first, though – keep to a simple system with a few straightforward enterprises.

## Notes

10. Often termed, ambiguously, a *complete budget*.
11 If they were transferred at cost of production, the 'selling' enterprise would look unduly poor and the 'buying' enterprise unrealistically good.
12. That is, enterprise budgets/accounts based on time periods identical to the production cycles of those enterprises (see Chapter 11).

## References


Barnard, C.S. & Nix, J.S. (1979) *Farm Planning and Control*, 2nd edn. pp. 45–47 (gross margins), 317–327 (complete budgets) and 537–539 (full costings). Cambridge University Press, London.

Berry, A. & Jarvis, R. (1991) *Accounting in a Business Context*. Chapman & Hall, London.

Broadbent, M. & Cullen, J. (1993) *Managing Financial Resources*. Butterworth Heinemann, Oxford.

Bull, R.J. (1990) *Accounting in Business*, 6th edn. Butterworth Heinemann, London.

Kerr, H.W.T. (1988) Management accounting and the marginal concept. *Farm Management*, **6**, 10.

Millichamp, A. (1995) *Finance for Non-financial Managers*, 2nd edn. DPP Publications, London.

Ministry of Agriculture, Fisheries and Food (1977) *Definition of Terms Used in Agricultural Business Management*. Section 4 (Margin Terms).

Warren, M.F. (ed.) (1993) *Improving your Profits*. Book 3 in series: Finance Matters for the Rural Business. ATB Landbase/Seale-Hayne Faculty of Agriculture, Food and Land Use, University of Plymouth.

| *Enterprise:* Spring barley | | | *Ha or head?* 12 | | | *Year* 199X/9Y *Version* 1 | | |
|---|---|---|---|---|---|---|---|---|
| *Products* | Grain | | Straw | | | | | |
| | Qty | £ | Qty | £ | Qty | £ | Qty | £ |
| Cash received<br>Closing debtors<br>–Opening debtors | | | | | | | | |
| **Sales** | | | | | | | | |
| Transfers out<br>Benefits in kind<br>Closing valuation<br>–Opening valuation | 42<br><br>48<br>–42 | 3 780<br><br>4 560<br>–3 780 | 26<br><br>24<br>–26 | 390<br><br>360<br>–390 | | | | |
| ***Output*** | 48 | 4 560 | 24 | 360 | | | | |

| *Inputs* | 20.10.10 | | Seed | | Sprays | | Casual labour | | Other | | | |
|---|---|---|---|---|---|---|---|---|---|---|---|---|
| | Qty | £ | Qty | £ | Qty | £ | Qty | £ | £ | £ | £ | £ |
| Cash paid<br>Closing creditors<br>–Opening creditors | 5 | 575 | 1.5 | 400 | 1 | 360 | 1 | 100 | 120 | | | |
| **Purchases** | 5 | 575 | 1.5 | 400 | 1 | 360 | 1 | 100 | 120 | | | |
| Transfers in<br>–Benefits in kind<br>Opening valuation<br>–Closing valuation | | | | | | | | | | | | |
| ***Variable cost*** | 5 | 575 | 1.5 | 400 | 1 | 360 | 1 | 100 | 120 | | | |

**Fig. 6.8** Home Farm barley gross margin.

| ***Miscellaneous valuation changes*** | Breeding livestock | | Crops in gd. | Total |
|---|---|---|---|---|
| | No. | £ | £ | £ |
| Closing valuation<br>–Opening valuation | | | | |
| **Valuation change** | | | | |

| *Spring barley* | | ***Gross margin*** | | | |
|---|---|---|---|---|---|
| *Ha or head*<br>*12* | *Total quantity* | *Qty/ha or /head* | *£ per unit quantity* | *£ Total* | *£/ha or /head* |
| Grain | 48.0 | 4.00 | 95 | 4 560 | 380 |
| Straw | 24.0 | 2.00 | 15 | 360 | 30 |
| Misc. valn. changes | xxxxxxxx | xxxxxxxxx | xxxxxxxxx | | |
| ***Output*** | | | | 4 920 | 410.00 |

| ***Variable costs*** | *Total quantity* | *Qty/ha or /head* | *£ per unit quantity* | *£ Total* | *£/ha or /head* |
|---|---|---|---|---|---|
| 20.10.10 | 5.0 | 0.42 | 115 | 575 | 48 |
| Seed | 1.5 | 0.12 | 267 | 400 | 33 |
| Sprays | | | | 360 | 30 |
| Casual labour | | | | 100 | 8 |
| Other | | | | 120 | 10 |
| ***Variable cost*** | | | | 1 555 | 129 |
| ***Gross margin*** | | | | 3 365 | 281 |

**Fig. 6.8** (cont'd)

| *Enterprise:* Fattening pigs | | *Ha or head?* 1000 | | | | *Year* 199X/9Y *Version* 1 | | |
|---|---|---|---|---|---|---|---|---|
| *Products* | Bacon pigs | | | | | | | |
| | Qty | £ | Qty | £ | Qty | £ | Qty | £ |
| Cash received | 1 060 | 73 140 | | | | | | |
| Closing debtors | 90 | 6 210 | | | | | | |
| –Opening debtors | –100 | –6 900 | | | | | | |
| **Sales** | 1 050 | 72 450 | | | | | | |
| Transfers out | | | | | | | | |
| Benefits in kind | | | | | | | | |
| Closing valuation | 500 | 22 500 | | | | | | |
| –Opening valuation | –550 | –24 750 | | | | | | |
| ***Output*** | 1 000 | 70 200 | | | | | | |

| *Inputs* | Bought weaners | | Conc. feed | | Transport | | Other | | Vet/med, etc. | | | |
|---|---|---|---|---|---|---|---|---|---|---|---|---|
| | Qty | £ | Qty | £ | Qty | £ | Qty | £ | £ | £ | £ | £ |
| Cash paid | 1 100 | 27 500 | 210 | 29 400 | 1 | 650 | 1 | 600 | 1 200 | | | |
| Closing creditors | | | 18 | 2 520 | | | | | | | | |
| –Opening creditors | | | –20 | –2 800 | | | | | | | | |
| **Purchases** | 1 100 | 27 500 | 208 | 29 120 | 1 | 650 | 1 | 600 | 1 200 | | | |
| Transfers in | | | | | | | | | | | | |
| –Benefits in kind | | | | | | | | | | | | |
| Opening valuation | | | | | | | | | | | | |
| –Closing valuation | | | | | | | | | | | | |
| ***Variable cost*** | 1 100 | 27 500 | 208 | 29 120 | 1 | 650 | 1 | 600 | 1 200 | | | |

**Fig. 6.9** Home Farm pig enterprise gross margin.

| *Miscellaneous valuation changes* | Breeding livestock | | Crops in gd. | Total |
|---|---|---|---|---|
| | No. | £ | £ | £ |
| Closing valuation<br>–Opening valuation | | | | |
| **Valuation change** | | | | |

| *Fattening pigs* | | | **Gross margin** | | |
|---|---|---|---|---|---|
| *Ha or head*<br>*1000* | *Total quantity* | *Qty/ha or /head* | *£ per unit quantity* | *£ Total* | *£/ha or /head* |
| Bacon pigs | 1 000 | 1.00 | 70.2 | 70 200 | 70.20 |
| Misc. valn. changes | xxxxxxxx | xxxxxxxxx | xxxxxxxxx | | |
| ***Output*** | | | | 70 200 | 70.20 |

| ***Variable costs*** | *Total quantity* | *Qty/ha or /head* | *£ per unit quantity* | *£ Total* | *£/ha or /head* |
|---|---|---|---|---|---|
| Bought weaners | 1 100 | 1.1 | 25 | 27 500 | 27.50 |
| Conc. feed | 208 | 0.208 | 140 | 29 120 | 29.12 |
| Transport | | | | 650 | 0.65 |
| Other | | | | 600 | 0.60 |
| Vet/med, etc. | | | | 1 200 | 1.20 |
| ***Variable cost*** | | | | 59 070 | 59.07 |
| ***Gross margin*** | | | | 11 130 | 11.13 |

**Fig. 6.9** (cont'd)

| *Enterprise:* Bed and Breakfast | | | *Beds* 6 | | | | *Year* 199X/9Y *Version* 1 | |
|---|---|---|---|---|---|---|---|---|
| *Products* | Night stays | | | | | | | |
| | Qty | £ | Qty | £ | Qty | £ | Qty | £ |
| Cash received<br>Closing debtors<br>–Opening debtors | 430 | 9 000 | | | | | | |
| **Sales** | 430 | 9 000 | | | | | | |
| Transfers out<br>Benefits in kind<br>Closing valuation<br>–Opening valuation | | | | | | | | |
| ***Output*** | 430 | 9 000 | | | | | | |

| *Inputs* | Food/consumables | | Advertising | | Postage | | Casual labour | | | | | |
|---|---|---|---|---|---|---|---|---|---|---|---|---|
| | Qty | £ | Qty | £ | Qty | £ | Qty | £ | £ | £ | £ | £ |
| Cash paid<br>Closing creditors<br>–Opening creditors | 1 | 1 680 | 1 | 700 | 1 | 50 | 1 | 1 500 | | | | |
| **Purchases** | 1 | 1 680 | 1 | 700 | 1 | 50 | 1 | 1 500 | | | | |
| Transfers in<br>–Benefits in kind<br>Opening valuation<br>–Closing valuation | | | | | | | | | | | | |
| ***Variable cost*** | 1 | 1 680 | 1 | 700 | 1 | 50 | 1 | 1 500 | | | | |

**Fig. 6.10** Home Farm bed and breakfast enterprise gross margin.

| ***Miscellaneous valuation changes*** | Breeding livestock | | Crops in gd. | Total |
|---|---|---|---|---|
| | No. | £ | £ | £ |
| Closing valuation<br>–Opening valuation | | | | |
| **Valuation change** | | | | |

| *Bed and Breakfast* | | ***Gross margin*** | | | |
|---|---|---|---|---|---|
| *Ha or head*<br>*6* | *Total quantity* | *Qty/bed* | *£ per unit quantity* | *£ Total* | *£/bed* |
| Night stays | 430 | | 21 | 9 000 | 1 500 |
| Misc. valn. changes | xxxxxxxx | xxxxxxxxx | xxxxxxxxx | | |
| ***Output*** | | | | 9 000 | 1 500 |

| ***Variable costs*** | *Total quantity* | *Qty/bed* | *£ per unit quantity* | *£ Total* | *£/bed* |
|---|---|---|---|---|---|
| Food/consumables | | | | 1 680 | 280 |
| Advertising | | | | 700 | 117 |
| Postage | | | | 50 | 8 |
| Casual labour | | | | 1 500 | 250 |
| ***Variable cost*** | | | | 3 930 | 655 |
| ***Gross margin*** | | | | 5 070 | 845 |

**Fig. 6.10** (cont'd)

*Enterprise:* cheese — *Cheese produced:* 600 — *Year* 199X/9Y, *Version* 1

| *Products* | Finished cheeses | | Maturing cheeses | | | | | |
|---|---|---|---|---|---|---|---|---|
| | Qty | £ | Qty | £ | Qty | £ | Qty | £ |
| Cash received<br>Closing debtors<br>–Opening debtors | 600 | 15 000 | | | | | | |
| **Sales** | 600 | 15 000 | | | | | | |
| Transfers out<br>Benefits in kind<br>Closing valuation<br>–Opening valuation | <br><br>50<br>–55 | <br><br>1 250<br>–1370 | <br><br>250<br>–248 | <br><br>3 750<br>–3 720 | | | | |
| ***Output*** | 595 | 14 880 | 2 | 30 | | | | |

| *Inputs* | Milk | | Materials | | Other costs | | | | | | | |
|---|---|---|---|---|---|---|---|---|---|---|---|---|
| | Qty | £ | Qty | £ | Qty | £ | Qty | £ | £ | £ | £ | £ |
| Cash paid<br>Closing creditors<br>–Opening creditors | | | 1 | 1 140 | 1 | 1 160 | | | | | | |
| **Purchases** | | | 1 | 1 140 | 1 | 1 160 | | | | | | |
| Transfers in<br>–Benefits in kind<br>Opening valuation<br>–Closing valuation | 20 000 | 4 000 | | | | | | | | | | |
| ***Variable cost*** | 20 000 | 4 000 | 1 | 1 140 | 1 | 1 160 | | | | | | |

**Fig. 6.11** Home Farm cheese enterprise gross margin.

| ***Miscellaneous valuation changes*** | Breeding livestock | | Crops in gd. | Total |
|---|---|---|---|---|
| | No. | £ | £ | £ |
| Closing valuation<br>–Opening valuation | | | | |
| **Valuation change** | | | | |

| *Cheese* | ***Gross margin*** | | | | |
|---|---|---|---|---|---|
| *Cheese sold*<br>*600* | *Total quantity* | *Qty/ cheese* | *£ per unit quantity* | *£ Total* | *£/cheese* |
| Finished cheeses | 595.0 | 0.99 | 25 | 14 880 | 24.80 |
| Maturing cheeses (valuation change) | | | | 30 | 0.05 |
| Misc. valn. changes | xxxxxxxx | xxxxxxxxx | xxxxxxxxx | | |
| ***Output*** | | | | 14 910 | 24.85 |

| ***Variable costs*** | *Total quantity* | *Qty/ cheese* | *£ per unit quantity* | *£ Total* | *£/cheese* |
|---|---|---|---|---|---|
| Milk | 20 000 | 33.33 | 0.20 | 4 000 | 6.67 |
| Materials | | | | 1 140 | 1.90 |
| Other costs | | | | 1 160 | 1.93 |
| ***Variable cost*** | | | | 6300 | 10.50 |
| ***Gross margin*** | | | | 8610 | 14.35 |

**Fig. 6.11** (cont'd)

| *Enterprise:* forage | | *Ha or head?* 42 | | | | *Year* 199X/9Y *Version* 1 | | |
|---|---|---|---|---|---|---|---|---|
| *Products* | Silage | | | | | | | |
| | Qty | £ | Qty | £ | Qty | £ | Qty | £ |
| Cash received<br>Closing debtors<br>–Opening debtors | | | | | | | | |
| **Sales** | | | | | | | | |
| Transfers out<br>Benefits in kind<br>Closing valuation<br>–Opening valuation | <br><br>10 000<br>–10 000 | | | | | | | |
| ***Output*** | | | | | | | | |

| *Inputs* | Ferts: 17:17:17 | | Ferts: 35% N | | Ferts: 20:10:10 | | Seed | | Sprays | | | |
|---|---|---|---|---|---|---|---|---|---|---|---|---|
| | Qty | £ | Qty | £ | Qty | £ | Qty | £ | £ | £ | £ | £ |
| Cash paid<br>Closing creditors<br>–Opening creditors | 21 | 2 835 | 13 | 1 495 | 7 | 805 | 0.4 | 200 | 420 | | | |
| **Purchases** | 21 | 2 835 | 13 | 1 495 | 7 | 805 | 0.4 | 200 | 420 | | | |
| Transfers in<br>–Benefits in kind<br>Opening valuation<br>–Closing valuation | | | | | | | | | | | | |
| ***Variable cost*** | 21 | 2 835 | 13 | 1 495 | 7 | 805 | 0.4 | 200 | 420 | | | |

**Fig. 6.12** Home Farm forage variable costs.

| *Miscellaneous valuation changes* | Breeding livestock | | Crops in gd. | Total |
|---|---|---|---|---|
| | No. | £ | £ | £ |
| Closing valuation<br>–Opening valuation | | | | |
| **Valuation change** | | | | |

| *Forage* | | **Gross margin** | | | |
|---|---|---|---|---|---|
| *Ha or head*<br>*42* | *Total quantity* | *Qty/ha or /head* | *£ per unit quantity* | *£ Total* | *£/ha or /head* |
| Silage | | | | | |
| Misc. valn. changes | xxxxxxxx | xxxxxxxxx | xxxxxxxxx | | |
| ***Output*** | | | | | |

There is no output in this case: the worksheet is used for calculation for variable costs only.

| ***Variable costs*** | *Total quantity* | *Qty/ha or /head* | *£ per unit quantity* | *£ Total* | *£/ha or /head* |
|---|---|---|---|---|---|
| Ferts: 17:17:17 | 21.0 | 0.50 | 135 | 2 835 | 67 |
| Ferts: 35% N | 13.0 | 0.31 | 115 | 1 495 | 36 |
| Ferts: 20:10:10 | 7.0 | 0.17 | 115 | 805 | 19 |
| Seed | 0.4 | 0.01 | 500 | 200 | 5 |
| Sprays | | | | 420 | 10 |
| ***Variable cost*** | | | | 5 755 | 137 |
| ***Gross margin*** | | | | –5 755 | –137 |

Transferred to the dairy enterprise as a variable cost.

**Fig. 6.12** (cont'd)

| | Per unit | Total | Total |
|---|---|---|---|
| *Gross margins* | £ | £ | £ |
| Pigs | 11.13 | 11 130 | |
| Dairy | 742 | 59322 | |
| Barley | 280 | 3365 | |
| Bed and breakfast | 845 | 5070 | |
| Cheese | 14.35 | 8610 | |
| | | | 87 497 |
| *Unallocated revenue items* | | | |
| Miscellaneous trading receipts | | | 2650 |
| *Miscellaneous valuation changes* | | | 848 |
| ***Total gross margin*** | | | 90 995 |
| *Fixed costs* | | | £ |
| Regular labour | | | 26 700 |
| Machinery and vehicle running costs | | | 15 400 |
| Buildings and property repairs | | | 200 |
| Rent and rates | | | 9500 |
| Miscellaneous fixed costs | | | 8600 |
| *less* benefits in kind, input items | | | –4 650 |
| ***Total fixed costs*** | | | 55 750 |
| | | £ | £ |
| ***Net profit*** before depreciation and finance | | | 35 245 |
| Depreciation: Farm machinery | | 9800 | |
| Vehicles | | 1800 | |
| Cheese-making equipment | | 800 | |
| Bed and breakfast equipment | | 400 | |
| Buildings (tenant's fixtures) | | 480 | 13 280 |
| ***Net profit*** before finance costs | | | 21 965 |
| Finance costs: Overdraft interest* | | 4 869 | |
| Loan interest | | 1 100 | |
| Leasing charges | | 2 100 | 8 069 |
| ***Net profit*** after finance costs | | | 13 896 |

* Calculated via the cash flow budget (Chapter 7).

**Fig. 6.13** Home Farm profit budget: financial summary.

# Chapter 7

# A Cash Flow Budget for the Whole Business

## 7.1 Basic principles

It is not enough for Adam to prepare a budget to show the likely profits of his business. In his situation as the new owner of a demanding business, and with little spare cash, it is vital that he should estimate the likely effect of his plans on the flow of cash into and out of the business. This he may do by means of a *cash flow budget*.

In compiling a cash flow budget for a given year, Adam must estimate all the likely cash receipts for that period (including capital and personal items) and deduct from them all the cash payments. This will give him an estimate of the *net cash flow* of the period or, in other words, it will show him how the bank balance of the business is likely to be affected during the year. You will remember from Chapter 2 that if the bank balance at the beginning of the year is known, or can be estimated, the bank balance at the end of the year can be calculated by adding the net cash flow to the opening balance.

It is normal to break up the year for the purpose of budgeting, so as to give a net cash flow and closing bank balance for each of the coming twelve months. Possible alternatives are bi-monthly or quarterly divisions, but the use of these longer intervals results in loss of precision and could lead, for example, to the overlooking of a severe fluctuation in the bank balance within a particular quarter. The monthly division has the further advantage that banks normally send statements to their customers at monthly intervals. The significance of this in the monitoring of cash flow is discussed in Chapter 10.

As well as a budget for each of the twelve months ahead, there should ideally be outline budgets for the following two or three years; quarterly budgets will usually suffice for this purpose. As with the profit budget, the cash flow budget should be 'rolling' (see Figure 7.1). In the example of the 'basic' twelve-monthly budget, as each half year passes, the equivalent half of the succeeding year should be budgeted. By this means, the manager always has at least six months' worth of budget to guide him. Where cash control is particularly important, quarterly re-budgeting may be necessary.

A simplified example of a twelve-monthly cash flow budget is shown in Figure 7.2. The net cash flow for each month is added to the opening bank balance for that

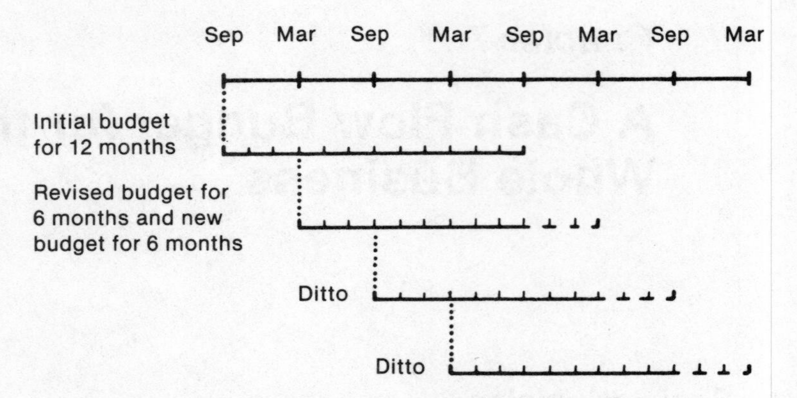


**Fig. 7.1** 'Rolling' cash-flow budget – graphical representation.

month to give the closing bank balance. This is then carried to the next column as the opening bank balance for the succeeding month.

An important feature is the 'total' column. This performs two functions; first it provides a summary of the whole year's cash flow, and second it enables checking of the arithmetic. The latter works through calculation of the annual net cash flow. This may be found both from the total column (taking total payments from total receipts) and by deducting the opening bank balance for the year from the closing bank balance (since any change in the bank balance can be attributed to the net flow of cash into or out of the business). If, when the annual cash flow is calculated in both ways, the two figures are identical, the arithmetic is likely to be correct. If they are not, a mistake has been made somewhere, and must be found. The value of this check may not be obvious with respect to such a simple example, but will be readily appreciated on first contact with a full-blown business budget.

The 'total' column in this example appears as the first of the columns. This ensures that the annual summary is adjacent to the category headings, and helps when reconciling with other budgets and interpreting the implications of the cash flow. Many prefer the 'total' column to be the last on the sheet: the choice is yours.

## 7.2 Value Added Tax and overdraft interest

The calculation of monthly net cash flows and bank balances is really very simple: the only possible difficulties are connected with Value Added Tax (VAT) and interest on overdraft, since calculation of each of these depends on the figures you put into the rest of the cash flow.

Value Added Tax is explained in more detail in Chapter 10. Briefly, certain types of product are taxed at a 'standard' rate of tax (17.5% at the time of writing) and others at 'zero rate' (meaning what it says – 0%). Others are 'exempt', or outside the scope of VAT. If a business buys standard-rated inputs to produce standard- or zero-rated outputs, it is allowed to set the VAT it has paid on the inputs against the VAT

| Cash flow budget | Year: 199X/9Y | Version: 1 | | | | | | | | | | |
|---|---|---|---|---|---|---|---|---|---|---|---|---|
| % interest rate: 15 | **Total** | October | November | December | January | February | March | April | May | June | July | August | September |
| *Receipts* | | | | | | | | | | | | | |
| Milk | 89 195 | 8 910 | 8 894 | 8 823 | 8 751 | 8 680 | 8 609 | 8 446 | 7 705 | 6 530 | 6 417 | 7 430 | 8 608 |
| Cottage let | 2 500 | | | | | | | 300 | 400 | 600 | 600 | 600 | 300 |
| Machinery sales | 5 000 | | | | | | | | | | | 5 000 | |
| **Total** | 96 695 | 8 910 | 8 894 | 8 823 | 8 751 | 8 680 | 8 609 | 8 746 | 8 105 | 7 130 | 7 017 | 13 030 | 8 908 |

| | **Total** | October | November | December | January | February | March | April | May | June | July | August | September |
|---|---|---|---|---|---|---|---|---|---|---|---|---|---|
| *Payments* | | | | | | | | | | | | | |
| Feed: cows | 9 200 | 3 000 | | 2 500 | | 1 000 | | 700 | | | | 2 000 | |
| Vet. and med. | 4 800 | 500 | 700 | 600 | 600 | 400 | 300 | 200 | 500 | 300 | 300 | 400 | 500 |
| Wages – permanent | 36 000 | 4 000 | 3 000 | 3 000 | 3 000 | 3 000 | 4 000 | 3 000 | 3 000 | 3 000 | 3 000 | 4 000 | 4 000 |
| Machinery purchase | 51 000 | | | | 30 000 | | | | 1 000 | | | 20 000 | |
| **Total** | 101 000 | 7 500 | 3 700 | 6 100 | 33 600 | 4 400 | 4 300 | 3 900 | 4 500 | 3 300 | 3 300 | 26 400 | 4 500 |
| Net cash flow | –4 305 | 1 410 | 5 194 | 2 723 | –24 849 | 4 280 | 4 309 | 4 846 | 3 605 | 3 830 | 3 717 | –13 370 | 4 408 |
| Opening balance | –8 000 | –8 000 | –6 590 | –1 396 | 1 327 | –23 522 | –19 242 | –14 933 | –10 087 | –6 482 | –2 652 | 1 065 | –12 305 |
| Closing balance | –12 305 | –6 590 | –1 396 | 1 327 | –23 522 | –19 242 | –14 933 | –10 087 | –6 482 | –2 652 | 1 065 | –12 305 | –7 897 |

**Fig. 7.2** Cash flow budget – outline example 1.

it has collected on its outputs: the balance is forwarded to the Customs and Excise. A business whose products are largely zero-rated, on the other hand, tends to pay out more VAT on inputs than it collects on outputs, and therefore periodically claims a refund of the balance from the Customs and Excise. Since most food products are zero-rated, this applies to the majority of farmers.

Thus the business acts as a collector of tax: VAT has no net effect on the cash flow of the business. In most cases, therefore, it is sensible to ignore VAT in cash flow budgeting, in the interests of simplicity. On the other hand, there is an inevitable delay between the payment of tax on inputs, and the refunding of that tax by the Customs and Excise. Where the business is a net reclaimer of VAT – a traditional farm business, for instance – this lag can have a serious negative effect on the bank balance.

In such cases, VAT can be incorporated by adding two further lines to the budget: in receipts, 'VAT charged on outputs' and 'VAT refunds'; and in payments, 'VAT paid on inputs' and 'VAT paid to Customs and Excise' (see Figure 7.3). Outputs and inputs are entered into the budget in the normal way, net of VAT. Taking the column for the first month, add up the value of all the outputs on which VAT will be charged (starred in Figure 7.3), multiply by the percentage tax rate and enter the result under 'VAT charged'. Do the same for the standard-rated input items, entering the result under 'VAT paid on inputs'.

If the VAT charged is more than the VAT paid, the difference between the two will be the amount payable for that month to the Customs and Excise. If, as is more likely in farming, the VAT charged is *less* than that paid, the difference represents the refund due from the Customs and Excise. Where the farmer is claiming refunds monthly, this refund should be shown under 'VAT refunds' in the next month's column. Where a farmer is claiming quarterly, the balance has to be accumulated for the three months before the business's usual VAT assessment date, and then shown in the next month's column under 'VAT refunds'. This is the method used in Figure 7.3. All other examples in this book, however, leave VAT out of the budgets.

One of the important functions of a cash flow budget is to give an estimate of likely overdraft interest payments – for use in calculating profit as well as cash flow. This must be done after all the other figures have been included in the budget. The first step is to estimate an interest rate for the year, based on forecasts in the business pages of the 'quality' press, phone calls to your accountant and other sources. This can be converted to a monthly rate by dividing by 12 (a crude estimate, since it ignores compound interest, but erring on the cautious side). Thus if you feel you are likely to be paying 12% interest per annum, the approximate monthly rate will be 1%.

If the opening bank balance is negative (as in Figure 7.3), multiply it by the monthly rate (e.g. 1%) and enter the result under 'overdraft interest' in the first month's column. Most banks charge interest monthly, one month in arrears, and this represents the interest due on the last month of the previous year. Now repeat the process using the closing balance of each month in turn: at the end of the year you will be left with the interest on the final month – this is a 'closing creditor' and will be paid in the first month of the following year.

| Cash flow budget | Year: 199X/9Y | Version: 1 | | | | | | | | | | | |
|---|---|---|---|---|---|---|---|---|---|---|---|---|---|
| % interest rate: 15 | **Total** | October | November | December | January | February | March | April | May | June | July | August | September |
| *Receipts* | | | | | | | | | | | | | |
| Milk | 97 803 | 8 910 | 8 894 | 8 823 | 8 751 | 8 680 | 8 609 | 8 446 | 7 705 | 6 530 | 6 417 | 7 430 | 8 608 |
| Cottage let* | 2 800 | | | | | | | 300 | 400 | 600 | 600 | 600 | 300 |
| Machinery sales* | 5 000 | | | | | | | | | | | 5 000 | |
| VAT charged on outputs | 1 365 | | | | | | | 53 | 70 | 105 | 105 | 980 | 53 |
| VAT refunds from C & E | 6 100 | | 325 | | | 5 583 | | | 105 | | | 88 | |
| **Total** | 113 068 | 8 910 | 9 219 | 8 823 | 8 751 | 14 263 | 8 609 | 8 799 | 8 280 | 7 235 | 7 122 | 14 098 | 8 961 |

| | **Total** | October | November | December | January | February | March | April | May | June | July | August | September |
|---|---|---|---|---|---|---|---|---|---|---|---|---|---|
| *Payments* | | | | | | | | | | | | | |
| Feed: cows | 9 200 | 3 000 | | 2 500 | | 1 000 | | 700 | | | | 2 000 | |
| Vet. and med.* | 5300 | 500 | 700 | 600 | 600 | 400 | 300 | 200 | 500 | 300 | 300 | 400 | 500 |
| Wages – permanent | 40 000 | 4 000 | 3 000 | 3 000 | 3 000 | 3 000 | 4 000 | 3 000 | 3 000 | 3 000 | 3 000 | 4 000 | 4 000 |
| Overdraft interest | 1 487 | 100 | 85 | 18 | | 363 | 246 | 195 | 137 | 95 | 47 | 1 | 199 |
| Machinery purchase* | 51 000 | | | | 30 000 | | | | 1 000 | | | 20 000 | |
| VAT paid on inputs | 9 853 | 88 | 123 | 105 | 5355 | 70 | 53 | 35 | 263 | 53 | 53 | 3 570 | 88 |
| VAT paid to C & E | | | | | | | | | | | | | |
| **Total** | 116 840 | 7 688 | 3 908 | 6 223 | 38 955 | 4 833 | 4 599 | 4 130 | 4 900 | 3 448 | 3 400 | 29 971 | 4 787 |
| Net cash flow | –3 772 | 1 222 | 5311 | 2600 | –30 204 | 9430 | 4 010 | 4 669 | 3 380 | 3 787 | 3 722 | –15 873 | 4 174 |
| Opening balance | –8 000 | –8 000 | –6 778 | –1 467 | 1 133 | –29 071 | –19 641 | –15 631 | –10 961 | –7 581 | –3 794 | –72 | –15 945 |
| Closing balance | –11 772 | –6 778 | –1 467 | 1 133 | –29 071 | –19 641 | –15 631 | –10 961 | –7 581 | –3 794 | –72 | –15 945 | –11 771 |

*Outputs on which VAT will be charged.

**Fig. 7.3** Cash flow budget – outline example 2.

| Cash flow budget | | Year: 199X/9Y | | Version: 1 (1st try) | | | | | | | | | |
|---|---|---|---|---|---|---|---|---|---|---|---|---|---|
| % interest rate: 12 | | **Total** | October | November | December | January | February | March | April | May | June | July | August | September |
| *Receipts* | | | | | | | | | | | | | | |
| Milk | 1 | 84 100 | 2 980 | 3 700 | 9350 | 11 450 | 10 760 | 9 050 | 7 445 | 6 600 | 6 485 | 6 480 | 5 800 | 4 000 |
| Calves | 2 | 7 400 | 500 | 3 500 | 2 000 | 1 000 | 400 | | | | | | | |
| Cull cows | 3 | 11 000 | | | | | | | | | | 6 600 | 4 400 | |
| | 4 | | | | | | | | | | | | | |
| Pigs | 5 | 73 140 | 6 900 | 6 900 | 3 680 | 6 900 | 6 900 | 6 900 | 6 900 | 6 900 | 3 680 | 6 900 | 6 900 | 3 680 |
| | 6 | | | | | | | | | | | | | |
| Bed and breakfast | 7 | 9 000 | | | | | | | 700 | 500 | 1 800 | 2 500 | 3 000 | 500 |
| | 8 | | | | | | | | | | | | | |
| Cheese sales | 9 | 15 000 | 1 300 | 1 300 | 1 300 | 1 300 | 1 300 | 1 300 | 1 300 | 1 300 | 1 300 | 1 300 | 1 300 | 700 |
| | 10 | | | | | | | | | | | | | |
| Sundries | 11 | 2 650 | | | | 220 | | | | 200 | | | | 2 230 |
| Capital: grants | 12 | | | | | | | | | | | | | |
| Machinery sales | 13 | 9 000 | | 9 000 | | | | | | | | | | |
| Loans received | 14 | | | | | | | | | | | | | |
| Personal receipts | 15 | 500 | | | | | | | | | | 500 | | |
| VAT charged | 16 | | | | | | | | | | | | | |
| VAT refunds | 17 | | | | | | | | | | | | | |
| **Total** | | 211 790 | 11 680 | 24 400 | 16 330 | 20 870 | 19 360 | 17 250 | 16 345 | 15 500 | 13 265 | 24 280 | 21 400 | 11 110 |

**Fig. 7.4** Cash flow budget (first try) for Home Farm.

| *Payments* | | | | | | | | | | | | | | |
|---|---|---|---|---|---|---|---|---|---|---|---|---|---|---|
| Dairy: l/stock purch. | 18 | 21 250 | | | | | | | | | 3 750 | 10 500 | 7 000 | |
| Feed | 19 | 12 900 | 3 750 | | 3 000 | 3 750 | | 1350 | | | | 1 050 | | |
| Vet and med. | 20 | 3 500 | 288 | 288 | 308 | 308 | 292 | 288 | 288 | 288 | 288 | 288 | 288 | 288 |
| Misc. costs | 21 | 4 000 | 333 | 333 | 333 | 333 | 333 | 333 | 333 | 333 | 333 | 333 | 333 | 337 |
| Pigs:l/stock purch. | 22 | 27 500 | 2 500 | 2 500 | 1 750 | 2 500 | 2 500 | 2 500 | 1 750 | 2 500 | 2 250 | 1 750 | 2 500 | 2 500 |
| Feed | 23 | 29 400 | 2 800 | 14 000 | | | | | 12 600 | | | | | |
| Vet and med. | 24 | 1 200 | 100 | 100 | 100 | 100 | 100 | 100 | 100 | 100 | 100 | 100 | 100 | 100 |
| Misc. costs | 25 | 1 250 | 104 | 104 | 104 | 104 | 104 | 104 | 104 | 104 | 104 | 106 | 104 | 104 |
| Seed | 26 | 600 | | | | | | 600 | | | | | | |
| Fertilizer | 27 | 5 710 | | 5 710 | | | | | | | | | | |
| Spray | 28 | 780 | | | | | | | | 780 | | | | |
| Misc. crop costs | 29 | 120 | | | | | | 30 | | | 40 | | | 50 |
| B & B: food | 30 | 1 680 | | | | | | | 200 | 100 | 300 | 400 | 500 | 180 |
| Other costs | 31 | 750 | 20 | 200 | 100 | 80 | 50 | 25 | 25 | 50 | 50 | 50 | 50 | 50 |
| Cheese: materials | 32 | 1 140 | 100 | 100 | 100 | 100 | 100 | 100 | 100 | 100 | 100 | 100 | 100 | 40 |
| Other costs | 33 | 1 160 | 100 | 100 | 100 | 100 | 100 | 100 | 100 | 100 | 100 | 100 | 100 | 60 |
| Wages: permanent | 34 | 25 500 | 2 125 | 2 125 | 2 125 | 2 125 | 2 125 | 2 125 | 2 125 | 2 125 | 2 125 | 2 125 | 2 125 | 2 125 |
| Casual | 35 | 2 800 | 100 | 100 | 100 | 100 | 100 | 100 | 200 | 200 | 500 | 500 | 500 | 300 |
| Power: fuel | 36 | 6 900 | 540 | 540 | 540 | 740 | 540 | 540 | 550 | 540 | 740 | 550 | 540 | 540 |
| Repairs | 37 | 7 500 | 625 | 625 | 625 | 625 | 625 | 625 | 625 | 625 | 625 | 625 | 625 | 625 |
| Contract hire | 38 | 1 000 | | | | 300 | | | | | | | 300 | 400 |
| Rent | 39 | 9 500 | | | | | | 4 750 | | | | | | 4 750 |
| Property repairs | 40 | 200 | | 200 | | | | | | | | | | |
| Overhead sundries | 41 | 8 600 | 625 | 1 725 | 625 | 625 | 625 | 625 | 625 | 625 | 625 | 625 | 625 | 625 |
| Overdraft interest | 42 | 4 870 | 210 | 241 | 637 | 583 | 504 | 401 | 400 | 443 | 398 | 394 | 351 | 308 |
| Loan interest | 43 | 1 100 | | | | | | 550 | | | | | | 550 |
| Hire purchase | 44 | | | | | | | | | | | | | |
| Leasing charges | 45 | 2 100 | | 525 | | | 525 | | | 525 | | | 525 | |
| | 46 | | | | | | | | | | | | | |
| Capital: buildings | 47 | 12 000 | | 12 000 | | | | | | | | | | |
| Machinery | 48 | 22 000 | | 22 000 | | | | | | | | | | |
| Capital repayment | 49 | 3 000 | | | | | | 1 500 | | | | | | 1 500 |
| Personal: drawings | 50 | 5 400 | 450 | 450 | 450 | 450 | 450 | 450 | 450 | 450 | 450 | 450 | 450 | 450 |
| Tax | 51 | 1 000 | | | | | | | | 1 000 | | | | |
| VAT paid | 52 | | | | | | | | | | | | | |
| **Total** | | 226 410 | 14 770 | 63 966 | 10 997 | 12 923 | 9 073 | 17 196 | 20 575 | 10 988 | 12 878 | 20 046 | 17 116 | 15 882 |
| Net cash flow | 53 | –14 619 | –3 090 | –39 566 | 5 333 | 7 947 | 10 287 | 54 | –4 230 | 4 512 | 387 | 4 234 | 4 284 | –4 772 |
| Opening balance | 54 | –21 000 | –21 000 | –24 090 | –63 656 | –58 323 | –50 376 | –40 089 | –40 035 | –44 265 | –39 753 | –39 366 | –35 132 | –30 848 |
| Closing balance | 55 | –35 619 | –24 090 | –63 656 | –58 323 | –50 376 | –40 089 | –40 035 | –44 265 | –39 753 | –39 366 | –35 132 | –30 848 | –35 620 |
| Monthly interest | | 210 | 241 | 637 | 583 | 504 | 401 | 400 | 443 | 398 | 394 | 351 | 308 | 356 |

**Fig. 7.4** (cont'd)

## 7.3 Cash flow budgets in practice

A typical cash flow budget (the one prepared by Adam Haynes for Home Farm) is shown in Figure 7.4. The most obvious difference between this and Figures 7.2 and 7.3 is the increase in detail and volume of figures: nevertheless, the calculation is exactly the same. To make things simpler, capital and personal receipts and payments have been separated from the trading items at the bottom of each of the relevant sections. This makes it easier to link the cash flow budget with those for profit and capital: the trading items have all appeared already within the profit budget.

The choice of headings needs care. Compiling the budget, and checking it subsequently, are made very much easier if the headings used in the trading section of the cash flow budget correspond to the categories used in the profit budget. This applies particularly to costs. Every effort should be made to ensure that variable costs and fixed costs are classified separately. Similarly the task of compiling a budgeted balance sheet and capital account is simplified by careful classification, so as to distinguish machine purchase from loan repayment, private drawings from tax payments, and so on.

An extra column headed 'actual' is often added for each month. This allows the budget to be used in 'control' of the business, by entry of the actual cash transactions for the month, and their comparison with the expected (i.e. budgeted) results (see Chapter 13).

There is no 'right' way to complete a cash flow budget form, but the following may help.

(1) A sensible starting point is a profit budget in gross margin form such as that described in Chapter 6. You can then go through each gross margin calculation in turn, and then the fixed costs, deciding when the various items of cost and output are likely to arise. Better still, a cash flow assumption form can be drawn up, such as the example in Figure 7.5: as well as helping to organize your budgeting process, it gives you a valuable source of reference when you come to use the budget throughout the year. Best of all, it reduces the inevitable frustration when you are trying to locate errors by cross-checking between gross margins and cash flow, since all the assumptions are spelt out in detail rather than left in your head.

(2) Deciding the timing of trading items is helped considerably by a sound knowledge of the production processes involved: agricultural husbandry with farm enterprises, hospitality management in a guest-house and restaurant, and so on. If you are unsure of these processes yourself, ask someone else: a neighbour, a consultant, or a Business Link advisor, for instance. Best of all, sign up for a full-time or part-time course to help you gain understanding.

(3) When deciding on the allocation of receipts and payments between months, it is important to remember that many of the transactions are likely to be made

***Cash flow assumptions*** Budget title: Home Farm – example entries
Year: 199X/9Y
Version No: 1

| Item | Amount bought/sold | Price per unit (£) | Total rec'd/paid (£) | Notes (including timings) |
|---|---|---|---|---|
| Milk | 420 500 l | 0.20 | 84 100 | Yields peaking Nov–Jan, very low Aug–Oct. One month delay |
| Calves | 74 | 100 | 7400 | 5 in Sep, 35 Oct, 20 Nov, 10 Jan, 4 Feb. One month delay |
| Cull cows | 20 | 550 | 11 000 | 12 in July, 8 Aug – paid same month |
| Pigs | 1 060 | 69 | 73 140 | 50 in Nov, April and May – rest 100/month. One month delay |
| Cheeses | 600 | 25 | 15 000 | Even through year except for September holiday. Cash sales |
| B & B | 430 | 21 | 9 000 | April–Sep: slow start peaking in August. Cash sales |
| . . . and so on | | | | |

**Fig. 7.5** Cash flow assumption sheet.

on a credit basis, i.e. the bill will not be paid until two, three or more weeks after the goods or services have changed hands. Since we are concerned with the effect of transactions on the bank balance of the business, an item which is bought or sold on credit should be entered in the month following that in which the goods are expected to be delivered, or the services provided. For instance, the cost of a delivery received in January, bought on normal trade credit terms, should be entered in the February column of the budget.

Occasionally a manager may be able to obtain longer periods of credit (as part of a special deal) or be forced to give more than a month's credit himself. If this is anticipated, it should of course be allowed for in the budget.

(4) Items bought or sold on credit in the last month of the financial year will not be paid for until the first month of the following year: these are the closing creditors and debtors referred to in Chapters 2 and 6. You can ensure consistency between the various budgets by noting these in the right-hand margin of the cash flow budget form.

Similarly, most businesses will have, at the beginning of the year, unpaid bills outstanding (i.e. opening creditors and debtors) which are likely to be settled in the first month of the year. Allowance must be made for this in the first column of the budget.

(5) Although it is inevitable that in practice a business will use a combination of cheques and cash-in-hand when buying and selling, it is not worth making the distinction between the two when budgeting; assume that all cash changes hands via the bank. The alternative is to have a separate budget for petty cash (see Section 10.3) which adds much work and potential confusion for little benefit.

(6) When calculating overdraft interest, remember to charge interest only on negative closing balances. Positive balances do not automatically attract interest, but if cash is siphoned off to a deposit account, it may be worth incorporating an 'interest received' heading in the 'receipts' section of the budget. The contents should be calculated in the same way as overdraft interest, but on positive closing balances and at a different (lower) monthly interest rate.

Interest on loans outstanding should be incorporated in the 'payments' section of the budget.

(7) Do not calculate to the nearest penny – use your judgement as to how far to round up or down. The larger the assumptions on which a figure is based, the more it should be rounded.

(8) The cash flow budget contains so many figures and involves so many additions and subtractions that it is almost inevitable that you will make mistakes (using a computer spreadsheet is even better, see Appendix A). Make sure you use pencil throughout the budget, and to keep an eraser handy. It is vital to check the budget at each stage, as described in the previous section, so that any slips are spotted and remedied.

## 7.4 Reconciling cash flow and profit

Reference has been made in both this and the preceding chapter to the need for checks *within* the profit and cash flow budgets. These checks can only verify the accuracy of calculation, however, and will not detect items that have been omitted, or wrongly included, in one or other of the budgets. One way of checking for such errors is to perform a *reconciliation* of cash flow with profit (see Figure 7.6).

| | | £ |
|---|---|---|
| Take | Annual net cash flow | –14 619 |
| Add (from cash flow budget) | Capital payments (such as machinery, loan repayments) | 37 000 |
| | Personal (such as drawings and tax) | 6 400 |
| and (from profit budget) | Benefits in kind | 4 797 |
| | Closing valuation of stock | 95 852 |
| | Closing debtors | 9 310 |
| | Opening creditors | 6 800 |
| *Sub-total* | | 145 540 |
| Then deduct (from cash flow budget) | Capital injections | –500 |
| | Capital receipts (such as machinery sales, receipt of loan) | –9 000 |
| and (from profit budget) | Opening valuation of stock | –93 344 |
| | Opening debtors | –10 000 |
| | Closing creditors | –5 520 |
| | Depreciation (on machinery, buildings, etc.) | –13 280 |
| Gives | Net profit after finance costs | 13 896 √ |

| | | £ |
|---|---|---|
| Net profit before finance costs (from profit budget) | | 21 965 |
| Finance costs | Overdraft interest (from Fig. 7.4) | 4 869 |
| | Loan interest | 1 100 |
| | Leasing charges | 2 100 |
| Net profit after finance costs | | 13 896 √ |

**Fig. 7.6** Reconciliation of cash flow and profit, Home Farm.

The aim of the reconciliation is to take the budgeted annual net cash flow and use it to arrive at a budgeted profit or loss. Reference to Chapter 2 will provide a reminder that profit differs from cash flow in including certain non-cash revenues and expenses, and excluding capital and personal receipts and payments. The reconciliation process may thus be summarized as:

*Take* annual net cash flow

*Include* non-cash revenue and expenses:

Add increase (or deduct decrease) in the value of stocks

Deduct depreciation

Add closing debtors and opening creditors
Add benefits in kind

*Exclude* capital, tax and personal items:
Add back capital, tax and personal payments
Deduct capital, tax and personal receipts

*Gives* budgeted net profit.

The result will be a budgeted net profit or loss which *should* be identical to that given by the profit budget. If it is not, the two budgets must be checked, checked and checked again until the errors are found. It is important to approach this task methodically; perhaps the best way is to take the profit budget and work through it checking that each relevant item is included in the cash flow budget, and ticking it off if the entry is correct. This can be a dispiriting process, but it is essential if you (and the bank manager) are to have any confidence in your budgets. Some small consolation can be derived from the knowledge that it is rare for even the best of us to get our budgets to reconcile first time round.

It would be a mistake, having completed the calculations and checked for accuracy, to breathe a sigh of relief, put it away and get on with other jobs. The budget is a means to an end – a way of assessing cash needs over the next year and planning accordingly. It is something to be manipulated to test various possible actions and their effects on bank balances and interest payments: thus Adam Haynes's budget in Figure 7.4 has to be regarded as a 'first try' rather than a finished article. A graphical cash flow profile helps put Adam's budget into context (see Figure 7.7). One of his objectives (Chapter 1) was to avoid an overdraft of more than £30 000: his bank manager has already indicated that he is not willing to offer a facility of more than £40 000, even with Adam's uncle acting as guarantor. Yet the budget shows a peak overdraft requirement of over £60 000, and only one month (the first) is less than £30 000. What can be done?

The cash flow budget can be used to identify opportunities for action as well as

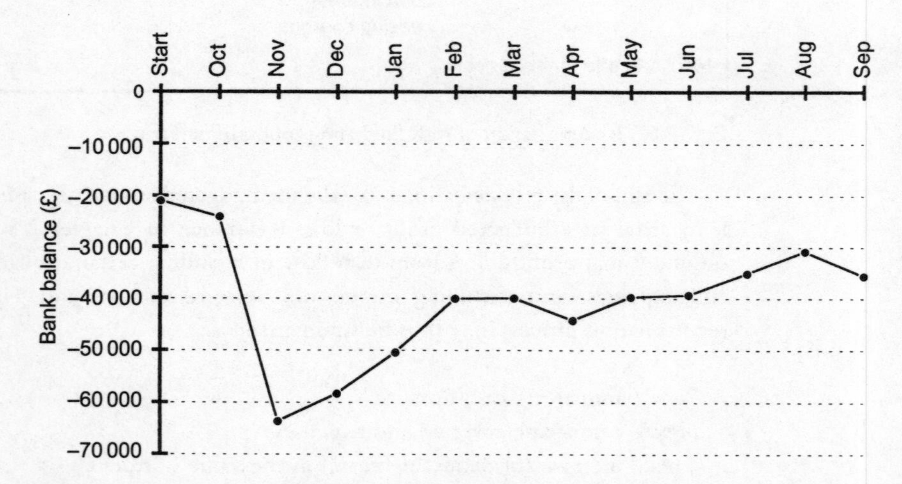


**Fig. 7.7** Cash flow graphical profile (based on Fig. 7.4).

highlighting problems. Those opportunities can be grouped under the following four headings.

(1) *Delaying payments*
It may be possible to delay *purchase* of goods and services (such as machinery bought months earlier than needed in order to claim a discount). Another possibility is delaying *payment* by taking credit, either by the simple expedient of not paying bills on time, or by an alternative form of borrowing such as loan, lease purchase or finance lease.

(2) *Advancing receipts*
Sales can often be brought forward, selling corn soon after harvest rather than storing it, for instance. Chasing slow payers will also help bring money in.

(3) *Boosting cash flow*
Increasing cash flow through trading is not easy; you will probably have investigated the obvious opportunities already when you compiled the gross margin budget. This may, however, be the time to reconsider some of the actions you rejected at that stage; cutting labour costs, for instance, or getting rid of a favourite, but cash-hungry beef herd. The most fruitful, if uncomfortable areas are often capital and personal items. Could private drawings be cut (the world cruise turned into a camping holiday in Cornwall), capital be injected from private funds, a pet investment project shelved? One of the biggest drains on many rural business families is private school fees: is the local comprehensive *really* that bad?

(4) *Renegotiating*
Many of the above suggestions have costs, financial and personal: lost discounts on early purchase, lost premium for out-of-season produce, credit charges, lost future rewards from long-term investment, reduced job satisfaction. If the cost of borrowing is outweighed by these other costs, the last alternative is to use the cash flow budget in helping to persuade the bank manager to raise the overdraft limit. The budget allows such persuasion to take place well before the money is actually needed (rather than after the manager has discovered that the limit has been exceeded without his permission).

The above comments are set in the context of an overdraft. Many businesses have bank accounts in surplus for all or most of the year. Although the need for a budget is not quite so pressing here, it can still be invaluable in identifying opportunities for using that surplus: short-term fattening of store lambs over the winter months, for instance, or reducing future tax liability by buying new machines and buildings before the end of the financial year.

To illustrate the potential for reducing cash flow, Figure 7.8 takes the Home Farm 'first try' budget and adjusts it as follows.

(1) Pig feed bought in smaller lots throughout the year, rather than in bulk and stored.

| Cash flow budget | | Year: 199X/9Y | | Version: 2 (revised) | | | | | | | | | |
|---|---|---|---|---|---|---|---|---|---|---|---|---|---|
| % interest rate: 12 | | Total | October | November | December | January | February | March | April | May | June | July | August | September |
| *Receipts* | | | | | | | | | | | | | | |
| Milk | 1 | 84 100 | 2 980 | 3 700 | 9 350 | 11 450 | 10 760 | 9 050 | 7 445 | 6 600 | 6 485 | 6 480 | 5 800 | 4 000 |
| Calves | 2 | 7 400 | 500 | 3 500 | 2 000 | 1 000 | 400 | | | | | | | |
| Cull cows | 3 | 11 000 | | | | | | | | | | 6 600 | 4 400 | |
| | 4 | | | | | | | | | | | | | |
| Pigs | 5 | 73 140 | 6 900 | 6 900 | 3 680 | 6 900 | 6 900 | 6 900 | 6 900 | 6 900 | 3 680 | 6 900 | 6 900 | 3 680 |
| | 6 | | | | | | | | | | | | | |
| Bed and breakfast | 7 | 9 000 | | | | | | | 700 | 500 | 1 800 | 2 500 | 3 000 | 500 |
| | 8 | | | | | | | | | | | | | |
| Cheese sales | 9 | 15 000 | 1300 | 1300 | 1300 | 1300 | 1300 | 1300 | 1300 | 1300 | 1300 | 1300 | 1300 | 700 |
| | 10 | | | | | | | | | | | | | |
| Sundries | 11 | 2 650 | | | | 220 | | | | 200 | | | | 2 230 |
| Capital: grants | 12 | | | | | | | | | | | | | |
| Machinery sales | 13 | 5 000 | | | | 5 000 | | | | | | | | |
| Loans received | 14 | | | | | | | | | | | | | |
| Personal receipts | 15 | 500 | | | | | | | | | | 500 | | |
| VAT charged | 16 | | | | | | | | | | | | | |
| VAT refunds | 17 | | | | | | | | | | | | | |
| **Total** | | 207 790 | 11 680 | 15 400 | 16 330 | 25 870 | 19 360 | 17 250 | 16 345 | 15 500 | 13 265 | 24 280 | 21 400 | 11 110 |

**Fig. 7.8** Cash flow budget (revised) for Home Farm.

| | | | | | | | | | | | | | | |
|---|---|---|---|---|---|---|---|---|---|---|---|---|---|---|
| *Payments* | | | | | | | | | | | | | | |
| Dairy: l/stock purch. | 18 | 21 250 | | | | | | | | | 3 750 | 10 500 | 7 000 | |
| Feed | 19 | 12 900 | 3 750 | | 3 000 | 3 750 | | 1 350 | | | | 1 050 | | |
| Vet and med. | 20 | 3 500 | 288 | 288 | 308 | 308 | 292 | 288 | 288 | 288 | 288 | 288 | 288 | 288 |
| Misc. costs | 21 | 4 000 | 333 | 333 | 333 | 333 | 333 | 333 | 333 | 333 | 333 | 333 | 333 | 337 |
| Pigs: l/stock purch. | 22 | 27 500 | 2 500 | 2 500 | 1 750 | 2 500 | 2 500 | 2 500 | 1 750 | 2 500 | 2 250 | 1 750 | 2 500 | 2 500 |
| Feed | 23 | 29 400 | 2 800 | 2 418 | 2 420 | 2 418 | 2 418 | 2 418 | 2 418 | 2 418 | 2 418 | 2 418 | 2 418 | 2 418 |
| Vet and med. | 24 | 1 200 | 100 | 100 | 100 | 100 | 100 | 100 | 100 | 100 | 100 | 100 | 100 | 100 |
| Misc. costs | 25 | 1 250 | 104 | 104 | 104 | 104 | 104 | 104 | 104 | 104 | 104 | 106 | 104 | 104 |
| Seed | 26 | 600 | | | | | | 600 | | | | | | |
| Fertilizer | 27 | 5 710 | | | | | | 5 710 | | | | | | |
| Spray | 28 | 780 | | | | | | | | 780 | | | | |
| Misc. crop costs | 29 | 120 | | | | | | 30 | | | 40 | | | 50 |
| B & B: food | 30 | 1 680 | | | | | | | 200 | 100 | 300 | 400 | 500 | 180 |
| Other costs | 31 | 750 | 20 | 200 | 100 | 80 | 50 | 25 | 25 | 50 | 50 | 50 | 50 | 50 |
| Cheese: materials | 32 | 1 140 | 100 | 100 | 100 | 100 | 100 | 100 | 100 | 100 | 100 | 100 | 100 | 40 |
| Other costs | 33 | 1 160 | 100 | 100 | 100 | 100 | 100 | 100 | 100 | 100 | 100 | 100 | 100 | 60 |
| Wages: permanent | 34 | 25 500 | 2 125 | 2 125 | 2 125 | 2 125 | 2 125 | 2 125 | 2 125 | 2 125 | 2 125 | 2 125 | 2 125 | 2 125 |
| Casual | 35 | 2 800 | 100 | 100 | 100 | 100 | 100 | 100 | 200 | 200 | 500 | 500 | 500 | 300 |
| Power: fuel | 36 | 6 900 | 540 | 540 | 540 | 740 | 540 | 540 | 550 | 540 | 740 | 550 | 540 | 540 |
| Repairs | 37 | 7 500 | 625 | 625 | 625 | 625 | 625 | 625 | 625 | 625 | 625 | 625 | 625 | 625 |
| Contract hire | 38 | 1 000 | | | | 300 | | | | | | | 300 | 400 |
| Rent | 39 | 9 500 | | | | | | 4 750 | | | | | | 4 750 |
| Property repairs | 40 | 200 | | 200 | | | | | | | | | | |
| Overhead sundries | 41 | 8 600 | 625 | 1 725 | 625 | 625 | 625 | 625 | 625 | 625 | 625 | 625 | 625 | 625 |
| Overdraft interest | 42 | 2 244 | 210 | 239 | 220 | 194 | 243 | 160 | 236 | 173 | 147 | 163 | 141 | 118 |
| Loan interest | 43 | 1 100 | | | | | | 550 | | | | | | 550 |
| Hire purchase | 44 | | | | | | | | | | | | | |
| Leasing charges | 45 | 2 100 | | 525 | | | 525 | | | 525 | | | 525 | |
| | 46 | | | | | | | | | | | | | |
| Capital: cubicles | 47 | 3 000 | | 1 000 | 1 000 | 1 000 | | | | | | | | |
| Machinery | 48 | 15 000 | | | | 15 000 | | | | | | | | |
| Capital repayment | 49 | 3 000 | | | | | | 1 500 | | | | | | 1 500 |
| Personal: drawings | 50 | 3 000 | 250 | 250 | 250 | 250 | 250 | 250 | 250 | 250 | 250 | 250 | 250 | 250 |
| Tax | 51 | 1 000 | | | | | | | | 1 000 | | | | |
| VAT paid | 52 | | | | | | | | | | | | | |
| **Total** | | 205 384 | 14 570 | 13 472 | 13 800 | 30 752 | 11 030 | 24 883 | 10 029 | 12 936 | 14 845 | 22 033 | 19 124 | 17910 |
| Net cash flow | 53 | 2 406 | −2 890 | 1 928 | 2 530 | −4 882 | 8 330 | −7 633 | 6 316 | 2 564 | −1 580 | 2 247 | 2 276 | −6 800 |
| Opening balance | 54 | −21 000 | −21 000 | −23 890 | −21 962 | −19 432 | −24 314 | −15 984 | −23 617 | −17 301 | −14 737 | −16 317 | −14 070 | −11 794 |
| Closing balance | 55 | −18 594 | −23 890 | −21 962 | −19 432 | −24 314 | −15 984 | −23 617 | −17 301 | −14 737 | −16 317 | −14 070 | −11 794 | −18 594 |
| Monthly interest | | 210 | 239 | 220 | 194 | 243 | 160 | 236 | 173 | 147 | 163 | 141 | 118 | 186 |

**Fig. 7.8** (cont'd)

(2) Fertilizer bought in March (when needed) rather than October. If it is *really* needed in October, the fertilizer company may be able to arrange long credit.
(3) Machinery purchase cut to £15 000 from £22 000 (and sales to £5000) by abandoning plans to buy a second-hand Land-Rover. Further savings might be possible by using contractors, and further leasing would spread the costs. Other machinery purchases delayed by two months.
(4) Capital demands for buildings reduced by deferring plans to buy new bulk feed storage for the pigs. Cost of cubicle improvements spread over three months.
(5) Private drawings cut by £200 per month. Adam will have to face the fact that at least at first he will have to relax his objective of £5000 drawings a year if he is to achieve the other objectives. He still won't be poverty-stricken, given that Evelyn is earning £6000 net per year, and their house, transport and some food come with the business.

The result is shown as a graph in Figure 7.9. The closing bank balance stays well within £30 000 overdraft, and the interest cost is considerably reduced (by over £2600): this has in turn improved the projected profit (Figure 7.10). The worries have not evaporated: the balance is heading downwards in the last month of the year, and some of the changes have merely deferred, rather than deleted expense. Adam will need outline budgets for the next two years for reassurance that the trend will reverse (more businesses fail in the second than in the first year of their lives).

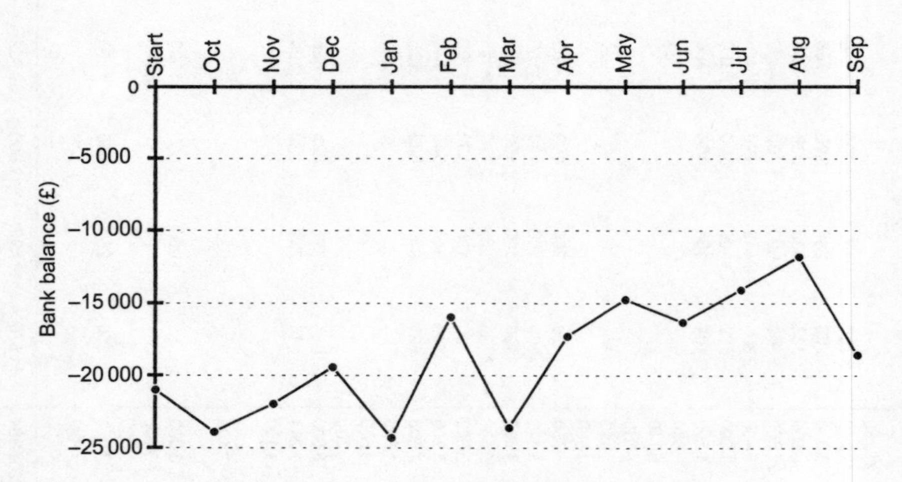


**Fig. 7.9** Cash flow graphical profile – revised budget.

In reality, many of these actions would incur costs such as loss of discounts. In order to maintain consistency with the gross margins in Chapter 6, these have not been built in to Figure 7.9. Partial budgets (see Chapter 9) can be used to test these effects in isolation, but in the end, both cash flow and profit budgets will have to be revised and reconciled. A computer spreadsheet (see Appendix A) is ideally suited to this process, but even if you have only pencil and paper the effort is well worth it.

| | | £ | £ |
|---|---|---|---|
| **Net profit** before depreciation and finance (Figure 6.13) | | | 35 245 |
| Depreciation: | Farm machinery | 9 800 | |
| | Vehicles | 1 200 | |
| | Cheese-making equipment | 800 | |
| | Bed and breakfast equipment | 400 | |
| | Buildings (tenant's fixtures) | 120 | 12 320 |
| **Net profit** before finance costs | | | 22 925 |
| Finance costs: | Overdraft interest (from Figure 7.8) | 2 244 | |
| | Loan interest | 1 100 | |
| | Leasing charges | 2 100 | 5 444 |
| **Net profit** after finance costs | | | 17 481 |

**Fig. 7.10** Net profit adjusted for revised depreciation and overdraft interest.

## References

Warren, M.F. (ed.) (1993) *Managing your Cash.* Book 4 in series: Finance Matters for the Rural Business. ATB Landbase/Seale-Hayne Faculty of Agriculture, Food and Land Use, University of Plymouth.

# Chapter 8

# A Budgeted Balance Sheet

## 8.1 Compiling the budget

The principles of construction and interpretation of a balance sheet have been explained in Chapters 2 and 3 of this book. All that remains is to apply those principles to budgeted information. Compared with other budgets, the budgeted balance sheet is usually easy to compile, since most, if not all of the information required is available from the profit and cash flow budgets, or the preliminary work for those budgets.

In budgeting a closing balance sheet for a particular year, the likely level of cash surplus or overdraft is taken from the cash flow budget. Debtors, creditors and stocks have been estimated in preparing the profit budget, and the written-down value of machinery and buildings derives from the depreciation calculations for that budget. Any repayments and/or extensions of hire purchase, loans or other credit will be shown in the cash flow budget, as will sales and purchases of land or investments.

Thus, as long as budgets for profit and cash flow are available for the year in question, budgeting a closing balance sheet is straightforward. If a detailed cash flow budget is not available, an alternative is to use a projected flow of funds statement (see Chapter 3) together with a simple capital budget, i.e. a statement of the initial capital required. This is very much a second-best, but it is better than nothing.

As always, it is essential to 'prove' the accuracy of the calculation by drawing up a capital account, and checking repeatedly for mistakes if capital account and balance sheet disagree. It is a waste of time and effort to perform these checks in a blind fashion. Certain aspects of a balance sheet are more prone to error than others, and should be checked first. Among these are trade debtors and creditors (often given different values in profit budget and budgeted balance sheet); loan repayments (often allowed for in the balance sheet but not in the cash flow budget, and vice versa); and written-down values of wasting assets, which can easily be miscalculated from the depreciation estimation.

It should also be checked that there has been no revaluation of assets without a corresponding allowance for the 'holding gain' (see Chapter 20). Other points, such as the valuation of stocks, level of bank balance, payments of tax and personal drawings, are easily checked with profit and cash flow budgets, respectively. If the cause of the discrepancy has still not been found, there is little alternative but to embark on a full-scale check of all three budgets. Two facts should always be remembered: (a) that until the error is traced you cannot have complete confidence in your budgets (after all, a discrepancy of £10 could be the result of an error of

£10 000 in one direction and an error of £10 010 in the other), and (b) that the error is not the result of some mysterious external influence, but lies somewhere within your calculations. It is up to you to find it.

## 8.2 A budgeted balance sheet for Home Farm

Figure 8.1 shows a budgeted balance sheet for Home Farm, based on the calculations of profit and cash flow in Chapter 6 and 7. The written-down values of

| 29/9/9X | Projected balance sheet at 28 September 199Y | | | |
|---|---|---|---|---|
| | *Fixed assets* | | | |
| 0 | Tenant's fixtures | | 2 880 | |
| 45 000 | Machinery and plant | | 41 600 | |
| 6 000 | Vehicles | | 7 200 | |
| 50 050 | Dairy herd | | 53 300 | |
| 101 050 | Fixed assets | | | 104 980 |
| | *Current assets* | | | |
| | Stocks | | | |
| 800 | Trading livestock: calves | 800 | | |
| 24 750 | Trading livestock: pigs | 22 500 | | |
| 3 780 | Crops in store: barley grain | 4 560 | | |
| 390 | Crops in store: barley straw | 360 | | |
| 10 000 | Crops in store: silage | 10 000 | | |
| 5 090 | Cheeses | 5 000 | | |
| 2 800 | Dairy feed | 2 800 | | |
| 5 684 | Miscellaneous stores | 6 532 | 52 552 | |
| 10 000 | Trade debtors | | 9310 | |
| 0 | Cash in bank | | 0 | |
| 0 | Cash in hand | | 0 | |
| 63 294 | Current assets | | 61 862 | |
| | *Current liabilities (due within one year)* | | | |
| 6 800 | Trade creditors | | 5 520 | |
| 21 000 | Bank overdraft | | 18 594 | |
| 27 800 | Current liabilities | | 24 114 | |
| 35 494 | Net current assets | | | 37 748 |
| 136 544 | Total assets less current liabilities | | | 142 728 |
| | *Creditors due after one year* | | | |
| 34 000 | Loan | | | 31 000 |
| | *Capital account* | | | |
| | Opening net capital | 102 544 | | |
| | + net profit | 17 481 | | |
| | + capital injections | 500 | | |
| | – private drawings | –3 000 | | |
| | – income and capital tax | –1 000 | | |
| | – benefits in kind | –4 797 | | |
| 102 544 | Closing net capital | 111 728 | | 111 728 |
| 136 544 | Capital employed | | | 142 728 |

**Fig. 8.1** Projected balance sheet for Home Farm.

vehicles and machinery have been adjusted to accommodate the reduced spending recommended in Chapter 7. Note that the capital account is shown here as part of the balance sheet: this is normal practice but remember that it is in fact a separate calculation to the balance sheet. The net capital given by the capital account matches that given by the balance sheet and is final confirmation that all the arithmetic is correct and that profit, cash flow and balance sheet budgets reconcile.

What does the balance sheet show? The techniques of balance sheet analysis have been described in some detail in Chapter 3, and it is enough here to refer to the summary of the analysis shown in Figure 8.2. On the positive side, the budgets anticipate a healthy return on capital of 16%, and a growth of 9% in net capital – considerably higher than the UK rate of inflation at the time of writing. The percentage owned is likely to improve marginally, and long-term debt to equity will fall, partly as a consequence of a loan repayment of £3000. Neither of these ratios is especially healthy, however, and the acid-test ratio is much less than 100%. This emphasizes even more the importance of planning and controlling cash flow in this business, and the need to increase profits to reduce borrowings.

| *Ratio analysis* | *Opening* | *Closing* |
|---|---|---|
| **Profitability ratios (%)** | | |
| Return on owner equity | | 15.6 |
| Return on capital employed | | 16.1 |
| Gross profit margin | | 45.6 |
| Net profit margin | | 8.8 |
| **Activity ratios (days)*** | | |
| Stock turnover period | | 191.2 |
| Debtor turnover period | | 17.7 |
| Creditor turnover period | | 11.3 |
| **Gearing ratios (%)** | | |
| Percentage owned | 62.4 | 67 |
| Long-term debt to equity | 33.2 | 27.7 |
| Interest cover | | 421.1 |
| **Liquidity ratios (%)** | | |
| Acid-test ratio | 36.0 | 38.6 |
| Current ratio | 227.7 | 256.5 |
| **Growth (%)** | | |
| Growth in net capital | | 9.0 |
| Growth in assets | | 1.6 |
| Growth in liabilities | | –12.1 |

* Calculated using average of opening and closing values.

**Fig. 8.2** Ratio analysis for revised profit and balance sheet budgets.

The flow of funds analysis (see Figure 8.3) shows trading providing nearly the whole of the cash flowing through the business during the year, with the overdraft providing the bulk of the rest. The resulting £22 406 is split between private uses, paying off part of the loan and overdraft, and investment in new assets. The latter

| | FLOW OF FUNDS ANALYSIS | | £ |
|---|---|---|---|
| **Sources** | *Net profit (loss)* | | 17481 |
| | Depreciation (+) | | 12320 |
| | Change in debtors (increase –, decrease +) | | 690 |
| | Change in creditors (increase +, decrease –) | | –1280 |
| | Valuations (increase –, decrease +) | | –2508 |
| | Benefits in kind (–) | | –4797 |
| **Cash from trading** | | | 21906 |
| **Add** | Capital introduced | | 500 |
| | Sale of land, property, investments | | 0 |
| | Increase in overdraft | | 0 |
| | Increase in loans | | 0 |
| | Reduction in cash balances | | 0 |
| **Cash from other sources** | | | 500 |
| **Total cash available** | | | 22406 |
| **Uses** | *Commitments* | | |
| | Private drawings | | 3000 |
| | Tax payments | | 1000 |
| | Reduction in loans | | 3000 |
| **Total commitments** | | | 7000 |
| **Add** | Purchase of land, property, investments | | 0 |
| | Net purchase of buildings | | 3000 |
| | Purchase of machinery | 15000 | |
| | Less sale of machinery | –5000 | 10000 |
| | Repayment of overdraft | | 2406 |
| | Increase in cash balances | | 0 |
| | | | 15406 |
| **Total cash used** | | | 22406 |

**Fig. 8.3** Home Farm – flow of funds analysis for revised budget.

should enable the business to be more profitable in future years, but in the short term Adam needs to look for other ways of increasing the cash generated by his business.

# References

Barnard, C.S. & Nix, J. S. (1979) *Farm Planning and Control*, 2nd edn. pp. 542–551. Cambridge University Press, London.

Berry, A. & Jarvis, R. (1991) *Accounting in a Business Context.* Chapman & Hall, London.

Broadbent, M. & Cullen, J. (1993) *Managing Financial Resources*. Butterworth Heinemann, Oxford.

Bull, R.J. (1990) *Accounting in Business*, 6th edn. Butterworth Heinemann, London.

Millichamp, A. (1995) *Finance for Non-financial Managers*, 2nd edn. DPP Publications, London.

Shepherd, C.G. (1991) *An analysis of performance by size, type and form of ownership of farms*. Unpublished Diploma in Farm Management project report, Seale-Hayne Faculty, Polytechnic South West.

Tyran, M.R. (1986) *Handbook of Business and Financial Ratios*. Prentice-Hall, Hemel Hempstead.

Warren, M.F. & Pierpoint, P.R. (1991) Vulnerability of farm businesses in the far South West of England. *Farm Management*, **7**, 12.

Warren, M.F. (ed.) (1993) *Using your Accounts*. Book 2 in series: Finance Matters for the Rural Business. ATB Landbase/Seale-Hayne Faculty of Agriculture, Food and Land Use, University of Plymouth.

# Chapter 9

# Budgeting for Incremental Change

## 9.1 Relevant costs and benefits

Up until now, we have been concerned with preparing budgets for the whole business. This is fine for the annual rebudgeting of a 'steady-state' business, or for major events such as acquisition of a new business or a significant upheaval in an existing one. It is not so useful, though, for investigating relatively small, incremental changes. Various types of incremental change can be identified. They include expansion of the business (such as the purchase of a few more hectares of land); changes within enterprises (such as changes in product, and changes in the type or quantity of inputs used); changes in the enterprise mix (such as the substitution of a riding stable for a beef herd); and changes in items of overhead cost. Related to the latter are changes in the financing of the business.

Despite the term 'incremental', such changes can have considerable effects on the financial health of the business. Therefore it is vital to have some reliable mechanism for assessing these effects. Conventional management accounting procedures can leave a lot to be desired, as the following example illustrates. It shows an abbreviated full cost budget of a business with three enterprises, A, B and C, with B showing a negative net margin.

| | Total (£) | A (£) | B (£) | C (£) |
|---|---|---|---|---|
| Revenue | 100 000 | 50 000 | 25 000 | 25 000 |
| Direct costs | 40 000 | 20 000 | 10 000 | 10 000 |
| Overhead costs | 46 000 | 20 000 | 16 000 | 10 000 |
| Net margin | 14 000 | 10 000 | –1 000 | 5 000 |

On the face of it, enterprise B should be scrapped, saving £1000 per year. Knowing some of the limitations of full costings, however (Section 4.3), and the importance of allowing for cost behaviour (Section 4.4) we might be cautious. In particular, we would want to know more about the allocation of general overheads, which often has to be made on the basis of a formula rather than factual data. For instance, the costs of office administration and the manager's vehicle could have been divided equally between the enterprises, even if in fact enterprise B makes little direct call on either.

More importantly, the judgement above makes no distinction between those costs which are relevant to the decision, and those which are not. *Relevant costs* and *relevant benefits* are those which will actually change as a result of the decision under consideration. Let us re-examine the example with a little more information:

- The manager is seeking a change which will improve profits within three months
- The direct costs include £1000 labour cost of a full-time regular worker shared between all three enterprises: no alternative revenue-earning use for that labour could be found within the timescale
- The overhead cost of B includes £6000 depreciation on a specialized building: a *sunk cost* relating to money that has already been committed
- That building could not easily be sold, but could be let for £3000 rent per year.

Thus within the timescale, £1000 direct costs and £6000 overhead costs would be unavoidable: they are not *relevant* to the decision. On the other hand, the potential rent from the building *is* relevant. It represents the *opportunity cost* of keeping the building in its present use: the benefit that is lost by using a resource in one activity rather than in the next most beneficial activity.

The relevant cost and benefit of keeping enterprise B would be:

| | £ |
|---|---|
| Relevant benefit | |
| Revenue | 25 000 |
| *less* opportunity cost of building | –3 000 |
| | 22 000 |
| Relevant cost | |
| Avoidable direct costs | 9 000 |
| Avoidable overhead costs allocated to B | 10 000 |
| Margin | 3 000 |

To put it another way, the relevant cost and benefit of *discontinuing* enterprise B would be:

| | £ |
|---|---|
| Relevant benefit | |
| Revenue | 3 000 |
| Relevant cost | |
| Unavoidable direct costs | 1 000 |
| Unavoidable overhead costs | 6 000 |
| Deficit | –4 000 |

Whichever way one looks at it, getting rid of B would worsen the situation by £4000 per year (although the story might be very different if the time horizon were longer, giving more opportunity for redeployment of labour, conversion of the building, and so on).

Neither full costing nor conventional gross margin accounting make allowance for variations in cost behaviour: the agricultural gross margin makes a brave stab at it but its compromise between variable and direct costs (Section 5.1.1) makes it a blunt instrument for most relevant cost decisions. What is needed is a simple but powerful variant of marginal costing which accounts specifically for relevant costs and benefits of a decision, i.e. those which:

- Relate to the future
- Are avoidable
- Include incremental costs/benefits
- Include opportunity costs/benefits

Such a device is the partial budget.

## 9.2 Partial budgets

### *9.2.1 Partial budget principles*

A partial budget is concerned only with those financial items which change as an outcome of a particular decision. Those items may be classified into four groups:

(1) Extra returns
(2) Costs saved
(3) Extra costs
(4) Returns lost

The partial budget is simply a device whereby groups (1) and (2) may be set against groups (3) and (4). If the resulting balance is in favour of groups (1) and (2), the proposed change is likely to be worthwhile. This is a direct application of the principle of opportunity cost: testing one course of action by comparing the profit it generates with that lost by not following an alternative course.

The partial budget may be used in two related ways. First, it may be used to test the effects on the whole-business budgets of an alternative to the plan embodied in those budgets. Several such alternatives may be under test, resulting in a 'central' whole-business budget with a series of 'satellite' partial budgets, thus avoiding the effort of preparing a new set of whole-business budgets for each alternative to the central plan. An example of this would be where the central budgets included forage maize in the cropping, but where the manager wished to check the potential of lucerne as an alternative.

The second way of using a partial budget is as a means of testing individual decisions in isolation, where there is insufficient time and/or need for the compilation of whole-business budgets. An example of this would be a decision on whether to sell a food product straightaway, or to put it into cold store in the hope of higher prices later. Another would be the decision on whether to employ a contractor for a particular job, or to use one's own labour and machinery. The partial budget's simplicity and suitability for rapid calculations make it ideal for such circumstances.

Ideally, both methods should be used – the latter as an initial screening device to see whether a change is worthwhile in itself; the former as a device for adjusting the central budgets once a change has passed the screening.

### 9.2.2 Partial profit budgets

The most common and most useful form of partial budget is the partial profit budget. Once this has been calculated, it is usually easy to see the potential effects of the decision on cash flow and capital. An example of the conventional layout of a partial profit budget is shown in Figure 9.1. It is important that the budget should begin with a clear statement of the decision under consideration, to ensure that one never loses sight of the main issues. As in the whole-business budget, all items should be specified in full physical and financial detail where possible.

*Decision* Eliminate 12 ha spring barley, introduce 180-ewe breeding flock. Housing alterations £2000.

| *Costs of change* | | *Benefits of change* | |
|---|---|---|---|
| *Revenue lost* | £ | *Extra revenue* | £ |
| Barley grain: 4 t/ha @ £95 | 4 560 | Lambs: 1.5/ewe @ £45/lamb | 12 150 |
| Barley straw: 2 t/ha @ £15 | 360 | Ewe premium @ £19/ewe | 3 420 |
| | | Wool: £3/ewe | 540 |
| | | Cull ewes @ 25% @ £30 | 1 350 |
| | | Cull rams: 2/year @ 35 | 70 |
| *Sub-total* | 4 920 | *Sub-total* | 17 530 |
| *Extra costs* | | *Costs saved* | |
| Replt. ewes: 30% @ £75 | 4 050 | Seed: 0.12 t/ha @ £267/t | 400 |
| Replt. rams: 2/year @ £250 | 500 | Fertilizer: 0.42 t/ha @ £115/t | 575 |
| Feed: 65 kg/head @ £150/t | 1 755 | Sprays: £30/ha | 360 |
| Vet. and med.: £4/ewe | 720 | Misc.: £10/ha | 120 |
| Misc.: £3/ewe | 540 | Machinery running costs | 900 |
| Extra forage costs | 500 | Casual labour, harvest | 100 |
| Depreciation of building £2000 over 10 years | 200 | Overtime | 200 |
| Extra labour (casual and overtime) | 700 | Combine depreciation (£3000 @ 20%) | 600 |
| *Sub-total* | 8 965 | *Sub-total* | 3 255 |
| **Total costs** | 13 885 | **Total benefits** | 20 785 |
| **Extra profit** before interest | 6 900 | | |

**Fig. 9.1** Partial budget for sheep flock, Home Farm – normalized year.

Items that will not be affected by the change should not be included. Particular care is needed with items of fixed cost, since in many circumstances a change in the intensity of use of these items does not result in a change in the cost to the business. In the example shown, depreciation of the grain store is not a cost saved, since the building is still owned by the business, even though it may be unused. Similar arguments often apply to regular labour and machinery depreciation; saving the labour of a quarter of a man (or tractor) may result in little cash saving if the man (or tractor) must still be employed to perform the other three-quarters of the work. Significant benefit will arise only if the released labour or machinery can be put to profitable use elsewhere (i.e. has a high opportunity cost).

The contents of the budget will depend to a certain extent on the way it is to be used. If it is to be used as a 'satellite' to the central budget, it will be concerned with the specific year covered by the latter. (The exception is where the central budget itself relates to a 'normalized' year.) In this case, the items in the partial budget should be calculated as in the central profit budget, using the same depreciation rates and methods, including the full cost of livestock purchases, and allowing for changes in stock values over the year. Interest charges can be estimated from a partial cash flow budget (see below) but if this is not available, they may be calculated on the initial extra capital (fixed and working) required for the proposed change. (If the change results in a system of lower capital intensity, this may be a cost saved rather than an extra cost.) Purchases of non-wasting assets, such as land, are not of course shown in a budget concerned with net profit, nor are changes in personal or tax items.

If, on the other hand, the partial budget is used as an initial screening device, it is more appropriate to look at a typical or normalized year. This implies using annual average extra costs and benefits. Thus an increase in the number of breeding livestock should be represented by an increase in herd depreciation rather than the full cost of purchase. Depreciation of wasting assets should be calculated by the straight-line method, since diminishing-balance depreciation relates to particular years. All other prices and quantities should be those you would expect in a typical year. Figures 9.1 and 9.2 use this form.

*Decision* Reduce spring barley by 4 ha for 50-van touring caravan site. Works cost £30 000

| *Costs of change* | | *Benefits of change* | |
|---|---|---|---|
| *Revenue lost* | £ | *Extra revenue* | £ |
| Barley grain: 4 t/ha @ £95 | 1 520 | 50 vans × 120 nights × 60% | |
| Barley straw: 2 t/ha @ £15 | 120 | × £4 per night | 14 400 |
| *Sub-total* | 1 640 | *Sub-total* | 14 400 |
| *Extra costs* | | *Costs saved* | |
| Running costs: casual labour | 800 | Seed: 0.13 t/ha @ £267/t | 139 |
| Publicity | 2 000 | Fertilizer: 0.42 t/ha @ £115/t | 193 |
| Other | 500 | Sprays: £30/ha | 120 |
| Depreciation on initial costs | | Misc.: £10/ha | 40 |
| over 10 years | 3 000 | Machinery running costs | 200 |
| | | Casual labour, harvest | 20 |
| | | Overtime | 80 |
| *Sub-total* | 6300 | *Sub-total* | 792 |
| **Total costs** | 7 940 | **Total benefits** | 15 192 |
| **Extra profit** before interest | 7 252 | | |

**Fig. 9.2** Partial budget for caravan site, Home Farm – normalized year.

One of three methods can be used to account for changes in the capital employed when compiling a normalized budget. The first is to charge interest on the borrowed capital employed, at the anticipated interest rate. This gives rise to problems of determining just how much of the extra capital is likely to be borrowed. More importantly, it does not cater for differences in the use of the owner's own capital, which has an opportunity cost.

The second alternative is to include a charge which represents the opportunity cost of *all* the extra capital employed (own and borrowed). The appropriate percentage rate is applied to the extra capital employed, including working capital. Estimation of the extra capital employed is not too difficult, but the estimation of the owner's opportunity cost of capital is more tricky.

The third, and perhaps neatest solution is to exclude charges for the use of capital from the partial budget, and then use the result to calculate the marginal rate of return on the capital invested. The result may then be compared to the potential rate of return from other uses of the capital (i.e. its opportunity cost) in order to decide whether or not to commit the capital.

All three methods are discussed in Chapter 18. Figure 9.3 illustrates the second

| *Introduction of sheep flock (Fig. 9.1)* | | | |
|---|---|---|---|
| *Initial capital* | | £ | |
| Fixed: | Building | 2 000 | |
| | Purchase of livestock | 14 500 | |
| | *less* sale of combine harvester | –3 000 | |
| Working: | 1 year sheep costs (excluding depreciation and livestock purchase) | 4 215 | |
| | *less* barley costs over half a year (excluding depreciation) £2655 ÷ 2 | –1 328 | |
| | **Total** | 16 387 | |
| Interest on initial capital @ 12% | | 1 966 | |
| Extra profit after interest (£6900–£1966) | | 4 934 | |
| Return on marginal capital (£6900 as a percentage of £16 387) | | | 42% |
| *Introduction of caravan site (Fig. 9.2)* | | | |
| *Initial capital* | | £ | |
| Fixed: | Building and works | 30 000 | |
| Working: | Running costs over 4 months (excluding depreciation) £3300 ÷ 3 | 1 100 | |
| | *less* barley costs over half a year | –396 | |
| | **Total** | 30 704 | |
| Interest on initial capital @ 12% | | 3 684 | |
| Extra profit after interest (£7252–£3684) | | 3 568 | |
| Return on marginal capital (£7252 as a percentage of £30 704) | | | 24% |

**Fig. 9.3** Allowing for capital effects in partial budgets.

and third approaches, taking the opportunity cost of borrowing as 12% per annum.

Net capital invested includes both fixed and working capital. Fixed capital is easy to understand, being the money invested in items such as buildings, machinery and livestock. Working capital is the money needed to pay the accumulating running costs until cash starts flowing in to offset them. Thus the business would have to finance the costs of sheep production for a year before seeing a return. In the case of the caravan site, Adam expects his first takings about four months after starting to incur running costs such as advertising. Set against the extra capital required is the capital that may be released by discontinuing the previous production: thus there is a combine harvester to sell when the barley enterprise is disposed of, and working capital set free when there are no barley running costs to finance any more.

The calculations use 'initial capital', in other words, the total amount invested at the start of the project. It could be argued that since the budgets deal with 'normalized' years, 'average capital' should be used, making allowance for the fact that borrowed money will be paid off steadily (interest calculation) or for the declining value of the assets over the years (return on capital). Apart from the assumption about paying back borrowed money often being fallacious, the use of average capital tends to give more optimistic results than other more sophisticated methods such as discounted cash flow (see Chapter 18). Use of initial capital errs on the pessimistic side, which is much safer and goes some way to counteract the over-optimism usually built in elsewhere in the budget.

### 9.2.3 Partial profit budgets for Home Farm

In Chapter 8, it was concluded that Adam should look for ways of improving profitability. The partial budget is ideal for this use, and Figures 9.1 to 9.3 investigate two such possibilities. The initial impression is that the caravan site would be the most profitable. When an estimate of the interest cost is included, though, the sheep enterprise emerges as the better of the two (Figure 9.3). Each has a return on capital greater than the interest rate, but that of the sheep flock is significantly greater.

The decision cannot rest solely on these budgets. Every budget is subject to the risk that the estimated figures used will turn out to be wrong, and this is especially so when investigating new projects. Adam thus needs to quantify this risk, and make sure that his business can withstand the worst outcome (see Chapter 19). Second, he needs to allow for the 'hassle factor'. The budgets deal with financial matters, and cannot accommodate the likely upheaval in the organization of the business, and the effect on the personal lives of himself, family and employees. A caravan site involves the intrusion of other people and their demands which, while many farmers enjoy the company, can cause considerable stress. A sheep flock creates seasonal demands, especially at lambing time, which can ruin family life. These and other similar effects, good and bad, need to be weighed against the financial consequences.

Finally, the change will affect cash flow and capital, particularly at the beginning when the project is getting under way. In this case, it needs little examination to realize that each of the projects would put unwarranted strain on the business. Potentially they are worthwhile developments, but require cash injections that would put the business at risk.

Other projects may not be so clear-cut, however. Figure 9.4 looks at the possibility of getting rid of the pig herd – a profitable opportunity, though with the unpleasant prospect of making a man redundant. Ideally, in order to see whether the change would release enough money to make a significant effect on the overdraft, the cash flow budget for the business should be reworked. If available, a computer spreadsheet makes this a relatively easy task: alternatively a useful short cut is a partial cash flow budget.

*Decision* Get rid of Home Farm pig herd. Use cottage for self-catering holiday lets.

| *Costs of change* | | *Benefits of change* | |
|---|---|---|---|
| *Gross margin lost* | £ | *Gross margin gained* | £ |
| 1000 fattening pigs @ £11.13 | 11 130 | Use of cottage for holiday accommodation (25 weeks @ £300/wk less maintenance and cleaning £800) | 6 700 |
| *Sub-total* | 11 130 | *Sub-total* | 6 700 |
| *Extra fixed costs* | | *Fixed costs saved* | |
| Depreciation of cottage furniture and fittings (£3000 over 5 years) | 600 | Labour: 1 man | 10 000 |
| | | Electricity and fuels | 1 000 |
| | | Misc. fixed costs | 1 000 |
| *Sub-total* | 600 | *Sub-total* | 12 000 |
| **Total** | 11 730 | **Total** | 18 700 |
| **Extra profit** before interest | 6 970 | | |

**Fig. 9.4** Partial budget for elimination of pig herd: normalized year.

### 9.2.4 *Partial cash flow budgets*

The principle of the partial cash flow budget is the same as that of the partial profit budget, except that only cash items are included, and that it is divided up to show the effect of the change on monthly (or quarterly) patterns of cash flow. Balancing extra cash receipts and cash payments saved against extra cash payments and cash receipts lost gives the effect of the change on net cash flow (before interest is taken into account). This can be related to the central cash flow budget to indicate the effect of the change on the closing bank balances for each month. If the results are favourable and the project adopted, the central cash flow budget should be amended accordingly.

In Figure 9.5, the effect of working capital released has an immediate effect through savings in livestock purchase, gathering momentum as more costs such as feed can be saved. The loss of sales from January onwards is partly balanced by

*Decision:* Get rid of pig herd plus one man, and use cottage for holiday lets.

| *Extra receipts* | *Total (£)* | *Oct* | *Nov* | *Dec* | *Jan* | *Feb* | *Mar* | *Apr* | *May* | *Jun* | *Jul* | *Aug* | *Sep* |
|---|---|---|---|---|---|---|---|---|---|---|---|---|---|
| Cottage rent | 3 000 | | | | | | | | | 500 | 1 000 | 1 200 | 300 |
| *Payments saved* | | | | | | | | | | | | | |
| Livestock purchase | 25 000 | | 2 500 | 1 750 | 2 500 | 2 500 | 2 500 | 1 750 | 2 500 | 2 250 | 1 750 | 2 500 | 2 500 |
| Concentrate feed | 19 344 | | | | | 2 418 | 2 418 | 2 418 | 2 418 | 2 418 | 2 418 | 2 418 | 2 418 |
| Misc. pig costs | 984 | | 20 | 50 | 80 | 104 | 104 | 104 | 104 | 104 | 106 | 104 | 104 |
| Vet. and med. | 950 | | 20 | 50 | 80 | 100 | 100 | 100 | 100 | 100 | 100 | 100 | 100 |
| Labour | 6 664 | | | | | 833 | 833 | 833 | 833 | 833 | 833 | 833 | 833 |
| Electricity and fuel | 664 | | | | | 83 | 83 | 83 | 83 | 83 | 83 | 83 | 83 |
| Misc. fixed costs | 664 | | | | | 83 | 83 | 83 | 83 | 83 | 83 | 83 | 83 |
| *Sub-total (a)* | 57 270 | | 2 540 | 1 850 | 2 660 | 6 121 | 6 121 | 5 371 | 6 121 | 6 371 | 6 373 | 7 321 | 6 421 |

| *Extra payments* | *Total (£)* | *Oct* | *Nov* | *Dec* | *Jan* | *Feb* | *Mar* | *Apr* | *May* | *Jun* | *Jul* | *Aug* | *Sep* |
|---|---|---|---|---|---|---|---|---|---|---|---|---|---|
| Cottage maintenance | 720 | | | | | | | | 150 | 100 | 170 | 220 | 80 |
| Cottage furnishing | 3 000 | | | | | | 1 000 | 1 000 | 1 000 | | | | |
| *Receipts lost* | | | | | | | | | | | | | |
| Sale of pigs | 48 760 | | | | | 6 900 | 6 900 | 6 900 | 6 900 | 3 680 | 6 900 | 6 900 | 3 680 |
| *Sub-total (b)* | 52 480 | | | | | 6 900 | 7 900 | 7 900 | 8 050 | 3 780 | 7 070 | 7 120 | 3 760 |
| *Effect on net cash flow before interest (a – b)* | 4 790 | | 2 540 | 1 850 | 2 660 | –779 | –1 779 | –2 529 | –1 929 | 2 591 | –697 | 201 | 2 661 |

**Fig. 9.5** Partial cash flow budget for initial year.

savings in wages. The net result in the first year is a saving of nearly £5000, with the prospect of more in subsequent years.

### 9.2.5 Effect of incremental change on capital

Partial budgeting principles are not appropriate to capital calculations, since one is here assessing a stock of wealth rather than a flow of income. Measuring the effect of an incremental change on the capital structure of a business is a matter of drawing up a new balance sheet in the light of the partial profit budget, and, if available, the partial cash flow budget. Such calculations are applicable only where budgets are used to examine the effects of change in a particular year, since a balance sheet shows net capital at a specific point in time.

## References

Attrill, P. & McLaney, E. (1994) *Management Accounting: an active learning approach.* Blackwell Business, Oxford.

Barnard, C.S. & Nix, J.S. (1979) *Farm Planning and Control*, 2nd edn. pp. 315–17. Cambridge University Press, London.

Broadbent, M. & Cullen, J. (1993) *Managing Financial Resources.* Butterworth Heinemann, Oxford.

Millichamp, A. (1995) *Finance for Non-financial Managers*, 2nd edn. DPP Publications, London.

Warren, M.F. (ed.) (1993) *Improving your Profits.* Book 3 in series: Finance Matters for the Rural Business. ATB Landbase/Seale-Hayne Faculty of Agriculture, Food and Land Use, University of Plymouth. Section 7: Making a partial budget.

# Part 3

# Financial History

So far we have been looking to the future. It would be most unwise to neglect the past, however, and every business needs a means of recording historical information.

A manager will wish to retain information for four reasons. The first is that history can be a valuable guide to the future – one can learn from past mistakes and make corresponding adjustments when budgeting. The second is that effective control of the business is impossible without records of actual performance to compare with the performance that was originally expected. Third, he is required by law to retain certain information, ranging from gun licences to VAT records. Finally, he may like to keep some information for pure enjoyment: a machinery or computer catalogue, for instance, can afford hours of pleasure even if the manager has no intention of buying any of the items shown.

This Part deals with the recording of the three main financial indicators, starting with cash flow (since the calculation of historic profit and net capital is based on records of cash transactions). A separate chapter is devoted to the supporting or 'back-up' records, in recognition of their importance both as a basis for the main financial records and as management tools in their own right.

# Chapter 10
# Recording Cash Flow

## 10.1 Designing a cash recording system

### 10.1.1 General considerations

A recording system of any sort, whether for cash, profit, capital or physical records, must have two characteristics, i.e. it should be as easy as possible to:

- Store the information
- Retrieve the information from its storage and use it effectively

These characteristics tend to work against each other, of course. The easiest way of storing information is to throw everything into a huge box, but this would make retrieval of a particular document somewhat difficult. It is worth a great deal of effort when setting up recording systems to find the 'right' compromise for the business concerned and for the person(s) who will be making use of them. As *Bluff your Way in Accountancy* (Anon., 1987) points out: 'Financial accounting is obligatory, and if you don't do it properly during the year someone will have to do it after the year end . . . Doing it after the year end, from a brown paper parcel full of vouchers, cheque stubs and wages sheets is . . . usually performed by accountants in training because their time is cheap, and a lot of effort is going to be wasted in finding the missing documents. There are *always* missing documents . . .'. What is relatively cheap where accountants' fees are concerned can still prove very expensive for you as a client.

Recording the cash flow of a business implies recording all cash transactions between the business and the 'outside world'. For such records to be used effectively in planning and control, they should be stored, ideally, in such a way that comparisons may be made easily between historic and budgeted cash flows, and that management accounts may be prepared easily. They should also provide the information required to meet the legal obligations of the business, especially in the UK, with respect to VAT and Income (or Corporation) Tax. A business registered for VAT is obliged to submit returns to the Customs and Excise to show the VAT paid and received, etc. All businesses are required to submit verifiable financial information to the Inland Revenue for Income/Corporation Tax purposes.

A method of recording is thus needed which is relatively easy to keep up to date and also allows the grouping of transactions in such a way as to ease the task of interpretation by the manager and accountant. The simplest form of grouping or classification is by the use of a code. Transactions are listed in a cash book, receipts on one page and payments on another. Each type of transaction is given a code, which is recorded in a column alongside each entry. The code 'C' might refer to cheese sales, and 'G' to grain sales, for instance. The main codes could be sub-divided: 'GW' might be wheat grain sales, 'GB' barley grain sales. A similar system would be used for payments.

Another method of classification is the use of columns. As well as the main lists of transactions and the amounts involved, subsidiary columns are used to collect together similar types of payment and receipt. Thus there could be a column for cheese sales, another for grain sales, and so on. On the payment page, columns could be used to distinguish fertilizer purchases from livestock purchases, wages, and other types of payment.

A third means of grouping transactions is by using separate account book pages, or even a separate file, for each type of goods or service.

### 10.1.2 Selecting the method

In choosing between these methods of book-keeping, the manager should bear in mind the two fundamental design criteria. Where *ease of entry* is concerned, he will like the simple coding best, since it involves only two lists of numbers, one for receipts and one for payments. The only slight complication is the need to look up the code for each entry.

The next favoured is likely to be the column analysis, since although the main amounts have to be split between columns, and there are several columns of numbers to add up on each page, they are at least all visible on the one page. The least popular for the small business, from the point of view of ease of entry, is likely to be the page-by-page analysis, where before an entry can be made, the appropriate page must be found. This can be a tedious process if several different types of transaction need recording.

When looking at *ease of retrieval* of the information, it is more difficult to predict opinions. Preferences will depend largely on the number of transactions, which in turn depends on the size and type of business. The simple coding system is least easy to use for recovery of information. Particular difficulties arise where a single payment is made in respect of several different items (such as settlement of a merchant's account in respect of fertilizers, feeds and baler twine). Great care is necessary when coding, since the scope for error is large. On the other hand, where the number of transactions is relatively small, these difficulties may be outweighed by the simplicity of the system.

If a very large volume of transactions is expected, classification by page provides the easiest retrieval of information. The most common form of book-keeping (the 'double-entry' method) is of this form. It can be expanded to cope with any volume of business, since the pages can be split between any number of clerks, each responsible for a particular set of pages, or 'accounts'. It incorporates an automatic record of debtors and creditors, which is valuable where the bulk of the trade is on a credit basis. It is also relatively easy to adapt to computer-based systems. The flexibility of the double-entry system has made it the first choice of book-keeping methods for commercial and manufacturing businesses for several hundred years. Where a business, because of size and/or type of production, generates a large volume of transactions, double-entry book-keeping will be essential.

The needs of most small rural businesses, however, are such that their owners find that the analysis column is the most suitable. While the volume of transactions in a typical farm business is too great to allow reliance on a simple coding system, it is insufficient to warrant the extra effort involved in keeping a double-entry system. Nor is the number of credit transactions likely to be large enough for the in-built debtor/creditor recording of the double-entry system to be much of an advantage.

### 10.1.3 Computerized book-keeping

The balance between effort and benefit is dramatically changed if a computer can be used for financial recording. A well-designed and properly set up accounts program can be easier to operate than a manual cash analysis book, producing a greater quantity of information in a more useful form. Few businesses now operate without a computer, and even in farming the growth in computer ownership has increased rapidly in the 1990s. Specialist programs are now widely available for particular types of business, including farming and tourism, but many prefer to use one of the general-purpose accounting packages available at lower cost. Even a simple spreadsheet can be used to very good effect with a column or coding system. Before finalizing a choice, it is important to research the field thoroughly, talking to existing users of the programs under review, making sure that training and documentation is adequate, and checking with your accountant that it will provide information in a form that is directly usable in authorization or auditing of year-end accounts.

There are still many situations, however, where a computer is not available: moreover, it is not sensible to use a computer program if you do not have some idea of the way in which information is being processed. Therefore this chapter concentrates on manual systems, and particularly the column, or *cash analysis* method.

## 10.2 The cash analysis system

### 10.2.1 Choosing column headings

The first step in establishing a cash analysis system is to buy a *cash analysis book* (CAB). Each page of a CAB includes, on the left-hand side, a broad column for a description of each transaction entered. To the right of this are two thin columns for recording the date of each transaction, and the number of the cheque or paying-in slip concerned. The rest of the page is taken up with a large number of analysis columns. Some publishers produce special account books where each of these columns has been given a heading. In most cases, however, it is better to choose your own headings to suit your own business and system of management accounting. A loose-leaf format adds versatility and allows pages to be sent to the accountant while the book itself is retained for continuing use.

In choosing the headings, it is essential to refer back to the way in which the information is to be used eventually. The legal obligations are fairly easily satisfied by ensuring that space is available for recording VAT (see Section 10.4) and that the classification is sufficiently clear for an accountant to follow when he prepares the financial accounts for the Inland Revenue. Perhaps more crucial is to design the classification to ease the preparation of management accounts, the completion of the 'actual' column in the cash-flow budget, and the compilation of a closing balance sheet. To this end, column headings should be chosen which separate, as far as possible, receipts and payments of one enterprise from those of another, and fixed costs from variable costs. To help in compiling the balance sheet and capital account, all capital, personal and tax items should be kept to the extreme right of the page, clearly separated from the trading items. The column headings should correspond throughout the headings used in the cash-flow budget; this is made much easier if the classifications used in the budget correspond to those used in the profit budget, as suggested in Chapter 7.

In addition to these analysis columns, two further columns are required on each page. One of these is the 'bank column', which is used to record the amount of cash changing hands. The other is the 'contra column', the use of which is explained later in this chapter. The column headings of the receipts and payments analysis pages for Home Farm might be as shown in Table 10.1. Adoption of any of the developments discussed in Chapter 9 would require the addition of further columns.

**Table 10.1** CAB column headings for Home Farm.

| *Receipts* | *Payments* |
|---|---|
| Bank | Bank |
| Contra | Contra |
| Milk | Cattle purchase |
| Cattle | Dairy feed |
| Pigs | Dairy miscellaneous costs |
| Cereals | Pig purchase |
| Bed and Breakfast | Pig feed |
| Cheese sales | Pig miscellaneous costs |
| Miscellaneous trading | Crop costs |
| | Bed and Breakfast costs |
| | Cheese costs |
| | Regular wages |
| | Casual wages |
| | Power and machinery |
| | Miscellaneous fixed costs |
| | Finance: interest and charges |
| *Capital receipts* | *Capital payments* |
| Personal and tax | Personal and tax |
| VAT charged on outputs | VAT paid on inputs |
| VAT refunds from Customs and Excise | VAT paid to Customs and Excise |

The number of columns per page limits the possible extent of classification by column. In Figure 10.1, for instance, 'crop costs' are allocated a single column, instead of the four (for fertilizer, seed, sprays and miscellaneous) which might be thought ideal. In the limited space available, the four columns could be accommodated only by allowing the payments analysis to spill over to another page. This should be avoided wherever possible, since it makes entry of transactions more difficult, and thus errors more likely. The problem may be overcome by using a coding system within the columns. The code 'BF' by an entry in the 'crop costs' column might be used to indicate barley fertilizer, and 'GS' to indicate grass seed. As with the 'all-coding' system described earlier in this chapter, it is essential to write an index to these columns in the front of the CAB.

Some managers prefer to classify inputs entirely by enterprise, resulting in payments analysis columns for dairy variable costs, bed and breakfast variable costs, and so on, with codes used to identify different types of input within the columns. This method tends to make retrieval of information slightly easier, at the expense of slightly more difficult entry. The choice between the two alternatives is largely a matter of personal preference, the priority being to ensure that both CAB and cash-flow budget use the same system.

### 10.2.2 Entry of transactions

For the time being, we will assume that all receipts are paid directly into the bank, and all payments are made by cheque. This simplifies the recording process and minimizes the scope for error (naturally some cash in hand or 'petty cash' will be required for small transactions – this is dealt with separately in Section 10.3). When starting a new account book, the first entry must be that of the opening bank balance; on the receipts page if a credit balance, and on the payments page if an overdraft, with the words 'Opening balance' written in the 'Details' column, the date in the 'Date' column and the amount in the 'Bank' column.

Whenever the cash changes hands, the date and the details of the transaction should be recorded on the stub of the cheque or paying-in book. The details should include the number and type of item, and the name of the other party to the transaction. At least once per month, the manager should go through the cheque book and paying-in book stubs, transferring the date, details and stub number into the relevant page of the CAB. The sum of money paid or received is entered into the relevant analysis column. Entries may be required in several columns if the cash payment or receipt covers a number of items. Figure 10.1 shows the entry of the first four payments made by Adam Haynes in December 199X, together with the entry of the opening balance of £26 302 overdraft.

### 10.2.3 Contra-transactions

Often a firm supplying goods or services to the farm will also purchase items from the farm. Examples are the merchant who buys corn as well as selling fertilizer, and the supplier who takes an old machine in part-exchange for the new. In such cases,

*Cheques paid*
926521 J. Bird: 53 weaner pigs £1325 (zero-rated for VAT)
926522 Cleverdon Feeds: 18 tonnes pig meal £2790 (zero-rated)
926523 Robert Martin: supply of fence posts £100 (supplier not registered for VAT)
926524 A. Jellings: vet services for dairy (£507) and pigs (£190) (at standard rate of 17.5% VAT)

| Date | Details and Supplier | Cheque No. | 1 Bank | 5 Dairy Misc. Costs | 6 Pig Purchase | 7 Pig Feed | 8 Pig Misc. Costs | 13 Misc. Fixed Costs | 14 Finance Interest and Charges | 15 Capital Payments | 16 Personal and Tax | 17 VAT Paid on Inputs |
|---|---|---|---|---|---|---|---|---|---|---|---|---|
| 1/12 | Opening balance (overdraft) | | 26302.00 | | | | | | | | | |
| 5/12 | J. Bird: 53 weaners | 521 | 1325.00 | | 1325.00 | | | | | | | |
| | Cleverdon Feeds: 18 t pig meal | 522 | 2790.00 | | | 2790.00 | | | | | | |
| 7/12 | R. Martin: fence posts | 523 | 100.00 | | | | | 100.00 | | | | |
| | Jellings: vet services | 524 | 818.98 | V 507.00 | | | V 190.00 | | | | | 121.98 |

**Fig. 10.1** Entry of transactions in cash analysis book – example.

the account will be reckoned up at a convenient time (probably the end of the month) and the business owing the most money will send a cheque for the balance to the other. This is known as a *contra-transaction*. Figure 10.2 gives a diagrammatic representation of an example where a farmer has, during a particular month, sold £1000-worth of wheat to a merchant, and has bought £600-worth of feed from the same firm.

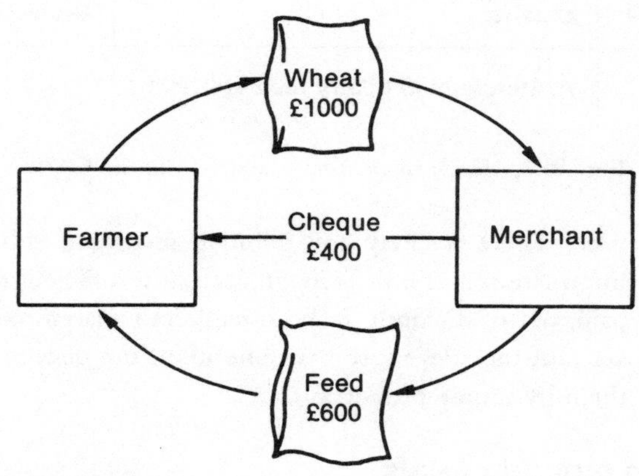


**Fig. 10.2** Contra-transaction.

A single cheque thus represents both a receipt and a payment. To maintain the accuracy of the classification, these two elements must be separated and recorded on the relevant pages. This is achieved through use of the *contra-column*. In Figure 10.2 the farmer receives a cheque for £400, which is entered in the 'Bank column' of the receipts page. The value of the wheat sold is £1000, so this is entered in the 'Wheat column'. The balance of £600 is entered in the 'Contra-column'; it is simply a way of explaining the discrepancy between the two previous figures. At the same time, the farmer buys feed to the value of £600. He thus enters £600 in the 'Feed column' of the payments page, but enters nothing in the 'Bank column' since no cheque has been sent. Again, the 'Contra-column' is used to explain the discrepancy between the amounts in the bank and analysis columns (see Figure 10.3).

The use of a contra-column allows contra-transactions to be recorded with the minimum disturbance to the cash analysis system, and allows easy checking of accuracy. In particular, it is worth noting that the figures in the contra-column of the receipts page are identical to those of the payments page. This provides a useful check on accuracy when the accounts are summarized at the end of the month (see Section 10.2.5).

### 10.2.4 Discounts

Discounts are of two types. The first is the *trade discount*, given as an incentive for bulk purchase, or to build up goodwill with a regular, even if not large, customer. The second is the *cash discount* given for settlement of an account within a stipulated

(a) Receipts page

| Date | Details | Stub | Bank | Contra | Wheat |
|---|---|---|---|---|---|
| 4/8 | A. Merchant (10 t wheat at £100/t) | 123 | 400 | 600 | 1000◎ |

(b) Payments page

| Date | Details | Stub | Bank | Contra | Feed |
|---|---|---|---|---|---|
| 4/8 | A. Merchant (5 t dairy feed at £120/t) | | | 600 | 600◎ |

**Fig. 10.3** Entry of a contra-transaction in the CAB.

time. There is rarely any point in recording either type of discount, since the information is of little value in itself. It is sufficient to record the amount of money paid, net of discount, in both bank and analysis columns of the CAB. Where an account includes more than one item, the discount should be allocated between them by simple proportion.

### 10.2.5 The monthly totals

At least once a month, time should be set aside for 'doing the books'. This process should include the totalling of each column, the finding of a closing bank balance, and the checking of the result by both internal tests and by reconciliation with the bank statement. The first step is to rule off each page below the last entry, and find the total of each column, including the bank and contra-columns. Two accuracy checks may then be made:

(1) The total of the receipts column should be identical to the total of the payments contra-column.
(2) When the bank column total on each page (excluding the opening balance) is added to the contra-column total, the result should be identical to the sum of the totals of the remaining columns (each receipt or payment is entered twice; once in the bank and/or contra-column, and again in one or more analysis columns).

If these checks do not work out, the entries for the month must be checked systematically until the error is found. Then the completeness of the recording can be checked by reconciling with the bank statement.

### 10.2.6 Reconciliation with the bank statement

Cash analysis is the manager's attempt to keep track of changes in his bank account. The bank keeps its own records, a copy of which may be sent, on request, to the manager at convenient intervals. This is the *bank statement*.

The existence of two sets of records of the same bank account provides a further opportunity to check the accuracy of recording. The first step is to ask the bank to send statements at a time in each month which is convenient for the manager to complete the accounts for that month (usually the end of the month). The cash analysis results and the bank statement may then be compared as soon as the former are complete. The chances are that the two will not agree, for one or more reasons:

(1) The bank may have made an error in recording (unlikely but possible).
(2) You may have made an error in recording (rather more likely).
(3) The bank may have omitted certain items, such as cheques which have not yet been presented by the recipient.
(4) You may have omitted certain items, especially those where cheques are not involved, such as standing orders, direct debits, and deduction of bank interest and charges.

The simplest way of identifying the reason for any discrepancy is to go through the bank statement, ticking off the entries which also appear in the CAB. Any entries which appear in the bank statement but not in the CAB may be checked, pencilled in the bank column below the bank balance, and a new balance found in the relevant analysis columns.

The balance can then be found between the receipts bank column total and the payments bank column total, and entered on the page with the smallest total. A credit balance is entered on the payments page, and an overdraft is shown on the receipts page. Any entries which appear in the CAB but not in the bank statement may similarly be checked, pencilled in on the statement, and a new balance found. The two balances should now agree. If they do not, they must be checked systematically until the error is found. The results of the cash analysis may then be used with reasonable confidence in cash flow control (Chapter 13) and as a basis for the preparation of profit and capital statements.

The month-end totalling and checking process is illustrated for Home Farm, December 199X, in Figures 10.4, 10.5 and 10.6.

After the closing balance has been reconciled with the bank statement, the accounts are ruled off and the balance 'carried down' to open the accounts for the following month. The opening balance is always entered on the opposite page to that on which the previous month's closing balance appears. Thus an opening overdraft is shown on the payments page, and an opening credit balance on the receipts page.

## 10.3 Petty cash

So far it has been assumed that all transactions take place via the bank. In most businesses, however, it is useful to have some cash in hand for small expenses, such as the purchase of half a dozen bolts from the local ironmonger's, or the employment

*Cheques paid*
926521 J. Bird: 53 weaner pigs £1325 (zero-rated for VAT)
926522 Cleverdon Feeds: 18 tonnes pig meal £2790 (zero-rated)
926523 R. Blackshaw: supply of fence posts £100 (supplier not registered for VAT)
926524 A. Jellings: vet services for dairy (£507) and pigs (£190) (at standard rate of 17.5% VAT)
926525 Treachers Ltd: dairy supplies (£345.36 + 17.5% VAT) and pig supplies (£136.65 + 17.5% VAT) less 3% cash discount
926526 Cash for wages: regular £680.00, casual £215.00
926527 Cash for wages: regular £685.00, casual £225.00
926528 Fisher Garage: diesel fuel £216.89 + 17.5% VAT less 3% cash discount
926529 Brassleys Agricultural Engineers: tractor gear-box and clutch repair £1050.00 less 5% trade discount + 17.5% VAT
926530 Cash for wages: £810.00 regular, £100 casual
926531 D. Jewett: purchase of tractor, £3545.89 + 17.5% VAT less 5% cash discount, less sale of tractor, £500 + 17.5% VAT
926532 A. Smith: cheese packaging materials £654 + 17.5% VAT

| Date | Details and Supplier | Cheque No. | 1 Bank | 2 Contra | 3 Cattle Purchase | 4 Dairy Feed | 5 Dairy Misc. Costs | 6 Pig Purchase | 7 Pig Feed | 8 Pig Misc. Costs |
|---|---|---|---|---|---|---|---|---|---|---|
| 1/12 | Opening balance (overdraft) | | 26 302.00 | | | | | | | |
| 5/12 | J. Bird: 53 weaners | 521 | 1 325.00 | | | | | 1 325.00 | | |
| | Cleverdon Feeds: 18 t pig meal | 522 | 2 790.00 | | | | | | 2 790.00 | |
| 7/12 | R. Blackshaw: fence posts | 523 | 100.00 | | | | | | | |
| | A. Jellings: vet services | 524 | 818.98 | | | | V 507.00 | | | V 190.00 |
| | Treachers: dairy & pig supplies | 525 | 551.90 | | | | M 335.00 | | | M 132.55 |
| | Cash for wages | 526 | 895.00 | | | | | | | |
| 13/12 | Cash for wages | 527 | 910.00 | | | | | | | |
| | Fisher Garage: diesel, etc. | 528 | 248.34 | | | | | | | |
| 20/12 | Brassleys: tractor repair | 529 | 1 172.06 | | | | | | | |
| | Cash for wages | 530 | 910.00 | | | | | | | |
| | Meredith auctions: commission | | | 89.07 | | | M 75.80 | | | |
| 30/12 | D. Jewett: purchase of tractor | 531 | 3 401.63 | 587.50 | | | | | | |
| | A. Smith: cheese packaging | 532 | 768.45 | | | | | | | |
| | Totals before reconciliation | | 40 193.36 | 676.57 | 0.00 | 0.00 | 917.80 | 1 325.00 | 2 790.00 | 322.55 |
| | Reconciliation adjustments | | | | | | | | | |
| | Kirks Dairy (fees) | | | 142.18 | | | M 121.00 | | | |
| | Personal drawings | | 260.00 | | | | | | | |
| | Bank interest and fees | | 254.98 | | | | | | | |
| | Adjusted totals | | 40 708.34 | 818.75 | 0.00 | 0.00 | 1 038.80 | 1 325.00 | 2 790.00 | 322.55 |

**Fig. 10.4** Home Farm – payments analysis for December 199X.

| 9 Crop Costs | 10 B & B Costs | 11 Cheese Costs | 12 Regular Wages and NI | 13 Casual Wages and NI | 14 Power and Machinery | 15 Misc. Fixed Costs | 16 Finance Interest and Charges | 17 | 18 Capital Payments | 19 Personal and Tax | 20 VAT Paid on Inputs |
|---|---|---|---|---|---|---|---|---|---|---|---|
| | | | | | | 100.00 | | | | | |
| | | | | | | | | | | | 121.98 |
| | | | | | | | | | | | 84.35 |
| | | | 680.00 | 215.00 | | | | | | | |
| | | | 685.00 | 225.00 | | | | | | | |
| | | | | | F 210.38 | | | | | | 37.96 |
| | | | | | R 997.50 | | | | | | 174.56 |
| | | | 810.00 | 100.00 | | | | | | | |
| | | | | | | | | | | | 13.27 |
| | | | | | | | | | 3 368.60 | | 620.53 |
| | | 654.00 | | | | | | | | | 114.45 |
| 0.00 | 0.00 | 654.00 | 2 175.00 | 540.00 | 1 207.88 | 100.00 | 0.00 | 0.00 | 3 368.60 | 0.00 | 1 167.10 |
| | | | | | | | | | | | 21.18 |
| | | | | | | | | | | 260.00 | |
| | | | | | | | 254.98 | | | | |
| 0.00 | 0.00 | 654.00 | 2 175.00 | 540.00 | 1 207.88 | 100.00 | 254.98 | | 3 368.60 | 260.00 | 1 188.28 |

**Fig. 10.4** (cont'd)

*Cheques received*

125629 Newington's, butchers: 52 cutter pigs £4269.00 (zero-rated)
125630 Meredith Auctions: sale of 19 calves (zero-rated), £1895.00) less commission £75.80 + 17.5% VAT
125631 Williams' Delicatessens: cheeses £1095.20 (zero-rated)
125632 Robin Clutterbuck: hedge-cutting, £340.00 + 17.5% VAT
125633 J. Bull: firewood, £50.00 + 17.5% VAT
125634 Customs and Excise: VAT refund, £1674.00

| Date | Details and Customer | Ref. No. | 1 Bank | 2 Contra | 3 Milk | 4 Cattle | 5 Pigs | 6 Cereals | 7 B & B Costs | 8 Cheese Sales | 9 Misc. Trading | 10 Capital Receipts | 11 Personal Receipts | 12 VAT Charged on Outputs | 13 VAT Refund from C & E |
|---|---|---|---|---|---|---|---|---|---|---|---|---|---|---|---|
| 1/12 | Newington's butchers | 629 | 4 269.00 | | | | 4 269.00 | | | | | | | | |
| 20/12 | Meredith Auctions: 19 calves | 630 | 1 805.94 | 89.07 | | 1 895.00 | | | | | | | | | |
| | Williams' Delicatessens: cheeses | 631 | 1 095.20 | | | | | | | 1 095.20 | | | | | |
| | Robin Clutterbuck: hedge-cutting | 632 | 399.50 | | | | | | | | 340.00 | | | 59.50 | |
| | J. Bull: firewood | 633 | 58.75 | | | | | | | | 50.00 | | | 8.75 | |
| 30/12 | Customs and Excise: VAT refund | 634 | 1 674.00 | | | | | | | | | | | | 1 674.00 |
| | D. Jewett: tractor (part-exchange) | | | 587.50 | | | | | | | | 500.00 | | 87.50 | |
| | Totals before reconciliation | | 9 302.39 | 676.57 | 0.00 | 1 895.00 | 4 269.00 | 0.00 | | 1 095.20 | 390.00 | 500.00 | 0.00 | 155.75 | 1 674.00 |
| | Reconciliation adjustments | | | | | | | | | | | | | | |
| | Kirks Dairy: 46 023 litres milk | | 9 111.82 | 142.18 | 9 254.00 | | | | | | | | | | |
| | Adjusted totals | | 18 414.21 | 818.75 | 9 254.00 | 1 895.00 | 4 269.00 | 0.00 | | 1 095.20 | 390.00 | 500.00 | 0.00 | 155.75 | 1 674.00 |
| | Closing bank balance (overdraft) | | 22 294.13 | | | | | | | | | | | | |
| | | | 40 708.34 | | | | | | | | | | | | |

**Fig. 10.5** Home Farm – receipts analysis, December 199X.

**MidWest Bank Plc**

Statement of Account

MR A HAYNES (FARM) A/C
HOME FARM
ANYWHERE

Account Number 867554321

| Date | Particulars | Payments | Receipts | Balance | |
|---|---|---|---|---|---|
| | Opening balance | | | 26302.00 | OD |
| 4-Dec | 125629 | | 4269.00 | 22033.00 | OD |
| 8-Dec | 926521 | 1325.00 | | 23358.00 | OD |
| | 926522 | 2790.00 | | 26148.00 | OD |
| 11-Dec | 926523 | 100.00 | | 26248.00 | OD |
| | 926526 | 895.00 | | 27143.00 | OD |
| | 926525 | 551.91 | | 27694.91 | OD |
| 19-Dec | 926524 | 818.98 | | 28513.89 | OD |
| | CHARGES TO 1 DEC | 32.00 | | 28545.89 | OD |
| | INTEREST TO 1 DEC | 222.98 | | 28768.87 | OD |
| | 926527 | 910.00 | | 29678.87 | OD |
| 20-Dec | KIRKS DAIRY | | 9111.82 | 20567.05 | OD |
| 21-Dec | 926528 | 248.34 | | 20815.39 | OD |
| 27-Dec | 125631 | | 1095.20 | 19720.19 | OD |
| | 125630 | | 1805.94 | 17914.25 | OD |
| | 125633 | | 58.75 | 17855.50 | OD |
| | 125632 | | 399.50 | 17456.00 | OD |
| 28-Dec | 926530 | 910.00 | | 18366.00 | OD |
| | 926529 | 1172.06 | | 19538.06 | OD |
| 31-Dec | A. HAYNES (PERS A/C) | 260.00 | | 19798.06 | OD |
| | *Reconciliation:* | | | | |
| | *less unpresented cheques rec'd* | | | | |
| | 634 | | | -1674.00 | |
| | *plus unpresented cheques paid* | | | | |
| | 531 | | | 3401.63 | |
| | 532 | | | 768.45 | |
| | *Adjusted balance* | | | 22294.14 | √ |

**Fig. 10.6** Home Farm bank statement for December 199X.

of a schoolboy on Saturday mornings. If such small or petty cash transactions do take place within a business, it is essential to record them. A number of small transactions can accumulate to a large sum and their omission will distort the picture shown by the accounts.

A basic requirement of any petty cash system is a petty cash box, the use of which helps avoid business money being mixed up with beer money, or being deposited in various forgettable places around the farmhouse. Where petty cash transactions are infrequent, it is possible to record them directly in the CAB, using a 'Petty cash' column in addition to 'Bank' and 'Contra', and analysing in the usual way. With a typical volume of petty cash transactions, however, it is more convenient to use a separate petty cash book; it avoids the CAB being cluttered with myriads of small transactions, and can be small enough to fit in the cash box.

It is best to restrict the use of a petty cash system to payments only. Receipts should ideally be kept separate and paid directly into the bank periodically. Enterprises which depend on high levels of cash trading, such as those in the retail or hospitality trades, should invest in an electronic till to ensure that an accurate and comprehensive record is kept. A petty cash book should have at least two columns per page; one for the total amount paid, and one for the VAT paid (see Figure 10.7). Given an accurate description of each transaction, and any VAT invoices, transfer of the details to the CAB at the end of the month is straightforward. It can be made easier by adding, if space permits, further columns to allow for more extensive classification.

| *Date* | *Details* | *Amount* | *Casual labour* | *Car* | *Other* | *VAT paid* |
|---|---|---|---|---|---|---|
| 21/8 | Bale carting | 20.00 | 20.00 | | | |
| 23/8 | Headlight unit (etc.) | 4.60 | | 4.00 | | 0.60 |
| 30/8 | **Totals** | 85.00 | 60.00 | 20.00 | 2.00 | 3.00 |
| 30/8 | Cheque drawn | 85.00 | | | | |

**Fig. 10.7** Petty cash book – example layout.

The most popular way of managing petty cash is the 'imprest' or 'float' method. The manager decides how much petty cash he is likely to need between visits to the bank. This amount, plus an allowance for contingencies, is drawn from the bank and deposited as a float in the cash box. The initial withdrawal from the bank (and any subsequent increase in the float) is recorded in the CAB as a capital payment, since it represents the establishment of a separate capital reserve. Before the next visit to the bank, the petty cash book is balanced and the cash in the box is counted. The two sums should tally: if they do not, there is something wrong with the accuracy of the manager's recording, the honesty of the employees, or his ability to keep his own spending money distinct from the business money.

Enough cash is drawn from the bank to bring the float up to the original level. This amount represents the petty cash paid out during the period. The cheque may thus be entered in the CAB payments page 'Details' column as 'Petty cash' and analysed out according to the information in the petty cash book (see Figure 10.8).

## 10.4 Statutory records

### 10.4.1 Value Added Tax

*Value Added Tax* is a tax on sales imposed throughout the European Union and elsewhere, but described here in the UK context. This is not the place to describe

| *Date* | *Details* | *Stub* | *Bank* | *Casual labour* | *Vehicle expenses* | *Misc.* | *VAT paid* |
|---|---|---|---|---|---|---|---|
| 30/8 | Petty cash | 123 | 85.00 | 60.00 | 20.00 | 2.00 | 3.00 |

**Fig. 10.8** Petty cash – entry in CAB.

the tax in great detail, but it is important to recognize certain characteristics of the VAT system. This tax distinguishes between four different types of goods and services:

(1) Standard-rated items
(2) Zero-rated items
(3) Exempt items
(4) Items which are outside the scope of VAT

Examples of each are shown in Table 10.2.

The producer of a standard-rated item must charge tax, at the relevant rate, on the sale of that item, but may, at the same time, reclaim any tax he has paid on the

**Table 10.2** Examples of VAT classifications (see *Scope and Coverage*. HM Customs and Excise Notice No. 701).

| Standard Rate | | Zero Rate | Exempt |
|---|---|---|---|
| *Fertilizers/sprays/manures*<br>*Machinery*<br>Including:<br>New plant, machines and implements<br>Car repairs<br>Machinery repairs<br>Workshop, tools, spares<br>Hire of machinery<br>Road fuels<br>Other fuels for business use (e.g. tractor diesel, gas, electricity, coal and coke)<br>*Non-food produce*<br>Including:<br>Flowers, flower seeds<br>Bulbs<br>Trees and wood<br>Wool<br>*Repairs*<br>Including:<br>Buildings (except protected buildings)<br>Roads<br>Fences<br>Electrical | *Professional services*<br>Including:<br>Solicitors, accountants<br>Consultants' fees<br>Auctioneers' commission<br>Vet. and med.<br>Leasing services<br>*General services*<br>Including:<br>Recording fees and AI<br>Contracting<br>Water and sewerage for business use<br>*Others*<br>Crop and livestock sundries<br>Holiday accommodation<br>Provision of meals<br>Hot take-away meals<br>Horses, sheepdogs<br>Petfood<br>Advertising<br>New building work<br>Building conversions<br>Building extensions | *Food production*<br>Including:<br>Produce (e.g. wheat, potatoes, beef, milk, fish, game birds)<br>Products (e.g. butter, cheese)<br>Livestock (e.g. cattle, sheep, pigs, poultry)<br>Feed (e.g. cattle feed, pig feed, poultry feed)<br>Seeds (e.g. wheat, potatoes, sugar beet)<br>*Others*<br>Grain and potato futures | *Bank charges and interest (paid or received)*<br>Building Society interest<br>Cottage rents (not holiday)<br>Finance charges<br>Insurance (business)<br>Land<br>Letting land for crops<br>Postages<br>Timber felling rights<br>Wayleaves<br>*Items outside scope (excluded from return)*<br>VAT payments and refunds<br>Wages, PAYE and NI<br>Insurance claims<br>Conmpensation<br>Life assurances/pensions<br>Private payments and income<br>Dividends<br>Grants and subsidies<br>Loans and repayments<br>Cars<br>Business rates<br>Levies<br>NFU subscriptions<br>Vehicle licences & MoT fees |

purchases of inputs used to produce the item. He pays the balance between receipts and payments of VAT to the Customs and Excise, who administers the tax. If receipts in a particular period are less than payments, he may claim a refund. Thus the net direct financial effect on the producer's business is zero (although the indirect effect is to increase the price of his products and thus depress sales). This procedure is followed at every stage of the production and marketing process, with the final consumer paying the accumulated tax (see Figure 10.9).

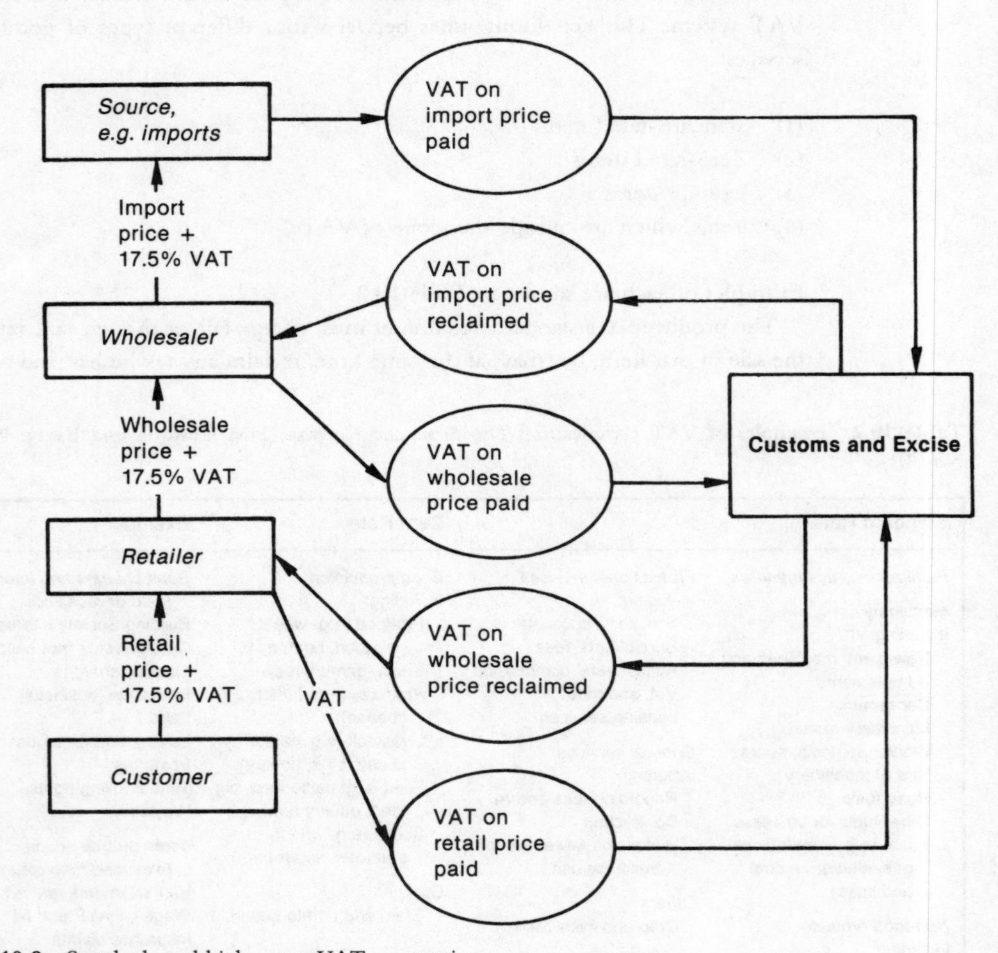


**Fig. 10.9** Standard- and higher-rate VAT – operation.

The producer of a zero-rated item is not required to charge VAT on sales of that item, but is allowed to reclaim the tax paid on inputs used in producing them[13]. The producer of an exempt item is not required to charge VAT on its sale, but may not reclaim VAT paid on inputs used in producing that item.

Producers with a taxable turnover of less than a stipulated amount (£47 000 per annum at the time of writing) are not required to register for VAT, i.e. they do not have to charge VAT on any sales, and are not able to reclaim tax paid on inputs. For businesses where most sales are taxable at the higher or standard rates, it pays to avoid registration if possible. Farmers, on the other hand, are concerned mainly with the production of food products, which are zero-rated. Since this implies that they will be paying more tax on inputs than they receive from sales, it is worth most farmers registering for VAT in order to claim the refunds.

The VAT system uses the businessman as an intermediate tax-collector. To prevent fraud, the law requires painstaking recording of each transaction made by a registered producer. This involves the issuing of special invoices, the storing of all invoices and related documents for three years, the regular completion of tax returns, and the ability to demonstrate that these returns are accurate. These requirements affect the manager's recording of transactions in two ways. First, the manager will wish to record the effect of VAT on his cash flow. Therefore he will require two VAT columns on each of the payments and receipts pages of his CAB. On the receipts page he will need columns for 'VAT received' and 'VAT refunds from Customs and Excise' and, on the payments side, columns for 'VAT paid on inputs' and 'VAT paid to Customs and Excise'. Second, he will wish to make the completion of the VAT return as easy as possible. The VAT return is a document that must be completed every three months and sent to the Customs and Excise[14]. The information required includes:

(1) VAT received on sales
(2) Deductible VAT paid on inputs (charged on the amount before cash discount)
(3) Net payment due to, or reclaimed from, Customs and Excise
(4) The value of all outputs before cash discounts given
(5) The value of all taxable inputs before cash discounts received

Items (1), (2) and (3) are catered for by the CAB columns described above; the additional information can be supplied by modification of the CAB. This entails adding further analysis columns to the receipts and payments pages. One column is added to the payments page, for recording the value of inputs charged at higher, standard and zero rates. Two columns are added to the receipts analysis, one to record the value of zero-rated sales, and the other for the value of exempt sales. These extra columns are not part of the cash analysis, and should be carefully ruled off from the other columns (see Figure 10.10).

Anyone faced with the practicalities of designing a financial recording system for a small rural business is strongly recommended to refer to Warren (1993), a step-by-step guide to cash analysis and VAT recording.

Various special schemes are available to help reduce the complexity of VAT recording for business of particular types. For instance, businesses with an annual turnover of less than £300 000 may register for an annual accounting scheme, making regular payments by direct debit during the year, and settling the balance at the end, thus avoiding the necessity for detailed quarterly or monthly returns (useful only for net payers of VAT, and thus not for most farmers). An agricultural flat rate scheme allows for farmers to charge 4% (which they can then keep) on all outputs sold to VAT-registered customers, rather than keeping the detailed records

*Recepits analysis*

| Date | Details and Customer | Ref. No. | 1 Bank | 2 Contra | 12 VAT charged on outputs | 13 VAT refund from C & E | 14 VAT RETURN ONLY | 15 |
|---|---|---|---|---|---|---|---|---|
| | | | | | | | Standard and zero-rated | Exempt and out of scope |
| 1/12 | Newington's, butchers | 629 | 4 269.00 | | | | 4 269.00 | |
| 20/12 | Meredith Auctions: 19 calves | 630 | 1 805.94 | 89.07 | | | 1 895.00 | |
| | Williams' Delicatessens: cheeses | 631 | 1 095.20 | | | | 1 095.20 | |
| | Robin Clutterbuck: hedge-cutting | 632 | 399.50 | | 59.50 | | 340.00 | |
| | J. Bull: firewood | 633 | 58.75 | | 8.75 | | 50.00 | |
| 30/12 | Customs and Excise: VAT refund | 634 | 1 674.00 | | | 1 674.00 | | 1 674.00 |
| | D. Jewett: tractor (part-exchange) | | | 587.50 | 87.50 | | 500.00 | |
| | Totals before reconciliation | | 9 302.39 | 676.57 | 155.75 | 1 674.00 | 8 149.20 | 1 674.00 |
| | Reconciliation adjustments<br>Kirks Dairy: 46 023 litres milk | | 9 111.82 | 142.18 | | | 9 254.00 | |
| | Adjusted totals | | 18 414.21 | 676.57 | 155.75 | 1 674.00 | 17 403.20 | 1 674.00 |
| | Closing bank balance (overdraft) | | 22 294.13 | | | | | |
| | | | 40 708.34 | | | | | |

**Fig. 10.10** Cash analysis book modified for VAT analysis.

*Payments analysis*

| Date | Details and Supplier | Cheque No. | 1 Bank | 2 Contra | 20 VAT paid on inputs | 21 VAT paid to C & E | 22 VAT RETURN ONLY: Standard and zero-rated | 23 VAT RETURN ONLY: Exempt and out of scope |
|---|---|---|---|---|---|---|---|---|
| 1/12 | Opening balance (overdraft) | | 26 302.00 | | | | | |
| 5/12 | J. Bird: 53 weaners | 521 | 1 325.00 | | | | 1 325.00 | |
| | Cleverdon Feeds: 18 t pig meal | 522 | 2 790.00 | | | | 2 790.00 | |
| 7/12 | R. Blackshaw: fence posts | 523 | 100.00 | | | | | 100.00 |
| | A. Jellings: vet services | 524 | 818.98 | | 121.98 | | 697.00 | |
| | Treachers: dairy & pig supplies | 525 | 551.90 | | 84.35 | | 525.00 | |
| | Cash for wages | 526 | 895.00 | | | | | 895.00 |
| 13/12 | Cash for wages | 527 | 910.00 | | | | | 910.00 |
| | Fisher Garage: diesel, etc. | 528 | 248.34 | | 37.96 | | 216.89 | |
| 20/12 | Brassleys: tractor repair | 529 | 1 172.06 | | 174.56 | | 997.50 | |
| | Cash for wages | 530 | 910.00 | | | | | 910.00 |
| | Meredith auctions: commission | | | 89.07 | 13.27 | | 75.80 | |
| 30/12 | D. Jewett: purchase of tractor | 531 | 3 401.63 | 587.50 | 620.53 | | 3 500.00 | |
| | A. Smith: cheese packaging | 532 | 768.45 | | 114.45 | | 654.00 | |
| | Totals before reconciliation | | 40 193.36 | 676.57 | 1 167.10 | | 10 781.19 | 2 815.00 |
| | Reconciliation adjustments: | | | | | | | |
| | Kirks Dairy (fees) | | | 142.18 | 21.18 | | 121.00 | |
| | Personal drawings | | 260.00 | | | | | 260.00 |
| | Bank interest and fees | | 254.98 | | | | | 254.98 |
| | Adjusted totals | | 40 708.34 | 818.75 | 1 188.28 | | 10 902.19 | 3 329.98 |

**Fig. 10.10** (cont'd)

necessary to reclaim input tax. (Since those who would gain by more than £3000 in a year compared to being VAT-registered are ineligible, this is likely to appeal only to the smallest, and most record-shy farm businesses). Other special schemes apply to retailers dealing directly with the public on a cash, rather than invoice basis, and dealers in second-hand goods.

### 10.4.2 Records for income tax

The Inland Revenue, in assessing an individual's liability for Income Tax in a given year, will ask him to complete a tax return. In the case of a businessman, it will require some verification of the accuracy of this return. One way of doing this would be for the Inland Revenue to exercise its right to investigate the manager's books of account, including the CAB. It is more convenient for both sides, however, if the manager makes available his profit and loss account and balance sheet for the year in question.

If these latter accounts are prepared by the manager himself, the Inland Revenue may still refuse to take them at face value. It is normal practice, therefore, to employ an 'outsider' to the business to undertake this task. This person should, preferably, be a member of a professional accountancy institute, such as the Institute of Chartered Accountants, which endeavours to ensure that its members adhere to a strict code of conduct. The fact that such a qualified outsider has prepared the accounts will give the Inland Revenue a degree of extra confidence in their accuracy.

It follows that the CAB must be as clear as possible to others, as well as to oneself – an accountant's time is expensive. This in turn implies the need for care in both the design of the CAB and the entry of transactions, and for the regular, i.e. at least monthly, reconciliation of the accounts.

## Notes

13. With the exception of tax paid on cars and business entertaining expenses.
14. A producer may elect to make monthly returns, as long as (a) he keeps it up for at least a year, and (b) he has not registered voluntarily. This can be quite an advantage for a farmer, since it enables earlier claiming of VAT refunds.

## References


### General

Anon. (1987) *Bluff your Way in Accountancy*. Ravette Books Ltd, Horsham.

### Cash analysis

Brown, B. (1991) *Practical Accounting for Farm and Rural Business*. Farming Press, Ipswich.

Critchley, R. (1982) *Understanding your Farm Accounts.* The North of Scotland Agricultural College Home Study Series. North of Scotland Agricultural College, Aberdeen.

Critchley, R. (1985) *Business Growth and Your Farm Accounts.* North of Scotland Agricultural College Home Study Series. North of Scotland Agricultural College, Aberdeen.

Critchley, R. (1986) *Financial Management Case Study I* and *Financial Management Case Study I Supporting Material.* North of Scotland Agricultural College Home Study Series. North of Scotland Agricultural College, Aberdeen.

Hosken, M. & Brown, D. (1991) *The Farm Office*, 4th edn. Farming Press, Ipswich.

Warren, M.F. (ed.) (1993) *Keeping the Books.* Book 1 in series: Finance Matters for the Rural Business. ATB Landbase/Seale-Hayne Faculty of Agriculture, Food and Land Use, University of Plymouth.

### *Value Added Tax*

Ball, A. & Marain, L. (1993, or most recent edition) *VAT: A Business by Business Guide*, 6th edn. Butterworths, London.

Foreman, A. (1996, published annually) *Allied Dunbar Tax Handbook 1996–97.* Pitman, London.

HM Customs and Excise (latest edition) *Value Added Tax: General Guide.* HMCE Notice No. 700. HMSO.

HM Customs and Excise (latest edition) *Value Added Tax: Scope and Coverage.* HMCE Notice No. 701. HMSO.

### *Double-entry book-keeping*

Bull, R.J. (1990) *Accounting in Business*, 6th edn. Butterworth Heinemann, London.

Etor, J. & Muspratt, M. (1982) *Keep Account: A Guide to Profitable Book-keeping.* Pan, London.

Wood, F. (1989) *Business Accounting 1*, 5th edn. Longman, London.

# Chapter 11
# Recording Profit and Capital

## 11.1 Preparing a profit and loss account

### *11.1.1 Adjustment for debtors and creditors*

Given a well-designed CAB, the compilation of a profit and loss account should be straightforward. In the Home Farm example used in Chapter 10, the 'trading' columns concerned with the productive activity of the business were kept together on the left-hand side of each analysis page, carefully segregated from the capital, personal and tax columns. In compiling a profit and loss account for a given period (say a year), the monthly totals in each of these trading columns are summed to give an annual total for each column. Each of these annual totals may then be transferred to the appropriate place in the profit and loss account.

It is convenient to make the necessary adjustment for opening and closing debtors and creditors (including prepayments, accruals, etc.) before this transfer takes place. Thus after the cash analysis has been ruled off for the year, two extra lines are included to allow for the deduction of opening debtors and creditors from the relevant columns, and the addition of closing debtors and creditors (these may be obtained from examination of invoices outstanding, but a separate monthly record of trade debts outstanding can save a lot of searching). The result of this calculation will be an adjusted total for each analysis column, ready for transfer to the profit and loss account (see Figure 11.1).

It may be that an opening debtor refuses or becomes unable to pay his debt. This bad debt must then be 'written off' by entering it in the profit and loss account as an expense, and omitting it from both the closing debtor adjustment in the CAB, and the 'trade debtors' item in the closing balance sheet.

### *11.1.2 Compiling a profit and loss account in conventional form*

The format of a conventional profit and loss account was discussed in Chapter 2. As a rule, it is insufficiently detailed for precise control, but it can be used to give a relatively quick check on profitability before more comprehensive enterprise accounts are prepared. To compile this form of account, the adjusted annual totals of the trading columns in the CAB are listed under their column headings. Depreciation is calculated as described in Chapter 5, and is entered as an expense

*Payments analysis*

| Date | Details and Supplier | Cheque No. | 1 Bank | 2 Contra | 3 Cattle Purchase | 4 Dairy Feed | 5 Dairy Misc. Costs | 6 Pig Purchase | 7 Pig Feed | 8 Pig Misc. Costs |
|---|---|---|---|---|---|---|---|---|---|---|
| | Opening balance for year (overdraft) | | 21 000 | | | | | | | |
| | ANNUAL PAYMENTS SUMMARY (sum of monthly totals, excluding opening balance) | | 203 760 | 3 514 | 18 260 | 11 362 | 2 912 | 23 514 | 28 950 | 1 348 |
| | Adjustments | | | | | | | | | |
| | – opening creditors | | | | | 4 000 | | | 2 800 | |
| | + closing creditors | | | | | 3 021 | 258 | 4 510 | 3 100 | |
| | Adjusted totals | | | | 18 260 | 10 383 | 3 170 | 28 024 | 29 250 | 1 348 |

*Receipts analysis*

| Date | Details and Customer | Ref. No. | 1 Bank | 2 Contra | 3 Milk | 4 Cattle | 5 Pigs | 7 B & B Costs | 8 Cheese | 9 Misc. Trading |
|---|---|---|---|---|---|---|---|---|---|---|
| | | | | | | | | | | |
| | ANNUAL RECEIPTS SUMMARY (sum of monthly totals) | | 224 760 | 3 514 | 84 239 | 18 089 | 75 231 | 8 250 | 13 950 | 4 897 |
| | Adjustments | | | | | | | | | |
| | – opening debtors | | | | 2 600 | 500 | 6 900 | | | |
| | + closing debtors | | | | 2 574 | | 6 336 | | | |
| | Adjusted totals | | | | 84 213 | 17 589 | 74 667 | 8 250 | 13 950 | 4 897 |

**Fig. 11.1** Home Farm – adjusting CAB for debtors and creditors.

(see Figure 2.3). The value of benefits in kind is taken from records (if they exist) or estimated. The opening valuation of stocks is given (it is the same as last year's closing valuation), and it only remains to estimate the closing value of stocks (see Section 11.2).

### 11.1.3 Compiling a profit and loss account in gross margin form

The particular value of the gross margin form of account is that it can provide a wealth of management information (see Chapter 4). This feature naturally adds to the difficulty of compiling the account. Nevertheless, if the CAB layout has been designed with gross margin accounting in mind, the task need not be daunting. In the CAB examples used in Chapter 10, the columns were arranged so as to keep fixed costs separate from variable costs. A judicious blend of classification by column and by code was recommended, to enable rapid identification of all the output and variable inputs relating to a particular enterprise.

The first step in the compilation of this type of account is to extract from the CAB all the receipts and variable cost payments relating to each enterprise. Where a column is highly specific (such as that for milk receipts), it will be sufficient to take the 'milk' column annual total, adjusted for opening and closing debtors and creditors. Where columns are shared, and codes are used within the columns, there is no alternative to searching the CAB records for the whole year, extracting all items identified by a particular code. The total thus derived must then be adjusted for those debtors and creditors which relate specifically to the subject of the code.

Once the picture of adjusted cash receipts and payments has been built up for each enterprise, adjustments may be made for transfers between enterprises. This will require reference to the physical 'back-up' records such as livestock movement records and stock records. If records of benefits in kind have been kept, these may be used to enter the home consumption of produce in the output of each enterprise. Usually these items are relatively so small that an estimate is sufficient. Finally, adjustments for changes in valuation may be made to enterprise output and variable inputs where necessary.

The gross margin for each enterprise is then calculated by deducting variable costs from enterprise output. As in the budget, all components of the gross margin should be expressed throughout in relation to some unit of scale, for instance per hectare of land used, per bed in an accommodation enterprise, or per kilogram of goods produced. All livestock enterprise gross margin accounts should show the final margin expressed both per head and per unit of area used by the enterprise (per forage hectare for grazing livestock; per square metre for intensive livestock). If capital, or even labour, are limiting resources, the gross margin may be expressed per £100 capital employed, or per 100 man-hours. The use of worksheets (such as those shown in Chapter 6) makes the process of drawing up an enterprise gross margin account considerably easier.

Most of the fixed costs can be taken from the relevant adjusted column totals in the CAB. The main exception is depreciation, which must be calculated by reference to the opening value of buildings and machinery, and to purchases and sales of such assets during the year (shown in the 'capital' columns in the CAB). A machinery register can be helpful in this calculation (see Chapter 12). Benefits in kind which relate to items of fixed cost should be deducted from the appropriate sub-total (usually miscellaneous fixed costs).

If the total fixed cost is now deducted from the total gross margin of the business, after allowance for any unallocated variable costs, a net profit or loss will result. This should be identical with that derived from the cruder conventional format. If it is not, checking for errors should begin with those revenues and expenses which are not recorded in the CAB, particularly changes in stock valuations, transfers between enterprises, and benefits in kind. It is very easy, when allocating these items between enterprises, to make an error or an omission. Inter-enterprise transfers are particularly prone to error: check that the transfers of a product out of one enterprise match the transfers of that product into other enterprises, in both quantity transferred and in unit price.

### 11.1.4 Compiling a profit and loss account in 'full costing' form

Chapter 4 indicated that there are some situations where the use of full cost absorption may be justified. The most common example in farming is that of intensive livestock production, where most of the costs are specific to the enterprise concerned.

Compilation of a full absorption costing is similar to that of a gross margin account, the calculation of the 'specific' fixed costs following the same pattern as the calculation of variable costs. There will be some costs, however, which cannot be allocated in this way – the 'general overheads' such as office expenses, use of the business truck, managerial salaries, and so on. A decision has to be made as to how those overheads are 'absorbed' by the various enterprises. The proportion of overheads absorbed by each enterprise may be determined by its financial contribution to farm output, its share of land occupied, its share of direct labour costs, or any other formula that the manager thinks appropriate. There is no 'right' basis of absorption; whatever method is used is likely to be fairly arbitrary. Provided that the general overheads are a small proportion of the total costs, this should not matter greatly.

While a full costing gives more precise detail than a gross margin account, and is useful for historic analysis, it is important to remember its limitations as a basis for decision-making (see Chapter 4) and to recognize the superior value of gross margin accounting in this respect.

### 11.1.5 Reconciliation of profit with cash flow

Whichever method is used to calculate the historic net profit, the result should be reconciled with the cash analysis results as described in Chapter 7. As with all such checks, if the figures do not reconcile, the two accounts must be checked and checked again until the error is found.

### 11.1.6 'Financial year' and 'production year' enterprise accounts

The above discussion is concerned entirely with the calculation of accounts for the financial year. The financial year calculation is important in many ways, not least in providing a vital link in the cash flow/profit/capital reconciliation chain, and in forming a basis for the estimation of likely tax liability for the year.

The financial year basis may not be the ideal for decision-making in some types of rural business, however, since the timing of the financial year can give rise to a distorted picture of the performance of an enterprise. This is especially the case where the production cycle is a year or more in length, and/or the production process is heavily influenced by the seasons, as is the case in extensive agriculture and horticulture, and tourism-dependent enterprises. An example is that of a breeding sheep flock on a farm where the financial year ends on 31 March. The financial year gross margin will include the sale of the previous year's crop of lambs, together with the variable costs of producing the current crop. The same applies to winter cereal crops. A September year-end might have similar effects for a tourism-related business whose season extends into the late autumn. Even in the absence of inflation, it can be very difficult to draw conclusions about the performance of the enterprise in such circumstances. A difference between years in climate, or in the size of the enterprise, will confuse the message of the account.

Where the financial year of the business finishes part-way through the 'production year' or 'crop year' of an enterprise, therefore, a production year account should be prepared in addition to the financial year version. The process can be summarized as follows:

(1) Take the two financial year accounts which span the production year in question.
(2) Remove from them all revenues and expenses relating to the production years before and after the production year in question.
(3) Add in any revenues and expenses arising in other financial years, but relating to the production year in question.

Adjustment (3) is not always necessary, but may arise as a result of advance buying, e.g. of fertilizer, or of long storage of output before sale.

In the interests of timely decision-making, it is desirable to prepare the production year gross margin as soon as possible after the end of that year. Where crops are stored beyond the end of the crop year, or where some livestock is still unsold, it will be necessary to estimate the likely value of the sales, using realistic, if conservative, valuations. The resulting gross margin should be corrected as soon as the sales or transfers are complete.

As an example of the preparation of a production year gross margin account, take a winter wheat crop in a business whose financial year ends in December. An appropriate 'production year' would run from 1 October to 30 September (see Figure 11.2).

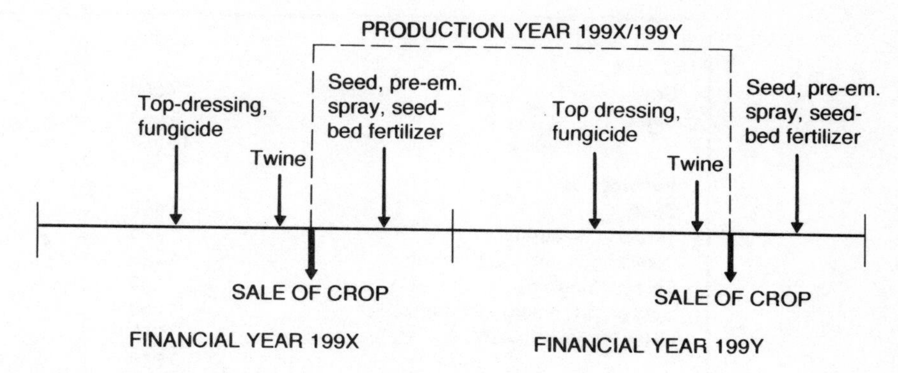


**Fig. 11.2** Illustration of financial and production years.

The gross margins for the financial years 199X and 199Y are shown in Figure 11.3, highly summarized. Note that it is impossible to express these sensibly in £ per hectare, since the various items of output and cost arising in a given financial year relate to different areas of land. To arrive at a production year gross margin, we can remove the sales of the 199X financial year, together with the top-dressing, the fungicide, and the baler twine. All relate to the previous production year. We can also remove the seed, seed-bed fertilizer and pre-emergence spray costs from the 199Y financial year, since these have been incurred in respect of the following crop. Assuming that there are no revenues or expenses relating to this crop and arising outside the two financial years shown, the production year gross margin would be as in Figure 11.4. The result is an account which can be used with ease in enterprise analysis (see Chapter 14). It is worth noting that the use of an October to September financial year minimizes the need for such adjustments.

| | *Year ending 31 December 199X* | *Year ending 31 December 199Y* |
|---|---|---|
| Hectares harvested | 30 | 35 |
| *Output (total)* | £ | £ |
| Grain | 21 000 | 17 500 |
| Straw | 1 500 | 1 050 |
| | 22 500 | 18 550 |
| *Variable costs (total)* | | |
| Seed | 1 400 | 1 120 |
| Fertilizer, seed-bed | 1 225 | 1 280 |
| Fertilizer, spring top-dressing | 900 | 1 225 |
| Sprays, fungicide | 1 500 | 2 100 |
| Sprays, pre-emergence herbicide | 1 050 | 1 280 |
| Miscellaneous (baler twine, etc.) | 300 | 350 |
| | 6375 | 7355 |
| ***Gross margin*** (total) | 16 125 | 11 195 |

**Fig. 11.3** Financial year gross margin for winter wheat.

| | £ total | £/ha |
|---|---|---|
| *Output* | | |
| Grain | 17 500 | 583 |
| Straw | 1 050 | 35 |
| | 18 550 | 618 |
| *Variable costs* | | |
| Seed | 1 400 | 47 |
| Fertilizer, seed-bed | 1 225 | 41 |
| Fertilizer, spring top-dressing | 1 225 | 41 |
| Sprays, fungicide | 2 100 | 70 |
| Sprays, pre-emergence herbicide | 1 050 | 35 |
| Miscellaneous (baler twine, etc.) | 350 | 12 |
| | 7350 | 246 |
| ***Gross margin*** | 11 200 | 372 |

**Fig. 11.4** Production year gross margin (30 ha winter wheat grown).

Whatever the financial year-end, crop-year adjustments are not necessary in 'continuous-production' livestock enterprises such as pigs, poultry and, for most purposes, dairy. Particular difficulties surround the calculation of enterprise accounts for trading livestock enterprises with long production cycles, such as rearing beef or dairy replacements. The easiest solution is to calculate the account on the basis of the financial year. For instance, in an established heifer-rearing enterprise, with 30 animals transferred per year to the dairy herd at three years old, one would expect the financial year account to show:

- Value of 30 heifers transferred out
- Cost of purchase or transfer in of 33 calves (allowing for mortality)
- Cost of rearing 33 heifers in their first year
- Cost of rearing 31 heifers in their second year
- Cost of rearing 30 heifers in their third year

If numbers stay more or less the same from year to year, this will give a reasonable approximation of the current cost of rearing heifers. If numbers change significantly from year to year, or one wants to determine what a batch of animals actually costs, the figures used should be those relating to an entire production cycle, rather than the cross-section described above. Whichever method is used, the components of the account should be expressed per finished animal sold or transferred, rather than per animal in the enterprise as a whole.

## 11.2 Valuing stocks

### 11.2.1 General principles

At the end of each financial year it is necessary for a valuation to be made of the stocks of the business. This valuation is required for the calculation of both net profit and net capital. You will remember from Chapter 2 that stocks include:

- Raw materials
- Work in progress
- Finished goods
- Goods purchased for resale

The values used for stocks will depend partly on the use to which the accounts are to be put. For accounts used in the assessment of income tax liability, the manager is allowed to use the lower of cost or net realizable value of the stocks, i.e. the anticipated market value after deduction of storage, transport and marketing costs. Furthermore, a farmer may use 75% of a 'reasonable' market value of home-bred sheep and pigs, 60% for cattle, and 85% for crops in store (Inland Revenue, 1993). Even if cost of production is used, this need only take into account the direct costs, as long as this method is used consistently from year to year. Breeding livestock may be given a cost of production value when they enter the flock or herd, and kept at that value until they are culled. An alternative is to remove them from the profit and loss account altogether by using the 'herd basis' method of valuation.

Price is not the only uncertain factor in valuing – it can often be difficult to determine quantities, especially of harvested crops stored in bulk. Where possible, such items should be weighed on storage; although some loss of weight is likely to occur during storage, weighing will at least provide the basis for a reliable estimate at the time of valuation. Where this is not possible, either because of the cost of weighing devices or the nature of the material, e.g. silage, it is necessary to determine the volume and density of the material and thus estimate the final weight. For relatively standard items, such as grain, specimen densities are published in various farm management handbooks. For other items, such as silage, sample weighings of measured volumes must be taken.

It is sound policy to employ a qualified valuer every two or three years, even if not specifically required to by the Inland Revenue, as a check on the accuracy of valuations.

### 11.2.2 Stock valuation for management accounts

Although beneficial from the point of view of minimizing the impact of taxation, the valuation methods described above can be unsatisfactory for management purposes. A manager needs a realistic indication of how his stocks have changed in value over the year, if he is to identify the relative contribution of each enterprise to net profit. This implies that, wherever possible, he should use cost or net realizable value, *whichever is higher* as the basis of valuation of finished goods for management accounts.

There is a danger, with this method, of valuing stocks at the peak of a very short-term price fluctuation. If the price slumps again before the stocks are sold in the next financial year, a distorted distribution of profit between years will result. To avoid this it is wise to err on the conservative side when valuing. Raw materials and goods purchased for resale should be valued at cost of purchase plus cost of conversion, where cost of purchase includes all costs borne by the buyer (e.g. duties, carriage, etc.) less trade discounts, subsidies, rebates, etc. Work in progress should be valued at cost of purchase plus cost of conversion, where cost of conversion is measured by the direct costs of transforming the raw materials into their present state.

Even this can be far from straightforward when the stocks contain more than one consignment of like items. For example, a farm guesthouse may make two purchases of the same brand of toilet rolls during the year: 10 boxes at £10 per box, and a later purchase of 20 boxes at £12 per box. At the end of the year there are two boxes in stock. We know, then, that 28 boxes of toilet paper have been used during the year, but what price should be used in valuing the boxes used and the boxes left in store: £10, £12 or a compromise figure?

Three valuation methods are commonly available (Millichamp, 1992, pp. 100–105). 'Last in, first out' or *LIFO*, values stocks used as if those last bought were first used, and values the remaining stocks as a residual. It is not very realistic, and is not generally accepted in the UK. As its name suggests, the 'average cost method', or *AVCO*, uses a weighted average of old and new prices for both use and stock values: more realistic than LIFO or FIFO, but has to be recalculated every time a new batch arrives. The most popular is 'first in, first out', or *FIFO*, which assumes that the older stocks are used before the newer consignment. Using this method on our example, the cost of the 28 boxes of toilet rolls used would be calculated as:

| | £ |
|---|---|
| 10 of first batch @ £10 | 100 |
| 18 of second batch @ £12 | 216 |
| cost: | 316 |

The value of remaining stock would be calculated as:

| | £ |
|---|---|
| 2 of second batch @ £12 | 24 |
| Total purchases | 340 |

As well as being intuitively reasonable, FIFO represents reality for many goods, especially perishables (food, chemicals, etc.), although it can underestimate the true expense of maintaining the business as a going concern.

### 11.2.3 Valuation of farm stocks

The valuation of farm stocks should follow the above principles, although their nature sometimes gives rise to specific difficulties. Farm stocks were earlier (Chapter 2) grouped in the categories of:

- Breeding livestock
- Trading livestock

- Crops in store
- Crops in ground, manurial residues, etc.
- Purchased materials

### *Breeding livestock*

Breeding livestock may be regarded as fixed assets, and thus there is much to be said for removing changes in their valuation from the profit and loss account altogether, accounting only for their net replacement cost or 'depreciation'[15]. The alternative is to value the animals at their net realizable value, which can be reasonably estimated as about half-way between the cull price and the price of a good replacement animal.

### *Trading livestock*

Trading livestock can be difficult to value, particularly those that are immature. Again they should be valued for management purposes at net realizable value, which will probably be determined by a mixture of the manager's experience of local markets, and intelligent guess-work. Livestock markets can be volatile, and it is wise to be conservative in times of high prices.

### *Crops in store*

Crops in store can be classified into saleable and unsaleable crops. The former include grain, potatoes, seeds, straw, hay, and so on. Pricing is rarely a difficulty here, as long as one remembers to make the appropriate allowances for storage, transport and marketing costs, and for possible deterioration in storage. The unsaleable crops in store include silage, for which there is no ready market due to its bulk. Without a reliable market value, one is forced back to using cost of production, or more specifically, the variable costs of production.

It can be argued that the use of opportunity cost would be more realistic, e.g. the profit obtainable from the best alternative use of the land conserved, less the cost of a ration composed of bought-in fodder and giving the same level of nutrients. While this principle should certainly be used for specific decisions in forage crop policy (via partial budgets), it is rarely likely to be worth the effort in valuation, since we are concerned here with changes in valuation over the year. The net effect of choice of method on profit is likely to be small compared with the effort involved in calculating the opportunity cost. The same argument applies to the inclusion of the fixed costs of production in the valuation.

### *Crops in ground, manurial residues, etc.*

Crops in ground are rarely saleable (exceptions being grass and fodder crops which can be let for keep, and cash crops at point of harvest), and are normally valued at their direct cost of production. Crops near to harvest may be valued at their net realizable value, less a substantial discount for the risk of calamity before harvest.

If it is thought that a significant change in manurial values has taken place over the previous 12 months, valuation tables may be used to calculate the net effect of the change. In view of the relatively small values involved, and the effort involved, it is

| *Enterprise:* Fattening pigs | | | *Ha or head?* 939 | | | | *Year* 199X/9Y *Version* 1 | |
|---|---|---|---|---|---|---|---|---|
| *Products* | Bacon pigs | | | | | | | |
| | Qty | £ | Qty | £ | Qty | £ | Qty | £ |
| Cash received | 1015 | 75 231 | | | | | | |
| Closing debtors | 93 | 6336 | | | | | | |
| –Opening debtors | –100 | –6900 | | | | | | |
| **Sales** | 1008 | 74 667 | | | | | | |
| Transfers out | | | | | | | | |
| Benefits in kind | | | | | | | | |
| Closing valuation | 481 | 24 050 | | | | | | |
| –Opening valuation | –550 | –24 750 | | | | | | |
| ***Output*** | 939 | 73 967 | | | | | | |

| *Inputs* | Bought weaners | | Conc. feed | | Transport | | Other | | Vet/med, etc. | | | |
|---|---|---|---|---|---|---|---|---|---|---|---|---|
| | Qty | £ | Qty | £ | Qty | £ | Qty | £ | £ | £ | £ | £ |
| Cash paid | 1016 | 23 514 | 225 | 28 950 | 1 | 625 | 1 | 480 | 1339 | | | |
| Closing creditors | 110 | 4510 | 18 | 3100 | | | | | | | | |
| –Opening creditors | | | –20 | –2800 | | | | | | | | |
| **Purchases** | 1126 | 28 024 | 223 | 29 250 | 1 | 625 | 1 | 480 | 1339 | | | |
| Transfers in | | | | | | | | | | | | |
| –Benefits in kind | | | | | | | | | | | | |
| Opening valuation | | | | | | | | | | | | |
| –Closing valuation | | | | | | | | | | | | |
| ***Variable cost*** | 1126 | 28 024 | 223 | 29 250 | 1 | 625 | 1 | 480 | 1339 | | | |

**Fig. 11.5** Actual gross margin of pig herd, Home Farm 199X/9Y.

| ***Miscellaneous valuation changes*** | Breeding livestock | | Crops in gd. | Total |
|---|---|---|---|---|
| | No. | £ | £ | £ |
| Closing valuation<br>–Opening valuation | | | | |
| **Valuation change** | | | | |

| *Fattening pigs* | | | **Gross margin** | | |
|---|---|---|---|---|---|
| *Ha or head*<br>*939* | *Total quantity* | *Qty/ha or /head* | *£ per unit quantity* | *£ Total* | *£/ha or /head* |
| Bacon pigs | 939 | 1.00 | 78.77 | 73 967 | 78.77 |
| Misc. valn. changes | xxxxxxxx | xxxxxxxxx | xxxxxxxxx | | |
| ***Output*** | | | | 73 967 | 78.77 |

| ***Variable costs*** | *Total quantity* | *Qty/ha or /head* | *£ per unit quantity* | *£ Total* | *£/ha or /head* |
|---|---|---|---|---|---|
| Bought weaners | 1 126 | 1.2 | 25 | 28 024 | 29.84 |
| Conc. feed | 223 | 0.237 | 131 | 29 250 | 31.15 |
| Transport | | | 625 | 625 | 0.67 |
| Other | | | 480 | 480 | 0.51 |
| Vet/med, etc. | | | | 1 339 | 1.43 |
| ***Variable cost*** | | | | 59 718 | 63.60 |
| ***Gross margin*** | | | | 14 249 | 15.17 |

**Fig. 11.5** (cont'd)

normally more sensible to keep the values the same from one end of the year to the other, leaving any revaluing to the valuer on one of his biennial or triennial visits. A modification of this is to revalue both opening and closing valuations by the percentage change in fertilizer price levels during the year. This will keep the values fairly realistic, while avoiding the appearance of 'paper' profit.

*Purchased materials*
Purchased materials should be valued as described in Section 11.2.2, using a LIFO, FIFO or AVCO method.

### 11.2.4 Home Farm profit statements

Figures 11.5 and 11.6 show the actual results for the Home Farm pig enterprise and for the farm as a whole – moving forward nine months from Chapter 10. To ensure easy comparison with the budgets in Chapters 6, 7 and 8, it is assumed here that Adam Haynes implemented none of the developments investigated in Chapter 9.

## 11.3 Recording capital

### 11.3.1 The balance sheet

The various formats of the balance sheet were fully described in Section 2.5. Once the format has been determined, it remains to complete the lists of assets and liabilities, and find the net capital. If the CAB and profit and loss accounts have been compiled as above, and rigorously checked at every stage, this should be easy.

Starting with the assets, the cash in hand is shown by the petty cash book, and the cash in bank (if any) by the closing balance in the CAB. Closing debtors and closing valuation of stocks have already been used in the calculation of profit. The written-down value of machinery, buildings and improvements may be derived from the depreciation calculations. The value of land and investments may be derived from the opening balance sheet, adjusted for any sales and purchases shown in the 'capital' section of the CAB.

The liabilities are even easier. Overdraft (if any) is taken from the CAB. Creditors, like debtors, have already been calculated. Loans may be calculated by taking the amount outstanding at the beginning of the year, and adjusting for any repayments or new receipts of loan finance, as recorded in the 'capital' section of the CAB. Total liabilities taken from total assets gives the closing net capital of the business. Figure 11.7 shows the actual balance sheet for Home Farm at the end of September 1 199Y.

### 11.3.2 The capital account

As always, it is vital to check the balance sheet through the capital account (see Chapter 2). Opening net capital is given by the opening balance sheet. Private

| | | Per unit | Total | Total |
|---|---|---|---|---|
| *Gross margins* | | £ | £ | £ |
| Pigs | | 14.25 | 14 249 | |
| Dairy | | 752 | 60 125 | |
| Barley | | 349 | 4 186 | |
| Bed and breakfast | | 821 | 4 925 | |
| Cheese | | 12.66 | 7 596 | |
| | | | | 91 081 |
| *Unallocated revenue items* | | | | |
| Miscellaneous trading receipts | | | | 4 897 |
| *Miscellaneous valuation changes* | | | | 1 234 |
| ***Total gross margin*** | | | | 97 212 |
| *Fixed costs* | | | | £ |
| Regular labour | | | | 26 485 |
| Machinery and vehicle running costs | | | | 18 502 |
| Buildings and property repairs | | | | 210 |
| Rent and rates | | | | 9 500 |
| Miscellaneous fixed costs | | | | 7 476 |
| *less* benefits in kind, input items | | | | –4 250 |
| ***Total fixed costs*** | | | | 57 923 |
| | | | £ | £ |
| ***Net profit*** before depreciation and finance | | | | 39 289 |
| Depreciation: | Farm machinery | | 9 500 | |
| | Vehicles | | 1 200 | |
| | Cheese-making equipment | | 800 | |
| | Bed and breakfast equipment | | 400 | |
| | Buildings (tenant's fixtures) | | 165 | 12 065 |
| ***Net profit*** before finance costs | | | | 27 224 |
| Finance costs: | Overdraft interest | | 2 175 | |
| | Loan interest | | 1 100 | |
| | Leasing charges | | 2 100 | 5375 |
| ***Net profit*** after finance costs | | | | 21 849 |

**Fig. 11.6** Home Farm financial summary – actual results 199X/9Y.

contributions are shown by the CAB column headed 'Personal receipts'. Grants will be shown in the 'Capital receipts' column, but be careful to avoid including grants which have already been shown in the balance sheet as deductions from the cost of buildings and machinery before depreciation. Private drawings and tax are easily obtained from the relevant CAB columns, and benefits in kind have already been calculated for the profit and loss account.

The result of the calculation should be identical with that of the balance sheet. If not, check again.

| 29/9/9X | Balance sheet at 28 September 199Y | | | |
|---|---|---|---|---|
| | *Fixed assets* | | | |
| 0 | Tenant's fixtures | | 3 972 | |
| 45 000 | Machinery and plant | | 42 798 | |
| 6 000 | Vehicles | | 4 800 | |
| 50 050 | Dairy herd | | 53 300 | |
| 101 050 | Fixed assets | | | 104 870 |
| | *Current assets* | | | |
| | Stocks | | | |
| 800 | Trading livestock: calves | 720 | | |
| 24 750 | Trading livestock: pigs | 24 050 | | |
| 3 780 | Crops in store: barley grain | 3 863 | | |
| 390 | Crops in store: barley straw | 520 | | |
| 10 000 | Crops in store: silage | 10 000 | | |
| 5 090 | Cheeses | 7 589 | | |
| 2 800 | Dairy feed | 3 150 | | |
| 5 684 | Miscellaneous stores | 7391 | 57 283 | |
| 10 000 | Trade debtors | | 8 910 | |
| 0 | Cash in bank | | 0 | |
| 0 | Cash in hand | | 0 | |
| 63 294 | Current assets | | 66 193 | |
| | *Current liabilities (due within one year)* | | | |
| 6 800 | Trade creditors | | 10 889 | |
| 21 000 | Bank overdraft | | 13 281 | |
| 27 800 | Current liabilities | | 24 170 | |
| 35 494 | Net current assets | | | 42 023 |
| 136 544 | Total assets less current liabilities | | | 146 893 |
| | *Creditors due after one year* | | | |
| 34 000 | Loan | | | 31 000 |
| | *Capital account* | | | |
| | Opening net capital | 102 544 | | |
| | + net profit | 21 849 | | |
| | + capital injections | 200 | | |
| | – private drawings | –3 520 | | |
| | – income and capital tax | – 856 | | |
| | – benefits in kind | –4 324 | | |
| 102 544 | Closing net capital | 115 893 | | 115 893 |
| 136 544 | Capital employed | | | 146 893 |

**Fig. 11.7** Actual balance sheet for Home Farm.

# Notes

15. This is the practice followed in the herd basis method of valuation mentioned earlier. Although it was designed to reduce the effect of inflation on tax liability, it is possible to use it for management accounts.

# References


### Recording profit and capital

Barnard, C.S. & Nix, J.S. (1979) *Farm Planning and Control*, 2nd edn. pp. 45–47 (gross margins, 317–327 (complete budgets) and 537–539 (full costings). Cambridge University Press, London.

Broadbent, M. & Cullen, J. (1993) *Managing Financial Resources*. Butterworth Heinemann, Oxford.

Kerr, H.W.T. (1988) Management accounting and the marginal concept. *Farm Management*, **6**, 10.

Ministry of Agriculture, Fisheries and Food (1977) *Definition of Terms Used in Agricultural Business Management*. Section 4 (Margin Terms).

Warren, M.F. (ed.) (1993) *Improving your Profits*. Book 3 in series: Finance Matters for the Rural Business. ATB Landbase/Seale-Hayne Faculty of Agriculture, Food and Land Use, University of Plymouth.

### Valuation of stocks: general

Berry, A. & Jarvis, R. (1991) *Accounting in a Business Context*. Chapman & Hall, London.

Bull, R.J. (1990) *Accounting in Business*, 6th edn. Butterworth Heinemann, London.

Millichamp, A. (1992) *Finance for Non-financial Managers*, 2nd edn. DPP Publications, London.

### Valuation of stocks: farming

Brown, B. (1991) *Practical Accounting for Farm and Rural Businesses*. Farming Press, Ipswich.

Gamble, B.T. (1986) *The Farming Business*. Institute of Chartered Accountants, London.

Hosken, M. & Brown, D. (1991) *The Farm Office*, 4th edn. Farming Press, Ipswich.

Inland Revenue (1993) *Farming – Stock Valuation for Income Tax Purposes*. Business Economic Notes 19.

Markham, G. (1996) *Farming – An Industry Accounting and Auditing Guide*. Accountancy Books (ICA), Milton Keynes.

# Chapter 12

# The 'Back-up' Records

## 12.1 Introduction

The financial records discussed in the preceding chapters are designed to give clear, simple summaries of the performance and state of the business. These summaries are based on raw information of considerable volume and variety: if they are to be reliable, the business needs a foolproof system of gathering and storing this raw or 'back-up' information.

For ease of discussion, this chapter makes a somewhat arbitrary distinction between three types of back-up record: transaction documents (such as invoices and statements); physical records (of matters relating to production); and 'background information' (the rest). An examination of methods of dealing with paperwork concludes the chapter.

## 12.2 Transaction records

A typical transaction between buyer and supplier, e.g. of fertilizer, may go through the following stages:

(1) The buyer seeks and receives a price quotation for the quantity and type of goods required.
(2) If the quotation is acceptable, the buyer places an order.
(3) The supplier, on receipt of the order, packs and despatches the goods to the farm.
(4) The supplier sends the buyer an invoice or bill.
(5) The supplier sends the buyer a statement of account at the end of the month, showing the total amount owed by the buyer (allowing for all goods supplied during the month, any credits to the account, and the amount owed at the beginning of the month).
(6) The buyer sends a cheque in full or part settlement of the account, some time after his receipt of the statement.

In most cases it is likely that stages (1) and (2) will be conducted by telephone, fax or e-mail. Where transactions are frequent, it is a sensible procedure to use order forms with carbon or pressure-sensitive paper, so that the buyer has a copy of the order to check against the goods delivered. Even where orders are made by telephone, they should ideally be confirmed in writing. It is not unknown for verbal messages to become distorted.

The supplier usually responds to the order by making out an invoice set. This is a bundle of forms, printed on pressure-sensitive paper. The details of the order are transcribed on to the top form (the invoice or bill) which is then given or sent to the customer to confirm the goods supplied and their cost. The second copy may be retained for the supplier's own records. The third and fourth copies may be sent with the goods as delivery notes, one to be retained by the buyer and the other, after being signed by the buyer as confirmation of the correct delivery of the goods, sent back with the lorry driver. Slightly different arrangements apply if the supplier uses a haulier, post, or public transport, but these make little effective difference to the documentation received by the buyer. An example of an invoice form is shown in Figure 12.1.

The statement of account shows any balance outstanding at the end of the previous month, followed by a list of the invoices issued within the current

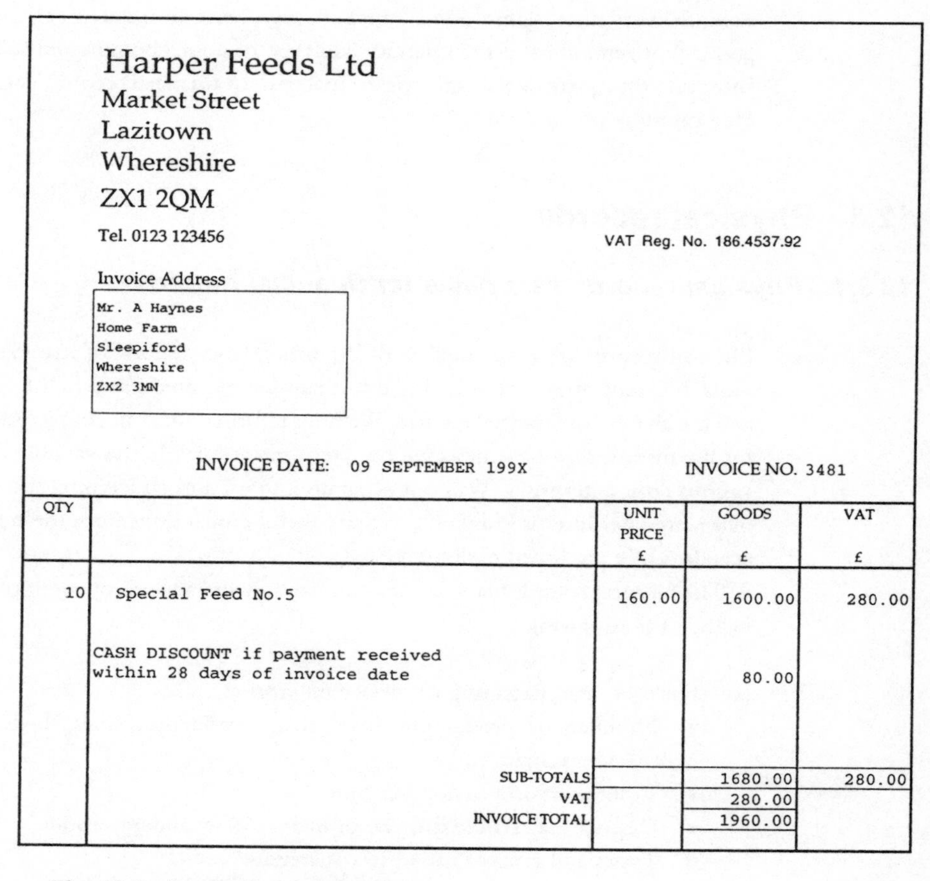

Harper Feeds Ltd
Market Street
Lazitown
Whereshire
ZX1 2QM

Tel. 0123 123456

VAT Reg. No. 186.4537.92

Invoice Address

Mr. A Haynes
Home Farm
Sleepiford
Whereshire
ZX2 3MN

INVOICE DATE: 09 SEPTEMBER 199X

INVOICE NO. 3481

| QTY | | UNIT PRICE £ | GOODS £ | VAT £ |
|---|---|---|---|---|
| 10 | Special Feed No.5 | 160.00 | 1600.00 | 280.00 |
| | CASH DISCOUNT if payment received within 28 days of invoice date | | 80.00 | |
| | SUB-TOTALS | | 1680.00 | 280.00 |
| | VAT | | 280.00 | |
| | INVOICE TOTAL | | 1960.00 | |

**Fig. 12.1** Invoice form – example.

month. If goods have been returned, e.g. faulty or wrongly supplied goods, or returnable containers and pallets, one or more credit notes (a sort of invoice in reverse) will also be recorded. The total of these figures represents money owed by the buyer to the supplier. The statement will also state the terms of trade, i.e. when payment is expected, and what discounts, if any, are given for prompt payment.

In principle the above applies whether you are the buyer or the supplier of goods. In practice it may be unnecessarily complex for the low-turnover business. In farming, in particular, outputs are often taken from the farm before the details of sale are fully known; milk, livestock (whether sold at auction or by deadweight), grain, sugar beet, and so on. In these cases, the farmer is notified of the details of sale by means of a *self-billed invoice*. The proceeds of the sale may even be paid directly into the farmer's bank account by direct debit. The typical farm manager thus has to issue invoices for a relatively small number of items. It is unlikely even here that he will want a complex set of documents; a two-copy set, one for the customer and one for himself, will probably suffice.

The documents that a manager will collect as a result of each transaction are useful both as short-term checks (e.g. order against invoice, invoice against delivery note, delivery note against goods supplied, invoices against statement) and as more permanent reminders of the quantity and type of goods bought and sold. In their latter role they provide valuable source material for financial records, and should be filed carefully (see Section 12.6).

## 12.3 Physical records

### *12.3.1 Physical records as a basis for financial records*

Physical records are concerned with the source and disposal of materials and services. A certain amount of physical information is required for both the compilation and the analysis of financial records. Without information concerning fertilizer rates, for instance, it may be impossible to determine accurately the variable costs of the various crop enterprises. Without accurate information concerning the number of pigs reared per litter it is difficult to draw useful conclusions from the management accounts of a pig breeding herd.

The physical records necessary for compilation and analysis of enterprise accounts include the following:

(a) For each manufacturing or service enterprise:
- Number of production units (e.g. production lines, beds, caravans, machines hired out)
- Volume of production per unit
- Disposal (sales and transfers) of main and secondary products
- Types and rates of use of raw materials
- Use of casual labour and contract hire of machinery

- Use of regular labour and own machinery (full costings only)
- Other useful information

(b) For each crop:
- Area grown of each variety
- Yield and quality of main product (by variety)
- Yield and quality of secondary product(s)
- Disposal (sales and transfers) of main and secondary products
- Types and rates of use of seeds, fertilizers, sprays, etc.
- Use of casual labour and contract hire of machinery
- Use of regular labour and own machinery (full costings only)
- Other useful information – such as soil analysis results, type and timing of cultivations, application of farmyard manure, etc.

(c) For each trading livestock enterprise:
- Sources (purchases and transfers) of young stock
- Weight of animals on entry to herd or flock
- Disposal (sales and transfers) of the finished animals
- Weight and quality of the finished animals (where sold for slaughter)
- Mortality
- Type and quantity of concentrate fed
- Type and quantity of forage fed
- Use of casual labour and contract hire of machinery
- Use of regular labour own machinery (full costings only)
- Other useful information – such as dates of housing and turning out, veterinary records, weaning dates, grazing days, pedigree documents, service dates and pregnancy diagnoses for replacement breeding stock

(d) For each breeding livestock enterprise:
- Number and type of progeny born and reared during the year
- Quantity and quality of other products, e.g. milk, eggs, wool
- Disposal (sales and transfers) of progeny and other produce
- Mortality of progeny
- Culls and deaths of breeding stock
- Sources (purchases and transfers) of replacement breeding stock
- Type and quantity of feeds (concentrate and forage)
- Use of casual labour and contract hire of machinery
- Use of regular labour and own machinery (full costings only)
- Other useful information – such as dates of service and parturition, dates of weaning, dates of housing and turning out, veterinary records, pregnancy diagnoses, grazing days, pedigree records, etc.

(e) For fixed cost analysis:
- Number of overtime hours worked by regular labour
- Number of hours of use of machinery, especially tractors

- Purchases and sale of machinery, equipment and buildings
- Service and repair records of machinery and equipment

Not all of the records mentioned above are applicable to every enterprise, of course, and it may be that particular enterprise types require records that do not appear in the lists given.

Many of the listed items may be obtained from the transaction documents described in Section 12.2. These include records of quantities and quality of produce sold, and quantities of inputs which are specific to a particular enterprise, such as a fertilizer used solely for winter wheat, or provisions purchased solely for the guest-house. How the remaining items are recorded depends largely on the needs of the particular enterprises, and is largely a matter of organization and commonsense. The golden rules of recording are:

- Make the records easy to keep and easy to use
- Build in checks on accuracy, where possible
- Never keep any record which is unlikely to be used effectively

Four types of farm physical record are sufficiently important to describe in detail. The first is a *field record*, which consists of a loose-leaf file or card index, with the details of crop inputs and outputs, cultivations, and other relevant matters recorded for each field on a separate page or card. Ideally, the record should also contain information relating to one or two previous years, since past experience is often a useful guide to present decisions. At the beginning of the production year for a crop, a new page or card can be inserted, and the oldest of the previous years' records removed and filed safely elsewhere.

For forage crops, the field record may include some items relevant to livestock, such as grazing days. Alternatively, this type of information may be kept in a separate *enterprise record* for each livestock enterprise, together with records of physical inputs (especially feed) and outputs and general information. An integral part of any livestock enterprise record should be a *livestock movement record (reconciliation)*. The movement of livestock on and off the farm, and between enterprises, is a source of much potential confusion. Where substantial numbers of various types of livestock are kept, it is worth keeping a monthly check; otherwise a quarterly or half-yearly check may suffice. Where the latter is the case, the record may include dates of transfer of stock between enterprises, as a guide to establishing appropriate transfer prices. An example of a livestock reconciliation form is shown in Figure 12.2; the farmer may of course design his own to suit his particular circumstances.

The use of reliable *barn records* helps to overcome two other potential sources of confusion: the disposal of stored crops and the allocation of feeds between livestock enterprises. Such a record should provide for the noting of all major items coming into store, particularly straw, bulk fodder, cash crops and fertilizer. It should also enable the recording of all disposals of these stored goods, particularly those used in growing next year's crops or in feeding of livestock. Where several livestock

| | Opening valuation | | Purchases | | Transfers in | | Sales | | Transfers out | | Deaths | Closing valuation | |
|---|---|---|---|---|---|---|---|---|---|---|---|---|---|
| | No. | £ | No. | £ | No. | £ | No. | £ | No. | £ | | No. | £ |
| Dairy cows | 89 | 44 500 | | | 24 | 14 400 | 19 | 7600 | | | | 94 | 47 000 |
| Calves | 4 | 400 | | | 92 | Born | 60 | 6000 | 30 | 3 000 | 2 | 4 | 400 |
| Dairy heifers i.c. | 24 | 14 400 | | | 33 | 14 850 | | | 24 | 14 400 | | 33 | 19 800 |
| Heifers (yearlings) | 13 | 3 900 | | | 27 | 5 400 | 3 | 750 | 33 | 14 850 | | 4 | 1 200 |
| Heifers under 1 yr | 23 | 4 140 | | | 30 | 3 000 | | | 27 | 5 400 | | 26 | 4 680 |
| Sub-total | | 67 340 | | | | | | | | | | | 73 080 |
| Ewes | 556 | 16 680 | 25 | 2 500 | 90 | 3 150 | 57 | 1 140 | | | 15 | 99 | 17 970 |
| Rams | 24 | 1 920 | 4 | 400 | | | 4 | 80 | | | 3 | 21 | 1 680 |
| Lambs | 764 | 19 100 | | | 868 | Born | 732 | 21 960 | 90 | 3 150 | 10 | 800 | 20 000 |
| Sub-total | | 37 700 | | | | | | | | | | | 39 650 |
| Sows | 46 | 4 600 | | | 30 | 3 300 | 14 | 980 | | | 1 | 61 | 6 100 |
| Boars | 3 | 300 | 1 | 100 | | | 1 | 40 | | | | 3 | 300 |
| Weaners | 39 | 975 | | | 1 020 | Born | | | 1 000 | 25 000 | 5 | 54 | 1 350 |
| Baconers | 388 | 17 460 | | | 970 | 24 250 | 998 | 59 845 | | | 4 | 356 | 16 020 |
| Malden gilts | 9 | 990 | 10 | 1 200 | 30 | 750 | 2 | 200 | 30 | 3 300 | | 17 | 1 870 |
| Sub-total | | 24 325 | | | | | | | | | | | 25 640 |
| **Total** | | 129 365 | | | | | | | | | | | 138 370 |

**Fig. 12.2** Livestock reconciliation – example for a farm with several livestock enterprises.

enterprises are fed with home-mixed rations, care must be taken to ensure that each batch of feed is recorded at the time of mixing.

The addition of simple records of opening and closing stocks allows regular checks on the accuracy of recording, by reconciliation with the field and enterprise records. This also helps in detecting waste and pilfering, and eases the task of valuing stock at the end of the year.

Where items are stored in bulk, e.g. grain, potatoes and concentrate feed, it is impossible to record accurately without some form of weighing mechanism, and the value of the barn records will suffer as a result. Up until now the adoption of such devices has been relatively slow, partly because of expense. Current developments, such as the electronic load cell, are reducing the real cost of weighing, and thus the validity of this excuse.

Two forms of record are particularly important in the context of fixed cost recording: the *wages book* and the machinery register. The wages book is the source of the single figure which appears in the CAB as a weekly or monthly expense. If the book is well set out, it can help the manager in the calculations of the various deductions from a worker's pay, both statutory (PAYE and National Insurance contributions) and voluntary (milk, eggs, cottage rent, Christmas Club, etc.). It provides the basis for the pay advice slip which should accompany the wages when paid, as well as keeping the cash analysis entries simple.

The *machinery register* records information relevant to the cost of ownership of machinery operated by the business, and eases the calculation of depreciation charges, written-down values, and profits or losses on sale. It may also be expanded to allow the recording of major repairs and maintenance, and hours worked over given periods. Although this information is often recorded in machine log-books, it is useful to have a central record in case a log-book is mislaid. Where the farm is large enough to justify a workshop, and even the employment of a fitter, more elaborate recording may be necessary. Such records should ideally be kept in the workshop rather than the office, with the fitter or head tractor-driver responsible for their upkeep.

### 12.3.2 Physical records in short-term control

Financial records are compiled at relatively wide time intervals. Where enterprise accounts (such as gross margins) are concerned, the calculations often cannot be made until the end of the production cycle, by which time it is too late to apply the lessons learnt to the current crop. Where production is continuous, it is possible to make more frequent calculations of margins, but at best this is likely to take place on a monthly basis. This limits the value of financial records in controlling the progress of an enterprise from day to day, week to week. Many operational decisions have to be made to very short deadlines. In such cases, control must be based on physical records.

Thus in a manufacturing enterprise, such as cheese-making, physical records of throughput and of quality are crucial. In a service enterprise, such as a tourist visitor attraction, head-counts of visitors provide opportunity for rapid monitoring. In farming, the use of physical records in short-term control is most highly developed in those enterprises which most nearly approximate to manufacturing, such as large-scale intensive livestock enterprises. This applies particularly to poultry, rabbits and pigs, and to a lesser extent to large dairy herds. In these circumstances, the scope for individual attention to animals or birds is limited by the sheer number in the care of the stockman. The use of physical records becomes a part substitute for the close relationship between man and beast which is possible in a smaller herd.

An extreme example of this is the daily recording of water consumption which is practised in many intensive poultry houses. Such recording would be of little value if used solely as a basis for enterprise accounts, since the value of the water used is so small that the effort of daily recording would not be justified. On the other hand, a significant deviation of water consumption from the norm gives the stockman a warning to look out for trouble (such as environmental problems, disease, or a leaking pipe). It also gives the vet or technical adviser valuable information for diagnosis. Other examples include the daily consumption of food, daily mortality, daily production of eggs and similar produce, and the quality of produce (such as the grading of eggs).

Where records are vital for short-term control, there is an extra need to make the job of keeping and using the records as easy as possible. By their nature, they will often be maintained by someone other than the manager. The 'other person' will almost certainly have less incentive than the manager to record clearly and accurately, and to take note of unexpected changes. The best way of increasing incentive is to keep the job simple.

To encourage accurate record-keeping, use should be made of clear, unambiguous record forms, displayed in a prominent position, well lit, and with a pencil provided (the occasional 'lost' pencil is a small price to pay for reliable records). To help in interpretation of records, graphs and other visual aids can be highly valuable, especially if they have the target performance marked on them. Colour can be used on record sheets to highlight the items that are really vital to check.

Increasingly computers are being used for physical record-keeping, with specialist industry-specific programs being developed (e.g. field record and mapping systems) and hand-held 'data-loggers' being used to collect information in electronic form in the work-place prior to linking up to the office computer.

## 12.4 Stock control

Stocks cost money. Holding stocks of raw materials, work in progress or finished goods incur costs related to storage (space, supervision, etc.), financing (working capital, overdraft), risk of damage or spoilage, and sometimes risk of obsolescence. On the other hand, the costs of not keeping *enough* stocks can be huge. A 'stockout' in materials or work in progress means that production stops, with machines and people idle, and overheads mounting. Running out of finished goods means immediate loss of sales (and cash flow), delay for the customer and possible loss of goodwill and trade in the longer term.

Maintaining up-to-date and accurate stock records is part of good stock control, using the principles of good recording outlined in Section 12.3. Another aspect, however, is *using* those records in order to balance the costs of holding stocks against the risk of running out. A useful tool in this is the calculation of the appropriate *reorder level* of a given stock:

$$\text{Reorder level} = \frac{\text{typical delivery time (weeks)} \times \text{use in year}}{\text{working weeks in year}}$$

Take the example of supplier 1 of a particular input, with a normal delivery time of 1 week. If you typically use 5000 units of the input in a year, and if your working year is 50 weeks, the reorder level is 100 units (1 week × 5000 units ÷ 50). Once your records of that particular input stock show a quantity in store of less than 100 units, you should put in an order for replacements.

If dealing with supplier 2 for the same input, whose normal delivery time is 10 weeks, the reorder level would be 1000 units (10 weeks × 5000 units ÷ 50). This is likely to involve a substantial extra cost in financing and provision of storage accommodation, and justifies paying a premium for purchases from supplier 1.

Formulas exist for the calculation of the *economic order quantity*, i.e. that quantity of input which you should reorder. They are often complex and rely on subjective assessments of some of the variables: while useful in bigger operations, with the assistance of purpose-designed computer programs, they are beyond the scope of most smaller businesses (Millichamp, 1992, p. 103).

'Just-in-time' systems represent the ultimate in reducing stock-holding. They involve a highly planned and controlled production process, with delivery specified to the hour, enabling stock-holding costs to be slashed or eliminated (by passing them back to the supplier). However, it is a risky procedure without a real hold over suppliers, and contingency plans for delivery failure. It is unlikely to be possible for the smaller business, as suppliers are likely to be reluctant to comply.

## 12.5 Background records

In addition to specific records of transactions and the technical performance of enterprises, there are a number of records which contribute in a general way to the financial management of the business. Some may be trivial, such as price lists and press cuttings: others may be essential, such as Integrated Administration and Control System (IACS) forms, insurance policies, loan agreements and the records needed to establish 'traceability' through the food chain (increasingly demanded by buyers of food products). The main requirement here is an orderly and secure method of storing them (see Section 12.5).

## 12.6 Organizing paperwork

It will be seen from the preceding sections that even a small business can generate a considerable volume of paperwork. Sooner or later all this paper will find its way into the files, into the post, or into the wastepaper basket. Two problems present themselves: how to file information so that it can be retrieved quickly and easily, and how to organize the flow of paper through the office so that nothing is lost or delayed.

The word 'filing' conjures up images of rows of filing cabinets, tended by secretaries. To file information simply means to store it, though. The 'files' may be anything from a cardboard box, through loose-leaf folder to microfilm; the method used depends on the size of business. For the small or medium-sized farm business, a system based on lever-arch loose-leaf files will probably be adequate. A three or four drawer filing cabinet will cost more initially, but will last longer and may be more convenient.

Whatever the container, some method of classification will be needed to permit the retrieval of stored documents. Ideally, this classification should relate in some way to the headings in the CAB. Files could thus be started for each of the various enterprises and/or types of input and output, and for the various categories of fixed costs. Retrieval can be further eased by the use of systematic indexing; popular methods are by alphabetical order of subject, and by numerical code.

Whatever indexing system is used, there will be times when a document relates to two or more file headings (such as an invoice which includes fungicide sprays for winter cereals as well as grassland fertilizer). In such cases a method of cross-referencing is required. In its simplest form this need consist only of a note in each of the related files, reminding the manager where the document may be found. More complex systems may be devised using card indexes.

It is of little use having a wonderful filing system if the documents never reach the file. Without a systematic processing of paperwork through the office, vital documents can easily be 'lost' in disorganized piles of paper. The consequences can be financially detrimental (such as the loss of a discount for prompt payment on a large order) or just embarrassing (such as the time when the adviser arrives to talk about the dairy and the milk record sheets cannot be found).

The first requirement of an office[16] system (see Figure 12.3) is a wastepaper basket. In addition at least three receptacles are needed. These can be standard office trays, but there are many alternatives; bulldog clips, cardboard boxes, or just tidy piles. The first tray (or other receptacle) is used to receive all incoming paper, such as the day's post. Another accumulates paper which has been dealt with, and which is

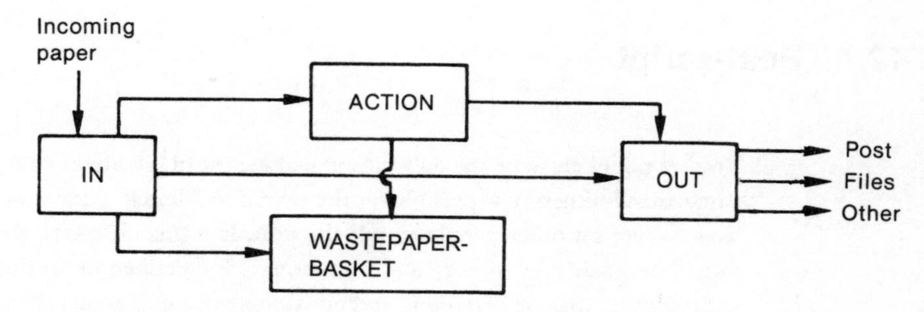


**Fig. 12.3** Paperwork system.

destined for the files or the post. The tray in between acts as a repository for all items which are awaiting action. Paper thus comes into the 'in' tray, stays there only until it is examined, after which it is transferred either to the wastepaper basket, or to the 'action' tray to await further developments, or to the 'out' tray for filing, posting, or other disposal.

A useful supplement to the 'action' tray is a loose-leaf diary. This has two advantages over the conventional form of diary. The first is that one can always keep a 'rolling' twelve months in the diary – as soon as one month has gone by, its pages can be removed and new pages inserted at the other end of the diary. The second, more important feature is the facility for clipping documents into the diary at certain dates. Thus if a large bill is due to be paid in twenty days' time, it can be filed so that on the payment day, the diary is opened to reveal that document. The same process can be used for amounts due to be received, licence renewal forms, and similar documents. This provides a more positive reminder than an action tray full of papers.

Because of their volume, and their financial significance, special care is needed in dealing with transaction documents. If the volume of transactions is sufficiently great, it may even be worthwhile installing a duplicate set of trays solely for such documents. The important points are first, to keep together all documents relating to a particular supplier or customer, and second, to avoid filing any documents before payment has been made or received. Thus, for purchases, the order copy should wait in the action tray for the invoice, delivery note, statement and other related documents; as these arrive they should be checked against each other and pinned to the order. When the cheque is written out, its number and amount can be recorded on the top sheet and the whole bundle of papers transferred to the 'out' tray ready for filing. Naturally the payment details will be recorded on the cheque stub, for transfer to the CAB at a later date. A similar procedure should be followed for sales documents.

As with any unwelcome work, it is advisable to make dealing with paperwork an integral part of the general routine. A suitable analogy relates to trips to the dentist – it is tempting to 'forget' about the six-monthly check-ups, but if you do so the consequences can be painful. Similarly, paperwork can be made less of a chore, and more efficient, if it is incorporated into the daily timetable (say half an hour after breakfast) instead of saving it all up for a rainy day.

## 12.7 Post-script

You are not likely to be the only person making use of records of your farm. Even in a one-man business, it is possible for the owner to fall sick or have an accident, and have to rely on other people to run the unit. It is therefore vital that the records should be made easy to keep and to interpret, as described in Section 12.6.

Employees may need training and initial supervision to cope with tasks that seem simple to you, but may be totally foreign to them. In order to encourage careful recording, it is good policy to explain why the various records are necessary, and to share with employees the lessons that you learn from those records.

A positive attitude to recording can be encouraged by providing a herdsman, say, with an 'office' area, where he can complete records in relative comfort, and which gives paperwork a higher status in the employees' eyes. It is a sad irony that far more attention has been given to creating warm and dry office space in farm businesses since the advent of the computer (perceived by its owner as more susceptible to adverse conditions than mere mortals, and whose repair costs have to be borne by the business).

## Notes

16. The term 'office' is used to denote any area set aside for paperwork, whether a purpose-built room or a desk in the corner of the spare bedroom. The essential features are comfort, convenience and quiet.

## References

### *General*

Hosken, M. & Brown, D. (1991) *The Farm Office*, 4th edn. Farming Press, Ipswich.

### *Physical records*

Barnard, C.S. & Nix, J.S. (1979) *Farm Planning and Control*, 2nd edn. pp. 516–519. Cambridge University Press, London.

Millichamp, A. (1992) *Finance for Non-financial Managers*, 2nd edn. DPP Publications, London.

Speedy, A.W. (1980) *Sheep Production*. Longman, London.

Whittemore, C.T. (1980) *Pig Production*. Longman, London.

# Part 4

# Controlling the Business

Earlier in this book it was suggested that the financial management of a business could be described in terms of 'planning and control'. Adam Haynes has used budgets in planning the financial affairs of Home Farm. As he puts those budgets into operation, he should start operating a system of financial control. 'Control' was earlier defined as the comparison of actual and expected performance, and the taking of any necessary corrective action. This process can, and should, be undertaken with respect to each of the three financial indicators: cash, profit and capital.

When using any of the control techniques described in the following chapters, it is essential to remember that one is seeking clues that might help in the solution of a problem, rather than the solution itself. As Rockley (1979) says in the context of performance ratios:

> [Performance comparisons] are but demonstrations of matters needing further scrutiny. They rarely give the final answers to any problem by themselves alone; it is the ensuing scrutiny of the underlying reasons for change which should present management with a plan of action to correct any undesirable developments.

## References

Rockley, L.E. (1979) *Finance for the Non-accountant.* Business Books, London.

# Chapter 13

# Monitoring Cash Flow

## 13.1 Why monitor cash flow?

Of the three financial indicators, the most sensitive, and probably most important to monitor is cash flow. Few rural businesses are likely to produce profit and capital statements at intervals of less than six months, if that. The cash flow budget and the CAB, on the other hand, are normally drawn up on a monthly basis. A monthly comparison of cash flow budget and cash analysis results can enable problems to be spotted much sooner than a six-monthly or yearly analysis of profit or capital. The earlier such potential problems are noticed, the sooner remedial action may be taken.

The penalty of *not* monitoring cash flow regularly and frequently could be at best the loss of an opportunity, and at worst the loss of a business. As an example of the first, take manager A, whose budgeted cash flow profile for 199X is shown in Figure 13.1. Manager A's results are a little better than he had budgeted at the end of month 8. If manager A does not take the trouble to monitor his cash flow, it will not be a disaster. His bank statement will show a larger cash surplus than expected, and he may then begin to think about doing something useful with that cash rather than let it lie in the current account. The chances are, though, that by the time realization dawns, many of the opportunities for short-term investment will have been missed.

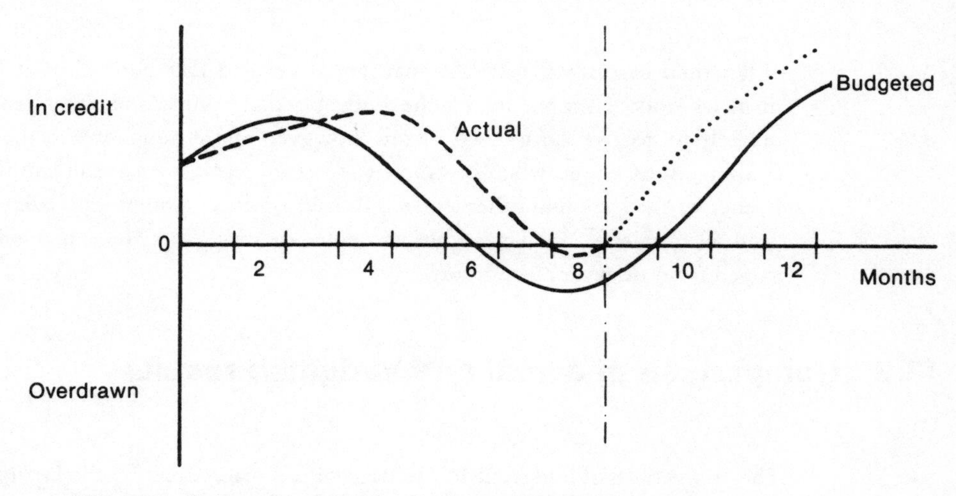


**Fig. 13.1** Budgeted and actual cash flow profiles, manager A.

If he had monitored his cash flow carefully, he might well have spotted the likelihood of such a discrepancy arising long before it showed up as an increased surplus. He would then have had plenty of time to make plans for short-term investments, storing produce that little bit longer, or diverting money to a deposit account, for instance.

As an example of a more extreme problem, take manager B, whose budgeted cash flow profile for 199X is shown in Figure 13.2. Whether because of bad luck, bad management, or a combination of the two, manager B is having a bad year. If he monitors cash flow carefully, he has a good chance of noticing danger signals at an early stage, and can have contingency plans ready in case his worst fears are justified. These contingency plans might include delaying expansion, bringing sales forward and/or delaying purchases, selling assets such as land or cottages, and making an early appointment to discuss the matter with his bank manager.

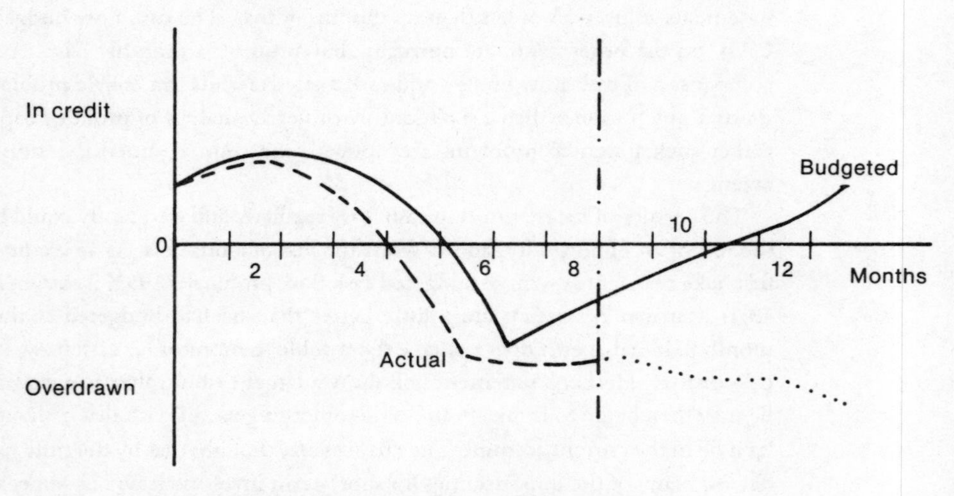


**Fig. 13.2** Budgeted and actual cash flow profiles, manager B.

It would be understandable if manager B decided that he could not face the financial realities shown up by the budget/actual comparison. The problem is unlikely to go away of its own accord, however, and if he ignores it, the crunch could come at a time when it is too late to change plans or to maintain the confidence of the bank and other lenders. Result: another business liquidation.

In all businesses, then, whether successful or struggling, there is much to be gained from monitoring cash flow.

## 13.2 Comparison of actual and budgeted results

The comparison of budgeted and actual results is made easier by the incorporation of 'actual' columns into the cash flow budget (Chapter 7). The figures needed to complete these 'actual' columns are derived from the monthly totals of the analysis columns in the CAB (Chapter 10). If these column headings have been chosen wisely they will correspond very closely to the headings used in the cash flow budget. The process then becomes one of a simple transfer of figures from cash analysis to budget. This transfer should ideally take place as soon as the monthly reconciliation of the CAB has taken place. Figure 13.3 shows this process completed for December 199X/9Y for Home Farm, using figures from the cash flow budget (Figure 7.8) and the reconciled cash analysis (Figures 10.4 and 10.5).

Once the 'actual' column for a particular month has been completed, it can be compared with the corresponding budget column. The golden rule applying to all financial comparisons is to start with the general and work back to the particular. In the case of cash flow control, this implies starting with a comparison of actual and budgeted closing bank balances for the month under consideration. The next stage is to examine the main factors which contribute to the bank balances, i.e. opening balance, total cash receipt and total cash payments. Then the individual items of receipts and payment should be investigated, concentrating on those which (a) are financially large, and (b) show a significant variance.

A systematic approach of this kind avoids becoming bogged down in minutiae and losing sight of the overall financial picture as a result. Figure 13.4 illustrates the process: analysis should start with the items at the top of the page and work down as far as the detail of the figures will allow. Memory is fallible, and it is useful to make a brief monthly record of events which led to discrepancies between budgeted and actual results; this can be jotted down on the back of the budget, or in a spare space in that month's column. This record is valuable not only for short-term control, but also in providing guidance for the preparation of next year's budgets.

| **Cash flow budget** | | 199X/9Y | | Version: 2 | | | |
|---|---|---|---|---|---|---|---|
| % interest rate | 12 | October | | November | | December | |
| | | Budget | Actual | Budget | Actual | Budget | Actual |
| *Receipts* | | | | | | | |
| Milk | 1 | 2 980 | 2 280 | 3 700 | 3 651 | 9 350 | 9 254 |
| Calves | 2 | 500 | 500 | 3 500 | 4 028 | 2 000 | 1 895 |
| Pigs | 5 | 6 900 | 6 300 | 6 900 | 5 214 | 3 680 | 4 269 |
| Bed and Breakfast | 7 | | | | | | |
| Cheese sales | 9 | 1 300 | | 1 300 | | 1 300 | 1 095 |
| Sundries | 11 | | | | 256 | | 390 |
| Capital: grants | 12 | | | | | | |
| machinery sale | 13 | | | | 500 | | 500 |
| VAT charged | 16 | | | | 132 | | 156 |
| VAT refunds | 17 | | | | | | 1 643 |
| **Total receipts** | | 11 680 | 9 080 | 15 400 | 13 649 | 16 330 | 17 403 |

**Fig. 13.3** Cash flow budget/actual comparison, Home Farm.

| | | October | | November | | December | |
|---|---|---|---|---|---|---|---|
| | | Budget | Actual | Budget | Actual | Budget | Actual |
| *Payments* | | | | | | | |
| Dairy: l/stock purch. | 18 | | | | | | |
| feed | 19 | 3 750 | 3 648 | | | 3 000 | |
| vet and med. | 20 | 288 | 440 | 288 | 390 | 308 | 507 |
| misc. costs | 21 | 333 | 320 | 333 | 180 | 333 | 532 |
| Pigs: l/stock purch. | 22 | 2 500 | 2 500 | 2 500 | 2 090 | 1 750 | 1 325 |
| feed | 23 | 2 800 | 2 800 | 2 418 | 2 295 | 2 420 | 2 790 |
| vet and med. | 24 | 100 | 120 | 100 | 105 | 100 | 190 |
| misc. costs | 25 | 104 | 115 | 104 | 115 | 104 | 133 |
| Misc. crop costs | 29 | | | | | | |
| B & B other costs | 30 | 20 | 100 | 200 | 100 | 100 | |
| Cheese: materials | 32 | 100 | 300 | 100 | 127 | 100 | |
| other | 33 | 100 | 50 | 100 | | 100 | 654 |
| Wages: permanent | 36 | 2 125 | 2 285 | 2 125 | 1 984 | 2 125 | 2 175 |
| casual | 37 | 100 | 1 026 | 100 | 829 | 100 | 540 |
| Power: fuel | 38 | 540 | 410 | 540 | 584 | 540 | 210 |
| repairs | 39 | 625 | 426 | 625 | 492 | 625 | 998 |
| Overhead sundries | 41 | 625 | 930 | 1 725 | 1 100 | 625 | 100 |
| Property repairs | 42 | | | 200 | 50 | | |
| Overdraft interest | 43 | 210 | | 239 | | 220 | 255 |
| Loan interest | 44 | | | | | | |
| Hire purchase | 45 | | | | | | |
| Leasing charges | 46 | | | 525 | 525 | | |
| Capital: cubicles | 49 | | | 1 000 | | 1 000 | |
| machinery purchase | 50 | | | | | | 3 369 |
| capital repayment | 51 | | | | | | |
| Personal: drawings | 52 | 250 | 250 | 250 | 300 | 250 | 260 |
| tax | 53 | | | | | | |
| VAT paid | 54 | | 483 | | 611 | | 1 188 |
| **Total payments** | | 14 570 | 16 203 | 13 472 | 11 877 | 13 800 | 15 226 |
| Net cash flow | 55 | −2 890 | −7 123 | 1 928 | 1 772 | 2 530 | 2 177 |
| Opening balance | 56 | −21 000 | −21 000 | −23 890 | −28 123 | −21 962 | −26 351 |

**Fig. 13.3** (cont'd)

## 13.3 Interpreting the results

Having examined the discrepancies, the manager must decide what action, if any, to take. If the various discrepancies more or less balance each other out, he may decide to keep the business on its present course. If he decides that corrective action is necessary, then he should prepare revised budgets to show the effect of the changes on cash flow in particular, but also on profit and capital. If the changes are small, outline or partial budgets will be sufficient; otherwise a complete revision will be necessary.

Even if there is no major change of policy during the year, the cumulative effect of

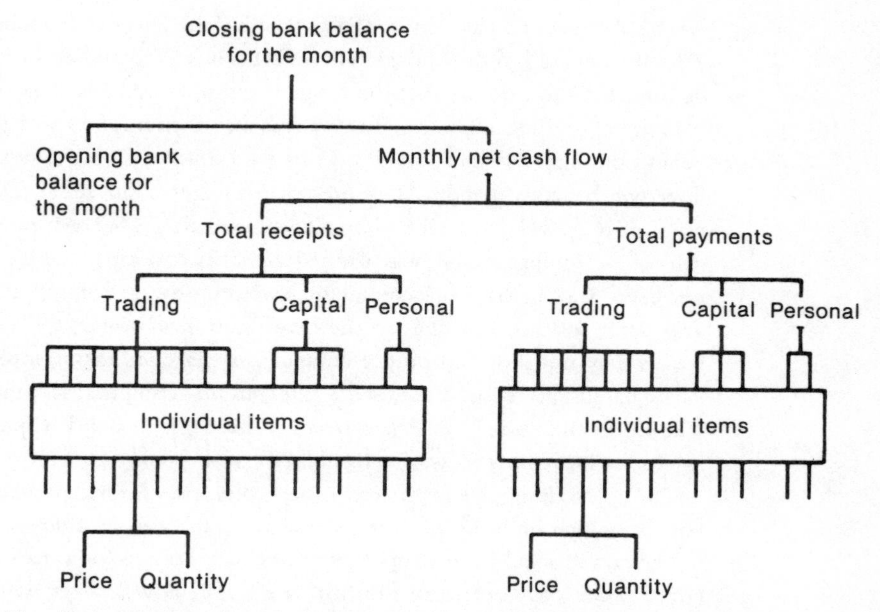


**Fig. 13.4** Working from the general to the specific in monitoring cash flow.

many small discrepancies over several months may result in the actual position bearing very little resemblance to the budget by half-way through the year. The closing bank balance may be very close to the target, but the way in which that balance has been derived may differ greatly from the budget. If this is the case, the budget should be revised to correct such anomalies and make the budget a more realistic and useful document. If the cash flow budget is treated as a 'rolling' plan as described in Chapter 7, this revision will be undertaken automatically at regular intervals.

In the example shown in Figure 13.3, relating to Adam Haynes' first three months' trading, the overdraft is £4742 higher than budgeted. This is disturbing enough in itself, but if it represents a trend that will continue for the rest of the year, it is a matter for serious concern. The difference reflects not only this month's activities, but also a carry-forward of variances in previous months: November's closing balance was already down by £4389.

Looking at the receipts analysis, it is possible to identify variations in individual items which more or less cancel each other out. Shortfalls on dairy and cheese receipts (around £400) are more than compensated by increases in pig receipts. The £1073 positive variance in total receipts arises primarily from VAT charged and refunded, and some unbudgeted sundry and capital income. The VAT charged and refunded, however, merely balance out VAT payments made over the last 3 months, leaving non-VAT receipts about £700 less than budgeted.

Total payments are up £1426 on budget in December – £238 excluding VAT. This was partly due to the unscheduled purchase of machinery. If this is merely bringing forward a purchase that was budgeted for a later month, the effect of this will be temporary. If, on the other hand, the purchase was not anticipated, it will adversely affect the results throughout the rest of the year.

The absence of a payment for dairy feed has the effect of reducing payments by £3000 compared to the budget. This effect is not likely to last, however. It would be unrealistic to suppose that the business did not need any dairy concentrates in November and December, and it is a reasonable assumption that the bill remains behind the kitchen clock, waiting to be paid. In that case, the next month's cash flow will be correspondingly poorer than budgeted: in effect, December's cash position is £7742 worse than budget, rather than £4742. Moreover, £1000 for dairy cubicle improvement was planned for both November and December, but has not been paid: perhaps the contractor has been slow in coming up with the bills. The latter still need paying, so the true position is nearer £9742 down. Other factors may not show up on the comparison, yet need taking into consideration. Adam has just placed an order for a £1500 business computer system, for instance. It was not in the budget, so Adam needs to bear this in mind when projecting the results for December forward over the rest of the year.

The £554 difference in 'cheese: other' costs arises from a mistake in the original budget, which omitted the costs of packaging the cheese. Budgeting mistakes are not always so small or so temporary: it has been known for a manager to omit the entire rental payment from the first year's cash flow budget, with painful consequences. Regular cash flow monitoring can help the manager to identify such problems at an early stage in the year, and he can embark on corrective action before it is too late.

Some differences are useful prompts to further investigation. The 'vet and med.' bill for the cows is consistently more than budgeted, for instance, a fact that might focus Adam's suspicions and encourage him to review the health status of his herd.

## 13.4 Annual cash flow monitoring

Examination of cash flow performance over a whole year can be valuable, although it is not as immediate as monthly monitoring, or as comprehensive as profit comparisons. Adam Haynes' first year cash flow is summarized in Figure 13.5, with variances shown in both money and percentage terms.

Since December 199Y (we are moving forward in time from the previous section) Adam has managed to keep a tight rein on cash flow, with the result that the closing balance is under that budgeted. This has partly been achieved by substantial economies on dairy, pig, and sundry costs, while increasing revenues in those areas. Cheese: other costs and power costs are considerably under-budgeted – this will need attention in revising the budgets. As so often happens, the capital payments are more than anticipated. The result should, nevertheless, reassure the bank manager at the impending annual review of Adam's account.

The comparison takes no account of money owed, or of goods in store: capital and personal transactions obscure the picture of trading performance. In other words, what is needed for an effective annual appraisal is an examination of net profit.

| | | Total | | Difference | |
|---|---|---|---|---|---|
| | | Budget | Actual | £ | % |
| *Receipts* | | | | | |
| Milk | 1 | 84 100 | 84 239 | 139 | 0.17 |
| Calves | 2 | 7 400 | 8 258 | 858 | 11.59 |
| Cull cows | 3 | 11 000 | 9 831 | –1 169 | –10.63 |
| | 4 | | | | |
| Pigs | 5 | 73 140 | 75 231 | 2 091 | 2.86 |
| | 6 | | | | |
| Bed and Breakfast | 7 | 9 000 | 8 250 | –750 | –8.33 |
| | 8 | | | | |
| Cheese sales | 9 | 15 000 | 13 950 | –1 050 | –7.00 |
| | 10 | | | | |
| Sundries | 11 | 2 650 | 4 897 | 2 247 | 84.79 |
| Capital: grants | 12 | | | | |
| machinery sales | 13 | 5 000 | 7 523 | 2 523 | 50.46 |
| loans received | 14 | | | | |
| Personal receipts | 15 | 500 | 200 | –300 | –60.00 |
| VAT charged | 16 | | 3 617 | 3 617 | N/a |
| VAT refunds | 17 | | 8 764 | 8 764 | N/a |
| **Total** | | 207 790 | 224 760 | 16 970 | 8.17 |

| | | Total | | Difference | |
|---|---|---|---|---|---|
| | | Budget | Actual | £ | % |
| *Payments* | | | | | |
| Dairy: l/stock purch. | 18 | 21 250 | 18 260 | –2 990 | –14.07 |
| feed | 19 | 12 900 | 11 362 | –1 538 | –11.92 |
| vet and med. | 20 | 3 500 | 3 346 | –154 | –4.40 |
| misc. costs | 21 | 4 000 | 2 912 | –1 088 | –27.20 |
| Pigs: l/stock purch. | 22 | 27 500 | 23 514 | –3 986 | –14.49 |
| feed | 23 | 29 400 | 28 950 | –450 | –1.53 |
| ved and med. | 24 | 1 200 | 1 096 | –104 | –8.67 |
| misc. costs | 25 | 1 250 | 1 348 | 98 | 7.84 |
| Seed | 26 | 600 | 895 | 295 | 49.17 |
| Fertilizer | 27 | 5 710 | 6 524 | 814 | 14.26 |
| Spray | 28 | 780 | 820 | 40 | 5.13 |
| Misc. crop costs | 29 | 120 | 153 | 33 | 27.50 |
| B & B: food | 30 | 1 680 | 1 865 | 185 | 11.01 |
| other costs | 31 | 750 | 951 | 201 | 26.80 |
| Cheese: materials | 32 | 1 140 | 1 893 | 753 | 66.05 |
| other costs | 33 | 1 160 | 3 257 | 2 097 | 180.78 |
| Wages: permanent | 34 | 25 500 | 26 485 | 985 | 3.86 |
| casual | 35 | 2 800 | 2 432 | –368 | –13.14 |
| Power: fuel | 36 | 6 900 | 7 502 | 602 | 8.72 |
| repairs | 37 | 7 500 | 9 900 | 2 400 | 32.00 |
| contract hire | 38 | 1 000 | 1 100 | 100 | 10.00 |
| Rent | 39 | 9 500 | 9 500 | | |
| Property repairs | 40 | 200 | 210 | 10 | 5.00 |
| Overhead sundries | 41 | 8 600 | 7 476 | –1 124 | –13.07 |
| Overdraft interest | 42 | 2 244 | 2 175 | –69 | –3.08 |
| Loan interest | 43 | 1 100 | 1 100 | | |
| Hire purchase | 44 | | | | |
| Leasing charges | 45 | 2 100 | 2 100 | | |
| | 46 | | | | |
| Capital: cubicles | 47 | 3 000 | 4 137 | 1 137 | 37.90 |
| machinery | 48 | 15 000 | 16 021 | 1 021 | 6.81 |
| capital repayment | 49 | 3 000 | 3 000 | | |
| Personal: drawings | 50 | 3 000 | 3 520 | 520 | 17.33 |
| tax | 51 | 1 000 | 856 | –144 | –14.40 |
| VAT paid | 52 | | 12381 | 12381 | N/a |
| **Total** | | 205384 | 217 041 | 11 657 | 5.68 |
| Net cash flow | 53 | 2 406 | 7 719 | 5 313 | 220.85 |
| Opening balance | 54 | –21 000 | –21 000 | | |
| Closing balance | 55 | –18 594 | –13 281 | 5 313 | –28.57 |

**Fig. 13.5** Budget/actual comparison, whole year.

# References

Warren, M.F. (ed.) (1993) *Managing your Cash*. Book 3 in series: Finance Matters for the Rural Business. ATB Landbase/Seale-Hayne Faculty of Agriculture, Food and Land Use, University of Plymouth.

# Chapter 14

# Monitoring Profit and Capital

## 14.1 The importance of monitoring profit

Although the monitoring of cash flow from month to month may be vital to the survival of the business, it is not enough on its own to ensure the continued prosperity of the firm. Without taking account of money owed to the business, for instance, or produce that is still unsold and in store, it is impossible to derive a clear impression of the efficiency of the business as a working unit. For this it is necessary to monitor profits.

The monitoring of profits is not as sensitive a control measure as the monitoring of cash flow. Although it is possible to draw up monthly profit and loss accounts (and common practice in manufacturing firms), the long production cycles of most agricultural products make it unlikely that the benefit will be worth the effort (possible exceptions are those items whose production most closely resembles a factory system, such as eggs, poultry and milk). A similar argument applies to other forms of rural business.

It is likely that the monitoring of profits will be an annual process for most enterprises. Ideally, it should take place as soon as the production cycle is complete (or as near as can be), so that any lessons learnt may be applied, without delay, to subsequent production. This will certainly mean the manager compiling his own management accounts, rather than waiting for the accountant's version. It may also mean anticipating the sale price of produce which is likely to be sold some time after the end of the production year. Broadly speaking, there are three complementary ways of monitoring profits, each relying on a comparison of some sort. The first is by comparison with the results of previous years; the second by comparison with other businesses; and the third by comparison with budgeted profits, i.e. budgetary control.

Certain comments may be made with respect to all three types of comparison, to avoid repetition. First, as with cash flow, the process should be systematic. The major summary figures should be investigated first, working progressively back to the smallest details in the account. The process is illustrated in Figure 14.1 using a profit and loss account in gross margin form as an example. Analysis should start with the items at the top of the page, and work down the page as far as the detail of the account will allow.

Second, since the analysis is intended as a foundation for decision-making, attention should be concentrated on those components of the account which:

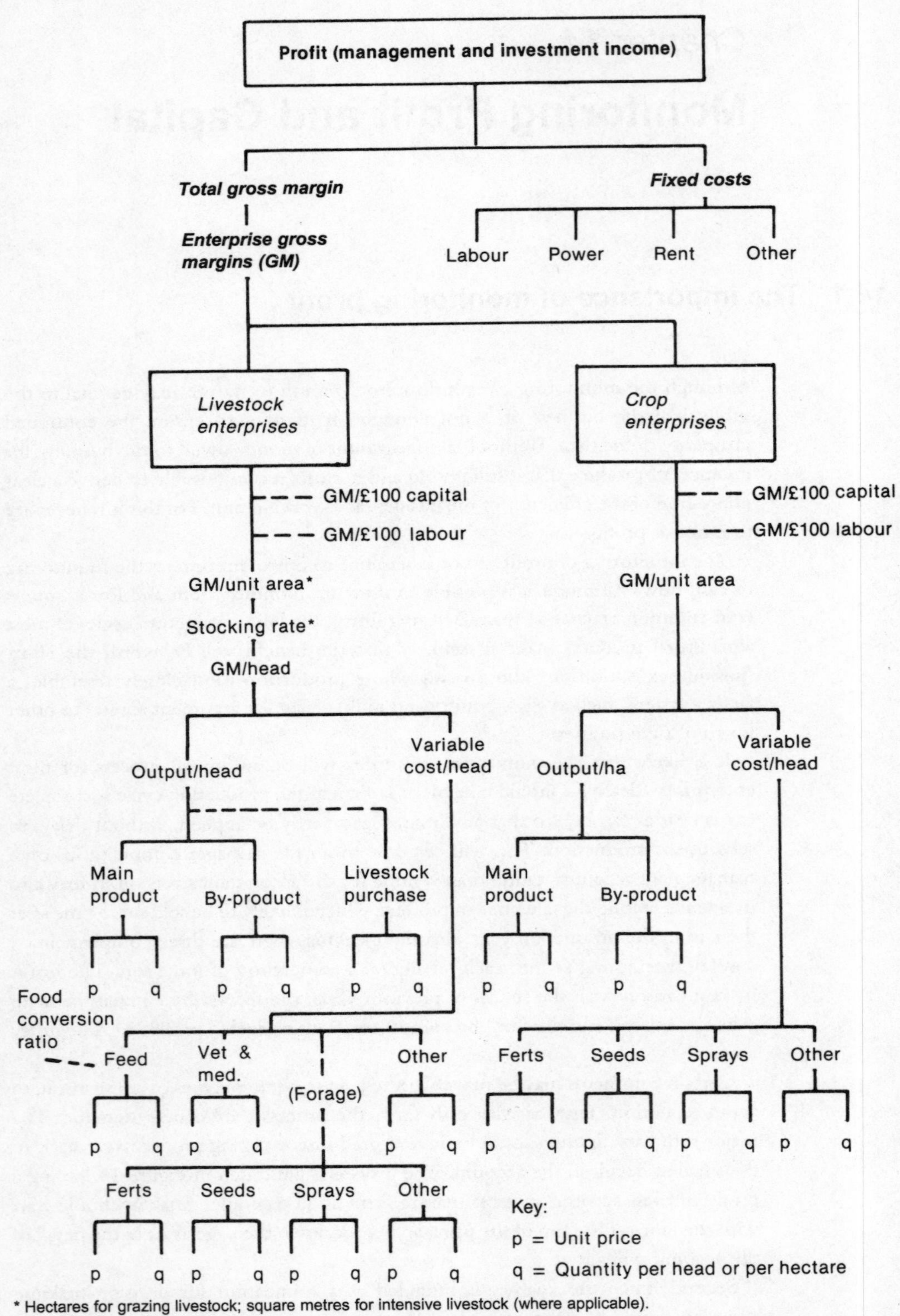


**Fig. 14.1** Working from the general to the specific in analysis of profit and loss accounts.

- Can be influenced by the manager within the time period concerned
- Are financially significant

Thus, in a given year, there may be little a manager can do to influence his rent bill, or even his regular labour costs. Most prices for standard farm commodities are set outside the business – the farmer is primarily a 'price-taker'. It is unwise to spend time analysing such items when that time could be spent more profitably in analysing items which *are* affected by the manager's decisions. It is similarly foolish to spend hours trying to find ways of reducing a comparatively minor cost when an equivalent time spent on a larger cost might bring substantially greater rewards.

This is not to say that other items should be completely ignored. Changes in minor costs can be valuable indicators of major problems, and information about the effect of factors outside the manager's influence (such as changes in price levels) can help him to adapt his strategy to the business environment. It is merely that such items should take second place in the analysis.

Finally, the opening remarks to Part Four should be borne in mind. The comparisons provide only clues – not whole answers. The answers arise only from combining these clues with a thorough knowledge of the business in question, and of the technical aspects of the production processes.

## 14.2 Comparison with previous years

### *14.2.1 The method of analysis*

Comparison of profits between years can be used with a profit and loss account of any form. Little needs to be said about the mechanics of the comparison, except to reiterate the need for a systematic approach. Where possible, the current year's results should be viewed in the light of those of several previous years, rather than just one. In this way the manager may avoid making judgements which are influenced by abnormal features of any one year.

### *14.2.2 Benefits and limitations*

Comparisons with previous years have an advantage over interfirm comparisons in that they relate specifically to the business in question rather than to a group of roughly similar businesses. They show the progress of the business between years, in terms of both major expansion and general improvements (or decline) in productivity. They also show the impact of price movements, both general (i.e. inflation), and specific.

Unfortunately, it is often difficult to distinguish between the two effects and to tell how much an improvement in a margin is due to increased efficiency, as opposed to increased product prices. In times of inflation the manager is likely to be forced back on to comparison of the physical components of the account, such as input and output quantities. A further limitation of this approach is that it is more backward-looking than either of the others. The past can be a valuable guide for future decisions, but it is dangerous to rely on it absolutely. Nevertheless, in view of the fact that it is specific to the business under scrutiny, measures progress and uses figures that are almost always readily available, inter-year comparison remains an important source of information.

## 14.3 Interfarm comparisons (1): conventional profit and loss account

### 14.3.1 Farm business analysis

The technique of interfirm comparison relies on comparing the manager's own profit and loss account with those of other, similar businesses. If this technique is to be used at all, it should ideally be based only on profit and loss accounts in enterprise account form. There are times, however, when the only profit information available is in the form of a conventional account.

The process of interfirm comparison was described in Chapter 3, and little needs to be added with respect to most service and manufacturing businesses. In farming, however, there are specific techniques meriting discussion. There are two influences on this: the tendency to be preoccupied with production efficiency in farming rather than sales efficiency (since the producer of conventional bulk farm commodities has little control over the amount he can sell or the price he receives); and the ready availability of comparison data in UK agriculture[17].

The usual source of comparison information in the UK is the *Farm Business Survey* (FBS). Each regional centre of the FBS (usually attached to the agricultural economics department of a university) collects financial information from a range of farms each year. The results are published annually in the form of a regional farm management handbook. Results are displayed for each of the main farming types in the region, and by farm size. In order to ensure that their statistics give a fair reflection of farming efficiency, the FBS researchers 'standardize' the profit information that they receive. Their intention is to avoid the results being distorted by factors which are not directly dependent on the standard of farm management. In order to make a fair comparison between his own profit and loss account and the FBS figures, the farmer must 'standardize' his account in the same way. The revised account will give the *Net Farm Income* (NFI) and the *Management and Investment Income* (MII) of the business.

The NFI is calculated as follows.

*Take* Net profit and *include* an extra expense to represent the value of any unpaid labour (not including that of the farmer and spouse)

*Exclude*
- Any non-farm revenue and expense
- Any interest charges paid
- Any revenue and expense connected with the ownership of land (e.g. repairs and maintenance, mortgage payments, receipt of rents). A

notional rent should take the place of these items for any land which is owned, rather than rented, by the farmer.

The MII is calculated by estimating the value of any physical work done on the farm by farmer and spouse, and deducting this from the NFI (see Figure 14.2). The MII per hectare may now be compared with that shown in the FBS handbook for similar farms in the same area. In attempting to explain deviations between his MII and that of the comparison, the farmer should work through the various items of revenue and expense in the same way. These will not be classified by enterprise, so the management information which can be derived is somewhat limited.

| | £ | £ |
|---|---|---|
| Net profit | | 10 000 |
| *less* Son's labour | –2 000 | |
| *less* Sundry external receipts | –500 | |
| *plus* Interest paid | 2 000 | |
| *plus* Land expenses | 500 | |
| *less* Rents received | –1 500 | |
| *less* Notional rent for land owned | –1 000 | –2 500 |
| **Net farm income** | | 7 500 |
| *less* Value of farmer's physical labour | | –2 000 |
| **Management and investment income** | | 5 500 |

**Fig. 14.2** Calculation of NFI and MII – example.

Various supplementary 'production standards' can be used in an attempt to remedy this deficiency, such as crop yields per hectare and livestock stocking rates. Those relating to fixed costs can be valuable, particularly the calculation of gross output or gross margin per £100 labour and/or power cost. The 'standards' used, and their calculation, can vary from one FBS region to another. Before calculating such standards, it is wise to check the glossary of terms in the FBS handbook.

Finally, MII may be compared with the capital employed. Since the standardization of the account removes all land-owning considerations, the appropriate measure of capital is taken to be 'tenant's capital'. Various definitions of tenant's capital are in use, the most common being the average of the opening and closing valuations of stocks, machinery and equipment. The MII and tenant's capital are then combined in the calculation of *return on tenant's capital*, using the formula:

$$\frac{\text{MII}}{\text{Tenant's capital}} \times 100$$

This can then be compared with the FBS results.

### 14.3.2 Benefits and limitations

The primary benefit of whole-farm account analysis is that it extracts decision clues from an otherwise unhelpful document. Where records are limited, this benefit can be substantial. Against this must be set the significant limitations of both the method and the comparison figures. The *method* is based on a whole-business approach. The result is that there is little scope for investigation of individual enterprises. Another criticism is that most of the comparison is aimed at determining the efficiency of one input, land, whereas the efficiency of use of other inputs, notably capital, is often more important.

The chief problem with the *comparison figures* themselves is their broad grouping of farm types and areas. Each FBS regional unit surveys a large number of farms. For convenience of presentation the results have to be classified into a relatively small number of groups. The 'average' results for each group thus arise from a wide variety of farms – some on good soil, others on poor; some with root crops, others with none, and so on. It follows that it is difficult for the farmer to be sure that he is comparing like with like. Further limitations arise from the inevitable delay between collection and publication of the data, and from doubts over the 'representativeness' of the sample of farms used in the survey.

In short, it is not worth bothering with this technique unless the only source of profit information is a conventional profit and loss account.

## 14.4 Interfarm comparisons (2): profit and loss accounts in enterprise account form

### 14.4.1 The method of analysis

In this form of analysis, enterprise accounts in either gross margin or full absorption costing form are compared with 'typical' accounts for similar enterprises. These 'typical' accounts will have been compiled from surveys of firms with similar enterprises, from the results of enterprise costing schemes, or by reference to standard performance figures.

A common source of agricultural survey data is the FBS; other sources are investigations carried out by commercial firms (such as feed and fertilizer suppliers); by the advisory and development services; and by researchers in universities, colleges, and research institutes (often producing reports for specialist producers in, for instance, horticulture, pigs and poultry).

Enterprise costing schemes are run by various organizations, such as the Meat and Livestock Commission (MLC) (beef, sheep and pigs); commercial firms (such as feed suppliers); Genus Management Services; the National Farmers' Union (NFU) (eggs and poultry); and various advisers and management consultants. The results may be made available only to subscribers to the scheme concerned, but are often widely published. A particularly useful type of scheme is that using comparisons between a relatively small group of farmers within the same locality.

'Standard' accounts are not built up directly from survey data, but from the opinions of experts as to the typical performance which might be expected from a reasonably well-run enterprise. They are particularly useful where survey data is out of date, insufficiently detailed, or in other ways unreliable. The various sources of standards include MLC publications, the Cambridge University Pig Management Scheme, the NFU poultry bulletins and, with forecast prices, farm management handbooks such as those produced by Wye College ( the 'Nix Pocketbook'), Agro Business Consultants, and the Scottish Agricultural Colleges.

For non-agricultural enterprises the search for comparison data might be more difficult, and is more likely to involve standards than straight survey results. It could start with enquiries to public and quasi-public organizations, such as regional tourist boards, chambers of commerce, professional institutes, and Business Link centres. Larger accounting and consulting firms are often able to provide comparison data of their own, based on information collated from a large number of clients. Although it is unlikely to be truly representative of the whole population of similar enterprises, such comparison can provide a valuable guide.

If a step-by-step analysis is followed, as recommended at the beginning of this chapter, few difficulties should arise. It is necessary occasionally to make judgements on a highly-summarized account, with little detail supplied. In such circumstances it is vital to ascertain the way in which the account was compiled. In a gross margin account, for instance, the inclusion of casual labour and/or contractors' charges in the account of one enterprise can make the margin look poor in comparison with another where all the labour and power is supplied from within the business. In comparing gross margin accounts it is easy to forget the existence of fixed costs, although there is often great scope for economy in labour and power costs. These should be investigated in the first place using the FBS comparisons described earlier, followed if necessary by the more detailed analysis described in Chapter 17.

For continuous-production farm enterprises such as milk and egg production, it is possible to make interim checks on performance by comparing such measures as the *margin over concentrates* (i.e. value of milk or egg sales minus cost of concentrate feed), or the quantity of concentrate feed used per unit of physical quantity of output (e.g. kilograms of concentrate feed per litre of milk or per dozen eggs). Although these are crude as measures of financial performance, they are quick and easy to calculate, and sensitive indicators of flock or herd health and management. The main benefit of this form of analysis is the ability to compare the economics of one's own enterprises with the performance that other businesses are currently achieving. As long as enough detail is provided in the comparison figures, a 'lower-than-typical' margin can be explained by careful examination of output and costs. This may, in turn, lead to corrective action being taken to prevent similar discrepancies occurring in future. Since each enterprise is investigated individually, the grosser limitations of the whole-account method are avoided.

On the other hand, it is difficult to avoid the problems inherent in the use of comparison with a 'typical' business. Survey and costing scheme data is likely to be drawn from a wide range of business types and, as with the whole-business comparisons, it is difficult for the farmer to be sure that he is comparing like with like.

There is inevitably a delay between collection of the data and its publication (often more of a problem with surveys than with costing services). Where collection of information depends on the willing co-operation of business owners, it is often difficult to be sure that the information is truly representative (more of a problem with costing services and accountant's data than with surveys). The wide range of types of contributing businesses may make difficulties in providing full physical detail in the comparison accounts.

The use of 'standard' accounts can overcome the problems of delay and detail. This is only gained, however, at the expense of reliability, since one is dependent on the compiler's opinion as to what level of performance is 'typical' or 'acceptable'.

The clues provided by this technique are of higher quantity and quality than those provided by the whole-business technique, but they must still be treated with considerable caution.

## 14.5 Budgetary comparisons

### *14.5.1 The method of comparison*

In attempting to control the finances of a business, it is far more relevant to make comparisons with targets set for that specific business than with the performance of a 'typical' business. This is the process of budgetary control which was discussed in Chapter 13 with respect to cash flow. The method is one of simple comparison between the profit and loss account and the profit and loss budget; it can be used with profit statements of either conventional or enterprise account form, although the latter will, of course, give a more comprehensive view. It is also possible to use summary enterprise checks (such as margin over concentrates) as interim control measures.

As with the other forms of comparison, it is essential to use a systematic approach (see Figure 14.1).

### *14.5.2 Benefits and limitations*

This method compares figures relating to the same business, in the same time period and with as much physical detail as the manager himself cares to supply. It thus avoids most of the problems associated with comparisons between years and/or between businesses. A more positive benefit is that, since the budget will have (or should have) been prepared with the manager's objectives firmly in mind, the budgetary comparison is a valuable check on the degree to which those aspirations have been achieved. Its contribution to the planning/control process is thus far more direct than that of the comparisons with a wide range of other businesses, each of which may have different objectives from those of one's own business.

The limitations of this technique are minor compared to the other forms of comparison. The most significant problem is likely to be insularity. If the manager relies entirely on comparisons within his own business, and never takes the trouble to compare with standard or survey data, there is a danger that he will be ignorant of opportunities to improve profits by bringing outputs and/or costs into line with those on comparable businesses. This can be avoided by use of a judicious blend of both interfirm and budgetary comparisons, the one helping the interpretation of the other.

## 14.6 Variance analysis

### 14.6.1 Principles and method

The primary aim in any form of account comparison must be to identify discrepancies between one account and another. It is often difficult to decide exactly *how* these discrepancies arise, however. Three factors are at work: volume, i.e. the size of the enterprise, as measured in number of head, hectares, square metres, beds, etc.; unit quantity, i.e. output (yield) or input per unit of volume; and price per unit produced or consumed.

As an example, the revenue from the sale of weaners in a pig breeding herd will vary according to the average number of sows in a herd during the year, the number of pigs weaned per sow, and the average price obtained for each weaner. Unless some attempt is made to isolate the effects of these on a comparison, valuable management clues may be lost. To use the above example, the actual revenue may be £5 more than the comparison. This £5 could be the result of a 'favourable variance' (say £1005) arising from better prices, and 'unfavourable variances' arising from poor sow performance (–£950) and low sow numbers (–£50). A serious management problem is thus masked by a fortunate increase in prices. This is an extreme example, but it demonstrates the danger of relying on the *measurement* of discrepancies, without any attempt to *explain* their origins.

Such explanations may be provided by the use of *variance analysis*. The principles of the technique can be illustrated, for a simple 'two-way' example (i.e. excluding the effect of volume variances), by means of a price/quantity diagram. Figure 14.3 relates to the revenue per hectare of a wheat crop; price is shown on the vertical axis, unit quantity on the horizontal axis, and revenue (price × quantity) is shown by the area of the resulting rectangle. Two revenue rectangles are shown on the same diagram, one representing the actual performance obtained, and the other the comparison (described here as the budget, but it could be from survey, standard, or previous-year data). It may be deduced that the effect of the unfavourable variance in quantity is represented by the small rectangle marked *U* (i.e. –1 t × £95). Similarly the effect of the favourable variance in price is given by the small rectangle marked *T* (i.e. £10 × 6 t). The measurement and explanation of variances may now be restated as:

| | | £ |
|---|---|---|
| Actual revenue: | 6 t at £105/t | 630 |
| Budget revenue: | 7 t at £95/t | 665 |
| *Difference* | | –35 |

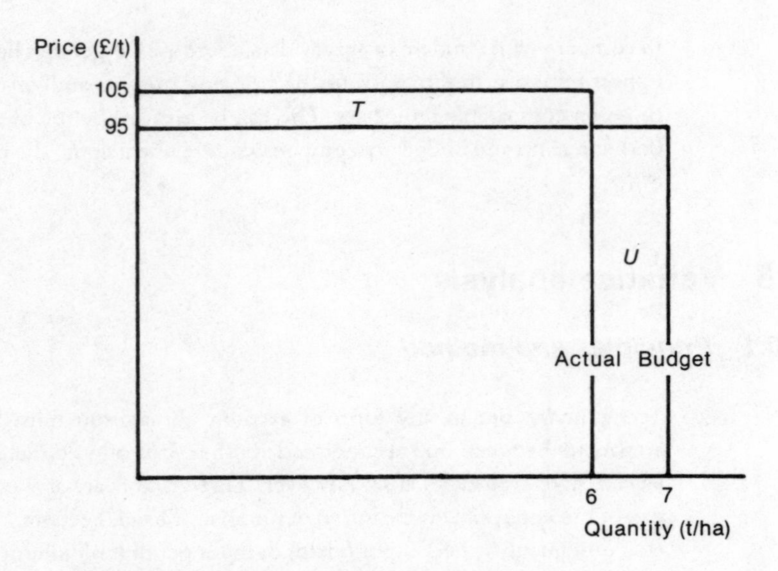


**Fig. 14.3** Price/quantity diagram for a wheat crop.

| | |
|---|---:|
| Variance due to unit quantity: −1 t × £95 | −95 |
| Variance due to price: + £10 × 6 t | +60 |
| *Total variance* | −35 |

From this may be derived a general formula for the calculation of two-way variances:

Variance due to unit quantity = difference in unit quantity x budget price
Variance due to price = difference in price × actual unit price

The technique may be extended to the analysis of three-way variances, so as to incorporate the effects of volume variances. If the wheat crop above was grown on only 30 ha instead of the 35 budgeted, the difference in total revenue would be:

(£630 × 30 ha) − (£665 × 35 ha) = −£4375

Variance due to volume = difference in numbers × budget unit quantity × budget price.
Variance due to unit quantity = difference in unit quantity × actual numbers × budget price.
Variance due to price = difference in price × actual numbers × actual unit quantity.
Thus the −£4375 revenue difference of the example may be explained by:

| | £ |
|---|---:|
| Variance due to volume (−5 ha × 7 t/ha × £95/t) | −3325 |
| Variance due to unit quantity (−1 t × 30 ha × £95/t) | −2850 |
| Variance due to price (+£10 × 30 ha × 6 t/ha) | +1800 |
| | −4375 |

The formula may be used in exactly the same way to explain variances in items of expense. It is important to remember, of course, that a negative variance in expense is a favourable variance.

### 14.6.2 Benefits and limitations

Little needs to be said about the benefits of using this technique. Its use enables the influences of volume, unit quantity and price to be identified separately. This enables the manager to see how much his profits have been affected by the quality of his management (usually reflected by volume and unit quantity variances) as opposed to chance (usually the main influence on prices, as far as the manager is concerned). Several items of revenue and/or expense may be analysed simultaneously, and the results related. In the example used for Figure 14.3, for instance, analyses could have been made of differences in fertilizer costs, and the results related to the yield variance. In this way the manager gains a comprehensive insight into the economics of his business.

Of the limitations, one will be immediately obvious; the time-consuming and complicated nature of the calculations. This need not be a great problem as long as the manager restricts his analysis to those items which are large enough, in money terms, to bother with, and which are well 'off target' with respect to volume and/or unit quantity.

A second problem lies in the application of a precise formula to data which is often rather imprecise. This applies particularly to interfirm comparisons. Much of the discussion in earlier sections was concerned with doubts over the validity of using survey or standard data as a guide to the performance of one's own business. Use of variance analysis in such circumstances has the danger that the results will attract an unjustified aura of precision. This does not mean that the technique should not be used; merely that the manager must continually remember that he is looking for clues rather than answers.

The final limitation arises from doubts over the formula itself. This formula has gained wide acceptance in management circles, and is described (if not explained) in a variety of publications (see, for instance, those listed under References). It works perfectly when a favourable price variance is combined with an unfavourable unit quantity variance. With any other combination of variances, however, it gives distorted results, as can be seen from Figures 14.4–14.6. Figure 14.4 shows the effect of favourable variances in both price and quantity. Areas $V$ and $X$ are clearly the result of variances in price and quantity, respectively, but area $W$ cannot be attributed to either, since it results from the interaction of price and quantity variances. The formula, however, attributes this area to price alone (try it and see).

In Figure 14.5, which shows the result of unfavourable variances in both price and unit quantity, a similar interaction is represented by area $Y$. This time the formula attributes it all to unit quantity variance.

Figure 14.6 shows a favourable variance in quantity combined with an unfavourable price variance. The diagram suggests that the price variance is given by difference in price × budget quantity, and the quantity variance by difference in quantity × actual price. The formula, however, states that these should be,

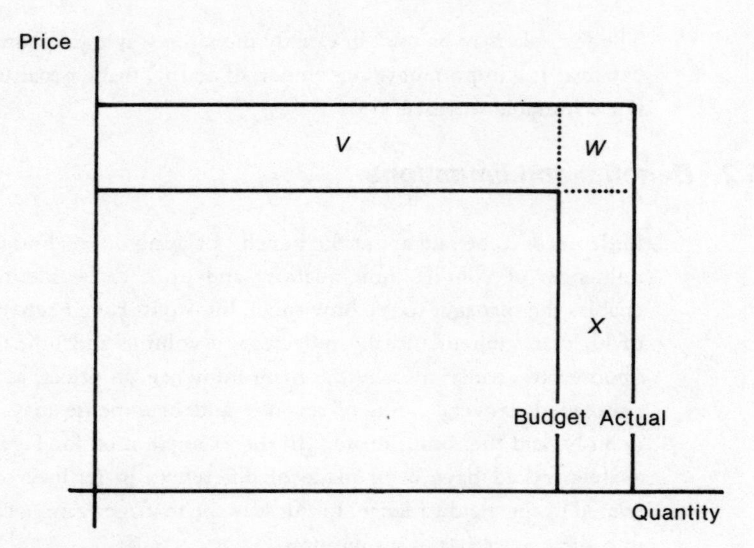


**Fig. 14.4** Favourable variances in both price and quantity.

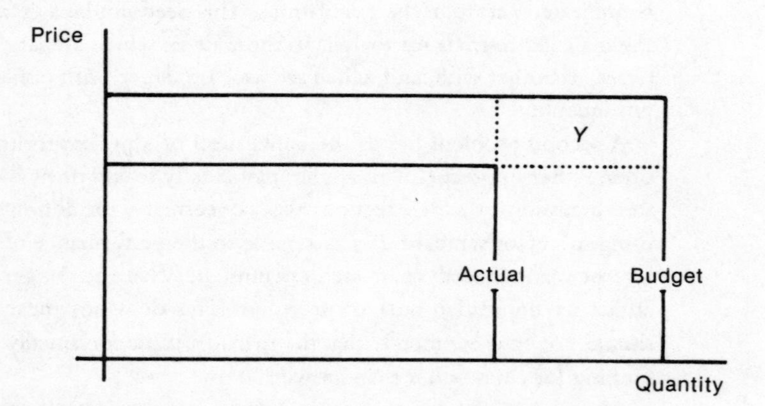


**Fig. 14.5** Unfavourable variances in both price and quantity.

respectively, difference in price × actual quantity, and the difference in quantity × budget price. The formula thus exaggerates the influence of both price and quantity by equal amounts, i.e. area *Z*. There are two sensible solutions to this difficulty. The first is to use the formula regardless, and assume that the resulting inaccuracies are likely to be small. The other is to refer to sketch diagrams rather than to use the formula. This has the advantage that one does not need to memorize the formula, but is only practicable for 'two-way' analysis. The irrational solution would be to avoid using the technique; despite its limitations it remains a highly valuable aid to the analysis of profit statements.

## 14.7 Profit monitoring for Home Farm

We have available the budgeted and actual financial summaries (Figures 6.13, 7.10 and 11.6) and the budgeted and actual pig gross margin accounts (Figures 6.9 and

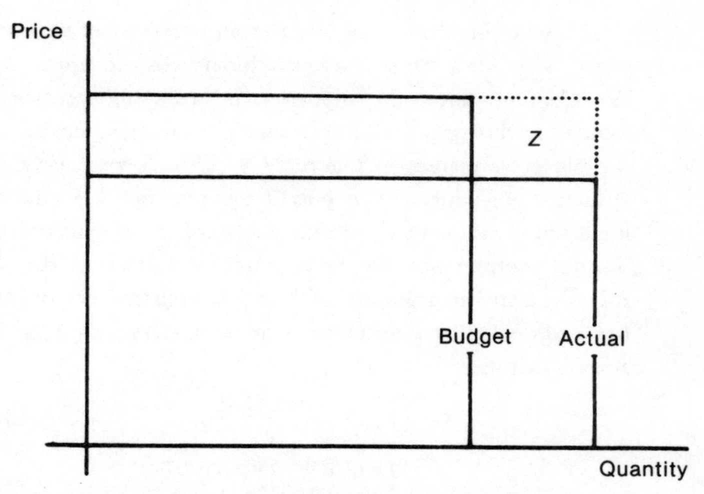


**Fig. 14.6** Favourable quantity variance combined with unfavourable price variance.

11.5) for Home Farm's first year of trading. This allows an illustration of the techniques described above, although in 'real life' it would be important to use previous-year and interfirm comparisons as well as budgetary control, and all the enterprises would come under scrutiny. It is important to note at this stage that the variances can just as easily arise from inaccurate budgeting in the first instance as from the quality of management during the year: another argument for continuous monitoring, particularly during the first years.

The comparison of results for the whole business is summarized in Figure 14.7. The diagram shows that net profit is up by £4368 compared with budget, this being brought about by substantially increased gross margins, partly offset by higher fixed costs. Increased machinery running costs have had the biggest effect on fixed costs: on the output side, large favourable variances in pigs and miscellaneous income compensated for poor performance on cheese.

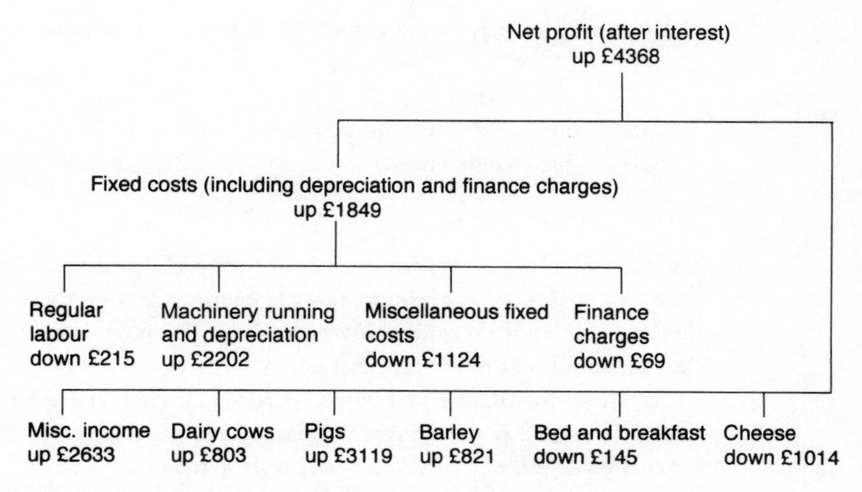


**Fig. 14.7** Breakdown of actual/budget comparison, financial summary.

The whole-business analysis gives an overview of performance, but begs questions. Why, for instance, were machinery costs so much higher than the budget? What brought about the improvement in the pig enterprise? For this we need to look more closely at the components of the budget, and this process is illustrated for the pig gross margin in Figure 14.8. This shows clearly that the main positive influence is an increase in output per pig, one big enough, even allowing for increased variable costs and reduced numbers, to improve the margin by £3119. Various interpretations can be put on these differences: they could have arisen from Adam's careful management of the pigs during the year, or his ineptitude could have been concealed by a fortunate increase in pigmeat prices. This is where variance analysis can help.

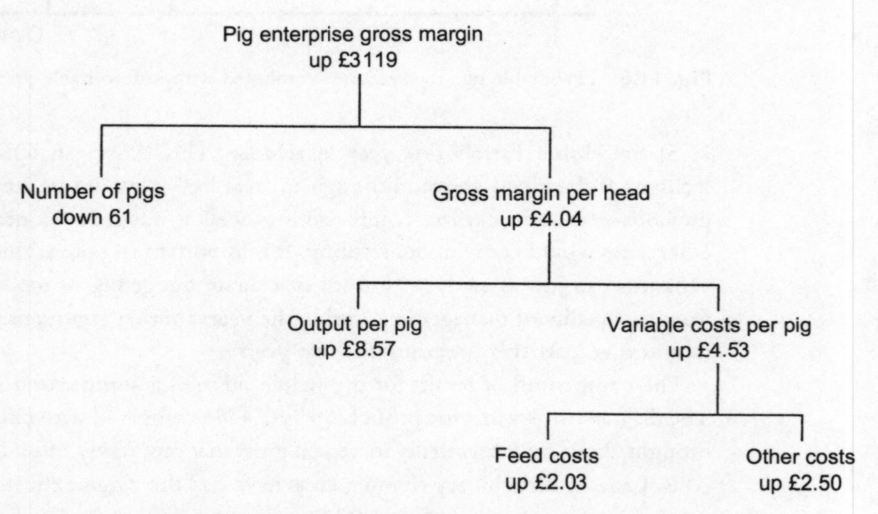


**Fig. 14.8** Breakdown of actual/budget comparison, pig enterprise.

A two-way analysis on pig output, for instance, is as follows:

| | | £ |
|---|---|---|
| Variance due to difference in numbers: | −61 × 70.20 | -4282 |
| Variance due to difference in price: | +£8.57 × 939 | 8047 |
| | | 3765 |

The full extent of the damage done by the shortfall in numbers can now be seen: over £4000. It may be, of course, that the improved price is partly a result of lower numbers and less stress through overcrowding. The point is that Adam now has a clear indication of the effect of that policy.

The most important cost factor is concentrate feed, and for this a three-way analysis is needed (a third factor of quantity per pig is involved here). Using the formula given earlier, this analysis would be as follows:

| | | £ |
|---|---|---|
| Variance due to difference in numbers: | −61 × 0.208 t × £140 | −1776 |
| Variance due to difference in quantity fed: | +0.0295 t × 939 × £140 | 3878 |
| Variance due to difference in price: | −£8.83 × 939 × 0.2375 | -1969 |
| | | 133 |

This analysis shows clearly that the drop in numbers (whose cost is explored in the two-way analysis above) and the drop in price (more likely to have arisen outside the business than from Adam's management) have obscured the full effect of an increase in the consumption of feed per pig of nearly £4000. Again, this may be part of Adam's policy to improve performance by producing a heavier, better quality animal, or it could be poor feed management, but we cannot draw any firm conclusions without talking to Adam and seeing the unit at first hand. At least Adam now has precise information with which to interpret the past results and to plan for the future.

As well as examining the other components of the budget in this way, Adam would be wise to compare his results with previous years and with local comparison figures in the manner described earlier.

## 14.8 Monitoring capital

There are two aspects to the monitoring of capital. The first is the examination of the actual balance sheet using ratio analysis, flow of funds analysis, etc. The techniques concerned have been fully explored in Chapter 3 and need no further discussion. The results of this examination are useful both in themselves and as a basis for the budget/actual comparison which forms the second aspect of the monitoring process.

One way of tackling this latter comparison is as follows. First the net capital figures are compared, budget against actual. Then the main ratios are compared: liquidity, gearing, activity. If significant discrepancies appear, the various groups of assets and liabilities should be examined. The flow of funds statements (budget and actual) will prove useful in detecting the root cause of any discrepancy.

Although the capital position of a business is, in most circumstances, a far less sensitive indicator of its financial health than are cash flow and profit, the process described above still merits the description 'control'. The time periods are much greater, but the logic is the same as that behind cash flow control. By careful monitoring, it is possible to spot unfavourable trends developing while there is still time to rescue the situation. Conversely, favourable movements may be detected early enough to capitalize on them.

In Figures 8.1 and 11.7 we have already seen the opening, budgeted closing and actual closing balance sheets for Home Farm, and Figure 14.9 shows a summary of the main features. The change in the main ratios has been generally positive, and the net capital has increased beyond expectations as a result of greater movement than anticipated in assets. This, together with the positive messages from the cash flow

| *Ratio analysis* | *Opening* | *Budget* | *Closing* |
|---|---|---|---|
| **Profitability ratios (%)** | | | |
| Return on owner equity | | 15.6 | 18.9 |
| Return on capital employed | | 16.1 | 18.5 |
| Gross profit margin | | 45.6 | 47.5 |
| Net profit margin | | 8.8 | 10.7 |
| **Activity ratios* (days)** | | | |
| Stock turnover period | | 191.2 | 190.8 |
| Debtor turnover period | | 17.7 | 16.9 |
| Creditor turnover period | | 11.3 | 15.8 |
| **Gearing ratios (%)** | | | |
| Percentage owned | 62.4 | 67.0 | 67.7 |
| Long-term debt to equity | 33.2 | 27.7 | 26.7 |
| Interest cover | | 421.1 | 506.5 |
| **Liquidity ratios (%)** | | | |
| Acid-test ratio | 36.0 | 38.6 | 36.9 |
| Current ratio | 227.7 | 256.5 | 273.9 |
| **Growth (%)** | | | |
| Growth in net capital | | 9.0 | 13.0 |
| Growth in assets | | 1.6 | 3.9 |
| Growth in liabilities | | –12.1 | –12.0 |

* Calculated using average of opening and closing values.

**Fig. 14.9** Ratio analysis for actual accounts, Home Farm.

and profit results, should enable Adam to proceed with cautious optimism – as long as he maintains the process of budgetary control for the following years.

Figure 14.10 completes the analysis with a flow of funds statement. Trading is seen to be the main source of cash flowing into the business, though it has needed a cash injection and increased overdraft in addition. Apart from the commitments represented by private drawings (could there still be scope for economy here?), tax payments and loan repayments, much of the cash has been spent on reinvestment in machinery and buildings, and in repaying borrowing.

# Notes

17. But not only in the UK: many other countries have comprehensive farm data collection and comparison systems. Some (see, for instance, Lund & Orens, 1997, concerning Denmark) are significantly more sophisticated.

# References

Agro Business Consultants (latest edition) *The Agricultural Budgeting and Costing Book.* Agro Business Consultants, Melton Mowbray (published half-yearly).

| | FLOW OF FUNDS ANALYSIS | | £ |
|---|---|---|---|
| **Sources** | *Net profit (loss)* | | 21 849 |
| | Depreciation (+) | | 12 065 |
| | Change in debtors (increase –, decrease +) | | 1 090 |
| | Change in creditors (increase +, decrease –) | | 4 089 |
| | Valuations (increase –, decrease +) | | –7 239 |
| | Benefits in kind (–) | | –4 324 |
| **Cash from trading** | | | 27 530 |
| **Add** | Capital introduced | | 200 |
| | Sale of land, property, investments | | 0 |
| | Increase in overdraft | | |
| | Increase in loans | | 0 |
| | Reduction in cash balances | | 0 |
| **Cash from other sources** | | | 200 |
| **Total cash available** | | | 27 730 |
| **Uses** | *Commitments* | | |
| | Private drawings | | 3 520 |
| | Tax payments | | 856 |
| | Reduction in loans | | 3 000 |
| **Total commitments** | | | 7376 |
| **Add** | Purchase of land, property, investments | | 0 |
| | Net purchase of buildings | | 4 137 |
| | Purchase of machinery | 16 021 | |
| | Less sale of machinery | –7 523 | 8 498 |
| | Repayment of overdraft | | 7 719 |
| | Increase in cash balances | | 0 |
| | | | 20 354 |
| **Total cash used** | | | 27 730 |

**Fig. 14.10** Home Farm – flow of funds analysis for 199X/9Y.

Barnard, C.S. & Nix, J.S. (1979) *Farm Planning and Control*, 2nd edn. pp. 530–542. Cambridge University Press, London.

Bright, G. (1996) An Exploration of Profit Measurement. *Farm Management*, **9**, 8, pp. 383–391.

Lund, M. & Orens, J.E. (1997) Computerised Efficiency Analysis in Farm Business Advice. *Farm Mangement*, **9**, 10.

Nix, J.S. (latest edition) *Farm Management Pocketbook*. Wye College, University of London, Wye, Kent (published annually).

Norman, L., Turner, R.A.E. & Wilson, K.R.S. (1985) *The Farm Business*, 2nd edn. Longman, London.

Scottish Agricultural College (latest edition) *Farm Management Handbook*. The Scottish Agricultural College, Edinburgh (published annually).

# Part 5

# More on Planning and Control

In Part 5 some of the possible complications are considered, starting with a look at planning in its widest sense, building up a business system from scratch. It then moves on to examine particular management problems, including the planning of livestock enterprises, the planning and control of farm labour and machinery, investment appraisal, and coping with risk and uncertainty.

*Chapter 15*

# The Planning Process – A Wider View

## 15.1 Tactics and strategy

In the interests of simplicity, the preceding chapters have viewed management, especially the planning process, in a very narrow way. Discussion of planning has been confined to the budgeting of a given system – little distinction has been made between the long and short terms; and it has been assumed that the objectives of a business can be described completely in financial terms.

It is now time to take a broader look at the planning process, with the aim of selecting and budgeting the *optimal* system for the business, rather than the uncritical rebudgeting of an existing business system.

A differentiation is often made between strategic and tactical plans. Strategic plans normally cover periods of several years, relating to the overall direction of business policy, while tactical plans typically refer to no more than 12 months at a time, and will ideally take their frame of reference from a pre-established strategic plan. Thus the battle tactics are played out in the context of the overall strategy for the war. As an example, the decision about what kind of caravan site Home Farm should have is a tactical decision: the decision about what kind of business Home Farm will be, and thus the decision to be involved in the tourism business, is one that relates more to the strategic planning of the business.

Ideally, strategic plans should be reviewed and revised at regular intervals – say every year. In this way they become 'rolling' plans, rather than being limited to a fixed period and, most important, becoming either cast in stone or ignored. Tactical plans should be formulated at the same time but are likely to need further revision as the year unfolds. A classic model of strategic planning is shown below.

- Determine the mission of the business
- Set aims and objectives
- Forecast likely changes in the external environment of the business
  - Assess the 'internal' characteristics of the business, i.e. strengths and weaknesses, including any constraints which are likely to limit the range of alternative actions open to the business, such as limited resources
  - Assess the external environment of the business, i.e. opportunities and threats, including forecasting likely changes in that environment

- Consider all the alternative plans which could be expected to achieve the objectives
- Select the plan which appears to have the greatest chance of achieving the objectives
- Implement that plan
- Periodically monitor, evaluate and revise plans

Other models exist, but this gives a good basis for considering the various components of the planning process.

## 15.2 The mission

The word *mission* has become grossly overused of late, but it is useful to consider what the overall purpose of the business *is*: it is very easy to take this for granted. The word 'vision' gives an alternative way of expressing this idea, implying a process whereby the owner(s) of the business, and ideally other stakeholders (such as employees) as well, attempt to identify what sort of business or organization they want to have. This could be defined in terms of size and type, but also by the values and principles they want to govern its operation. Adam Haynes' 'vision' of Home Farm, for instance, might be of a business which will provide a good level of living for him and his family, an opportunity for his children to succeed him, employment for local people, while operating in harmony with the natural environment. The values underlying this mission might include those relating to people (employees, customers, neighbours).

## 15.3 Setting aims and objectives

An *aim* is a general statement of the desired benefits for the owners (and possibly the other 'stakeholders') from running the business, for the period to which the plan applies. An example for Home Farm, for a three-year plan, might be to generate sufficient cash flow for Adam and his family to live, without further recourse to bank borrowing, and allowing essential reinvestment. In future periods this modest aim will, we hope, be superseded by more ambitious aspirations, in order to achieve the overall mission, but this might be as far as it is sensible to aim in the medium term.

*Objectives* are more specific goals that need to be achieved in order to attain the aim. Ideally they should be expressed in measurable outcomes, and relate to a specific time period, so that one can test whether or not one has achieved them. This is relatively easy where financial objectives are concerned: examples were given in Chapter 1. Not all objectives are financial, though, particularly where, as in most rural businesses, the owners of a business also manage it (and thus will have a freedom to exercise personal preference in a way that a salaried manager rarely has).

Non-financial objectives come in a host of varieties. A farmer may have a strong desire to own land, to have the best Hereford bull in the county, to be highly regarded in the local community, to conserve the flora and fauna of his holding. He may want a business that occupies him 24 hours a day, or one that will tick over happily while leaving him with half the week for fishing, going to market, or visiting his mistress. He may place importance on bringing his son into the business early so that he can retire, but on the other hand he could be determined to keep hold of the reins for as long as he is able. He may aspire to the chairmanship of the county NFU, or to a seat on the local Bench. The list is endless, and can easily be adapted to owners of other forms of business.

Non-financial objectives will always be subservient in some degree to the financial objectives, since if the business collapses financially, *none* of the objectives can be achieved. On the other hand, if such non-financial objectives are not considered, there is a risk of planning a business which proves to give the owner little or no satisfaction – and what is the point in possessing such a business? It can be very difficult to state such objectives in quantifiable terms, but nevertheless every effort should be made to do so. An example might be the intention to maintain good relations with the neighbouring managers and villagers (with implications for methods of slurry handling and the siting of intensive livestock units). Although it would be difficult to measure positive aspects of 'good relations', it would be possible to set an objective of 'zero complaints', albeit a rather negative one.

Setting objectives of any kind is rarely easy. Among the difficulties are those of:

- Ensuring that *all* objectives are explicitly considered
- Ensuring that objectives are realistic
- Resolving conflict between objectives

The *explicit consideration* of all objectives requires a methodical approach if the manager is to avoid the danger of omitting an objective which, though crucial to the business, escapes his mind at the planning stage. One way of avoiding this is to run through a checklist such as the one below:

(1) Personal:
  - Leisure and pleasure
  - Family considerations
  - Prestige
  - Others

(2) Financial:
  - Profitability
  - Stability
  - Adaptability

(3) Physical:
  - Innovation (techniques and products)
  - Productivity (of land, labour and capital)

(4) Responsibility to:
- Employees
- Local community
- Society in general

This list is by no means exhaustive, but provides a foundation on which the manager may build his own, more detailed schedule.

Even a checklist may fail to ensure the explicit consideration of all objectives, one of the principal reasons being lack of honesty. It may seem peculiar to refer to honesty in describing one person's attempt to decide what he wants from the business, but in fact lack of such honesty is common. It is often linked with the fact that some objectives are disturbing or distressing to consider.

An example of this might arise where a young manager has taken over a business from an ageing parent. A major consideration in this situation must be to minimize the threat to the business imposed by the likely impact of capital taxation after the parent's death. The prospect of this event may be so distressing that both parties blinker themselves to its inevitability. They ignore it and hope that everything will work out fine when the time comes. But such issues *must* be faced squarely, despite the emotional discomforts involved, if a policy for preserving the business is to be formulated.

The problem of *realism* is that of setting targets which are capable of achievement, without being too easily achieved. The manager who sets himself impossible targets risks constant self-dissatisfaction and frustration. This leads to loss of interest in the performance of the business, and probably a deterioration in the standard of management as a result. Objectives which are too easily attained, on the other hand, are of little value in the management of a business. Once the objective is achieved, what then? The business is floundering again through want of direction.

It is almost inevitable that *conflict* will arise between the various objectives of a business. Two common forms of conflict are those between long- and short-term objectives, and between financial and non-financial objectives. An example of the former might be provided by the farmer mentioned earlier, who was anxious to set his sons up in farming. If his own business has a succession of bad seasons, there is a chance that he will be faced with the choice between this long-term objective and the shorter-term one of keeping his own business intact.

A familiar example of the conflict between financial and non-financial objectives is that of the young manager who has a strong wish to spend at least one day a week with his family. He quickly finds that he has to work all day, every day, to keep the business going. The alternative is to hire additional labour, which will push the overdraft well beyond the limits of safety. He has to hold one of the objectives in abeyance – the financial or non-financial. If he wants to stay in business it will be the latter: if he wants to keep his family together, he may have to rethink from scratch.

Finally, it is important to remember that objectives and their priorities change with time. For this reason, a critical review of objectives should be undertaken as an integral part of the annual planning procedure.

## 15.4 Assessing 'internal' characteristics and external environment

In any plan, it is vital to take explicit account of both the internal resources of the business and of the external environment in which it operates. A simple but powerful device for doing this is SWOT analysis. The 'SW' stands for 'Strengths and Weaknesses', the aim being here to identify all the positive and negative aspects of the business as it stands. The primary focus will be the physical and financial resources available (quantity and quality of land, labour, cash reserves, stocks, etc.) but it should also take in other characteristics such as location, management expertise and experience, goodwill, and so on. The weaknesses often represent the reverse side of the strengths: for instance a good location for a tourist accommodation enterprise, with a degree of seclusion, may be a weakness as far as a farm shop is concerned, reliant on passing trade.

The 'OT' stands for 'Opportunities and Threats': considering these should force explicit consideration of the environment, in terms of political, economic, social and technological factors. Anticipating changes in these areas may, for instance, require forming views about possible reform of the Common Agricultural Policy, about forthcoming environmental legislation, about demand for cheese two years hence, and about social trends in the area. This would be a tall order if there were not so much assistance available in newspapers, conferences, journals, television programmes, discussion groups, teletext and the Internet, Business Link, MAFF documents ... the list seems endless. Even then, however, the manager must be prepared to devote a considerable amount of time and expertise to absorbing and evaluating the large amount of information available. A vital skill in the 'Information Age' is the ability to sift for the 'need-to-know' material without being distracted or daunted by the huge volumes of 'nice-to-know'.

## 15.5 Generating alternative plans

One of the most common faults in management is 'tunnel vision'; the inability or unwillingness to consider any action other than the one first thought of. It is closely connected with another common problem, the 'emotional decision'. This is a decision made entirely on the grounds of a 'gut feeling'. The feeling might be that you must have XYZ tractors, even if they are the most expensive available, or that you must buy that 20 ha parcel of land next door, as it may never come up for sale again. It is not that such gut feelings should be ignored; merely that they should be viewed in the light of the other possibilities. A vital stage of the planning process is thus the consideration of a range of courses of action.

It is easy to talk of considering a wide range of alternatives, but less easy to do it. This is partly a matter of lack of knowledge; partly a matter of lack of imagination. It is impossible for one person (especially a busy manager) to be fully informed about all the possible opportunities for development of his business. It can be very difficult (especially after a day's hard physical work) for him to stretch his mind to make maximum use of the knowledge that is available.

The only remedy for the first problem is to make every effort to keep well informed, as discussed in Section 15.4, and be prepared to pick the brains of others. Neither is there an easy way of solving the second problem. The main requirement is to keep as open a mind as possible. One way of doing this is to adopt the suggestion of Humble (1980) and ask the classic question, 'What business am I really in?' An attempt to find an honest answer to this deceptively simple query can often provide the idea that transforms an adequate business into a highly successful one. For example, the farmer who has always thought of himself as first and foremost a stock farmer (perhaps because his father and grandfather were stock farmers) may thus come to realize that his crop production is of a higher quality than his livestock production. This may in turn break mental barriers to the transfer of resources (such as land and capital) away from the stock towards the crops.

Similarly, the person who thinks of his business as a farm with a caravan site attached may come to realize that it is in fact a leisure business with a farm attached, and concentrate on developing the former as a more effective way of achieving his objectives than further development of the farming enterprises.

Another way of generating a wide range of ideas is to gather a group of other people, e.g. family, friends or neighbours, round a pot of tea or a crate of beer, tell them the problem in hand, and ask for as many suggestions as possible within a given time. No suggestion should be commented on, nor should apparently wild or fanciful ideas be discouraged – everything should be noted down. When the flow of suggestions has dried up, each of the ideas noted should be discussed at greater length. Often ideas which at first seemed crazy will, on consideration, begin to look interesting. The resulting list of possible plans will almost certainly be more comprehensive than one person could have produced on his own.

## 15.6 Selecting the optimal plan

Finding the optimal plan requires a process of sifting through all the credible alternatives arising from the ideas-generating stage, and selecting that which is most likely to meet the objectives. This is easier said than done, of course. For a start, we know from the above discussion that objectives will often conflict, and although some objectives will be obvious imperatives (especially financial ones) it can be difficult to compare alternatives. Another problem is caused by the sheer number of possible combinations and permutations possible in all but the very simplest of one-product businesses, and the enormity of the task of quantifying their likely outcomes.

Various techniques have been developed over the years in attempts to systematize the process of arriving at 'the' optimal plan for the business. Previous editions of this book have outlined methods such as gross margin planning, programme planning and linear programming, for instance. These methods have rarely proved satisfactory in planning for the small business level, however. The first two take no account of anything but the simplest of objectives and the most limited financial information. Linear programming struggles to take account of all the intricacies of even the smallest business, and in attempting to do so results in a complex and unwieldy procedure which requires too much to be taken on trust by the decision-maker. Linear programming is best left as a useful research and, in some cases, advisory tool.

The most common – though less systematic – process in selection is the *short-list and budget method*. This relies on the manager sifting through the wide range of alternative plans and subjectively choosing a selection of options that looks credible in terms of the objectives and the SWOT analysis. The selection should always include a plan based on the continuation of the present system, for comparison. The chosen few are then subjected to closer scrutiny. Given the importance of financial objectives, this scrutiny must include the preparation and comparison of budgets for each of the short-listed plans, based on normalized figures. The non-financial aspects of each plan should of course be examined before a final decision is reached.

This method has its limitations, of course. The selection of the initial short list relies on subjective, and often arbitrary judgement. Moreover, the work involved in preparing budgets for even a small range of plans can be considerable, although this can be reduced by the use of partial budgets round a 'central' whole-business budget (see Chapter 9). More importantly, the widespread adoption of the personal computer has considerably eased the burden of budget generation. Once a set of figures has been entered into a budgeting program (whether this is part of a specialized accounts package or just a spreadsheet), variations of these figures can be entered and the likely results seen immediately. This is probably the single most important development in small business planning in the last decade (see Appendix A).

## 15.7 Implementing and monitoring the selected plan

Implementation of the plan should be as systematic as its development. The selection process should have involved preparing detailed budgets: what is needed now is an action plan, setting out all the tasks likely to be involved. From this can be developed a set of targets, with deadline dates specified for each. This establishes the basis for a continual process of review and adaptation, to guide the implementation through the inevitable departures of reality from expectations as time goes on[18].

It goes without saying – since that is the main message of this book – that the budgets will form the basis of a budgetary control process, linked to the action plan, allowing comparison of budgets with accounts, and providing the impetus for periodic replanning.

## 15.8 Involving other members of the organization

A business is run by and for people. The manager who is unmarried, a sole proprietor, and employs no other labour but his own is lucky in the sense that he or she has only him- or herself to organize and motivate. As soon as other people – family, employees, partners – become involved, the problems arise of ensuring that the objectives of the business reflect those of *all* concerned with the ownership of the business, and that the efforts of all the members of the organization contribute toward these business objectives.

One of the most effective ways of coping with these problems is to make sure that those 'other people' are involved in the planning process. Objectives should be agreed (not just discussed) with those who are going to have to achieve them, and the resulting plans should be the result of not just the manager's efforts, but also those of the people who are responsible for their implementation. This process can take place at various levels; the stockperson need only be concerned with a narrow range of plans, relating to physical performance, while the assistant manager should probably be involved in the formulation of objectives and plans for the business as a whole.

This aspect of management deserves more space than can be afforded in a book concerned primarily with financial management techniques. Its best-known manifestation is 'management by objectives', and the writings of Humble (1980), and Giles and Stansfield (1990), are strongly recommended.

## Notes

18. Various project planning computer programs are available that simplify this process, producing charts and tables of deadlines, resources needs, cash flow, and so on. For the larger planning exercise they can be a valuable investment, particularly as they allow the consequences of a variety of alternative plans to be investigated quickly and easily.

## References

Anderson, A.H. & Barker, D. (1996) *Effective Enterprise and Change Management*. Blackwell Business, Oxford.

Argenti, J. (1980) *Practical Corporate Planning*. Allen & Unwin, London.

Bowman, C. & Asch, C. (1987) *Strategic Planning*. Macmillan, Basingstoke.

Giles, A.K. & Stansfield, J.M. (1990) *The Farmer as Manager*, 2nd edn. CAB International, Oxford.

Humble, J.W. (1980) *Improving Business Results*. Pan Books, London.

# Chapter 16

# Planning for Livestock Enterprises

## 16.1 Introduction

Crop enterprises are fairly simple to budget for, and the basic principles have been covered in previous chapters. The biggest single difficulty is probably formulating appropriate rotations, and determining the value of including certain break crops in the cropping system; see Barnard and Nix (1979).

Livestock enterprises are subject to various additional complicating factors, including the dependence of grazing livestock on crop enterprises, the continuous nature of some forms of livestock production, and the long production cycles of some types of animal. Factors such as these have implications for the choice of feeding policy, the allocation of forage costs, the estimation of potential production levels, and the choice of replacement policy for breeding livestock, etc. Some of these topics (feeding, for instance) cannot be discussed in detail here; although they are vital to the planning process, they are more the province of a textbook on animal production. Where this is the case, the appropriate methods are described in outline, and references provided for further reading.

## 16.2 Feeding livestock

### 16.2.1 Selection of feeding policy

The aim in selecting a feeding policy for a given group of livestock is to find the lowest-cost combination of feedstuffs which will provide the nutrients required by the animals. In non-ruminant livestock it is possible to avoid this process by buying complete rations, or supplements for use with home-grown barley, from a feed manufacturer. In ruminant livestock enterprises, however, concentrate feeds, whether home-grown or purchased, must be carefully balanced with the bulk fodder which is available. Even in a non-ruminant livestock enterprise, the farmer may prefer to design his own rations.

Selection of rations has some similarity to the enterprise-selection problem discussed in the previous chapter. The difference is that here the objective is minimum cost rather than minimum profit, and the constraints tend to be minima, e.g. 'the ration must contain at least $x$% digestible crude protein', rather than the maxima of the enterprise-mix problem, e.g. 'no more than $y$ ha of sugar beet may be grown in one year'. One exception is the constraint of appetite; it is obviously no use selecting a ration which, though providing the necessary nutrients at the least cost, is so bulky that the animal cannot consume it all. A second is the quantity available of any ingredient which is in short supply, particularly forage. Other maximum constraints concern palatability, toxicity and similar factors.

The 'minimum' constraints are imposed by the nutrient needs of the animals concerned. Among these are protein, vitamins and minerals of various types, but the most critical is energy, needed not only to maintain the animal's bodily condition, but also to enable production of milk, wool, liveweight gain, and so on. This energy is measured in terms of *metabolizable energy* (ME) for ruminants, and *digestible energy* (DE) for non-ruminants, i.e. pigs and poultry. Tables of the nutrient requirements of different classes of livestock, and the nutrient content of various types of feed ingredient, are published in specialist publications such as MAFF (1984) and Whittemore (1993) on pig production.

The choices of selection method in planning a feeding policy are similar to those in planning an overall farm policy. At the most sophisticated level, a version of linear programming (LP) can be used, requiring the use of a computer. No modern feed compounder would dream of formulating feeds without the aid of this technique. The farmer who wishes to mix his own rations, and who does not have access to a suitable computer, has two alternatives. He can either use the technical advisory services of ADAS (Agricultural Development and Advisory Service) or the feed firm that sells him the feed supplement, or he can use a trial-and-error process. The latter is akin to the budgeting procedure described earlier in the book. The first task is to work out various rations which will give the required levels of nutrients without exceeding the appetite limit. The cost of each ration is then calculated, and the cheapest alternative chosen.

This procedure does not have the precision and speed of the LP method, but as long as a number of alternatives are considered, the chances of finding a ration that comes near to the least cost are fairly high. Unfortunately, since lengthy calculations are involved, the busy farmer is likely to be tempted to take a short cut and limit his investigation to one or two possibilities only. Software is available that allows use of on-farm microcomputers to simplify and speed up the process. This enables the farmer to examine a wide range of alternative rations in a short time, and provides a reasonable compromise between the extremes of LP and hand calculation.

A detailed, yet easily-followed guide to the selection of a feeding policy is provided by Nelson (1979).

### 16.2.2 The cost of feed ingredients

A key factor in the choice of feedstuffs is that of cost. Bought-in ingredients should be valued at their purchase price (historic or anticipated); home-grown, saleable crops at their net realizable value at the time of use (sale price less an allowance for marketing and transport costs).

Certain crops may be included in the farm system solely to provide feed for livestock. This applies to all forage crops, and often cash crops such as barley. The produce from these crops should be valued at a price which includes both the variable costs of production and the opportunity cost incurred by giving up the chance of growing an alternative crop.

Growing grass for silage on a mixed farm may, for instance, limit the area available for winter wheat. In this case, the cost per hectare of growing silage is the variable cost of production plus the gross margin lost from a hectare of winter wheat. On a good arable farm, it might be worth cutting down on silage production, buying in barley and hay, and putting the saved land down to cash crops. For a large proportion of farms, however, the area of cash crops is limited by husbandry considerations and by contracts, quotas, etc. Thus grass is forced into the rotation; as may low-value crops such as barley. In such circumstances, the silage should be valued at the variable cost of production (making some allowance for the variable elements of labour and machinery costs – see Chapter 17).

Some farms will lie between the two above extremes – the grass area could be cut down by $x$ ha, for instance, in order to make way for roots or a cereal crop. In such cases, the cost of the $x$ ha of grass should include an opportunity cost element as described above. Only if this is done will the farmer realize the full marginal cost of growing his own feed.

## 16.3 Allocation of forage costs

The budgeting of forage costs for the first year of Adam Haynes' tenancy of Home Farm was straightforward. There was only one grazing livestock enterprise, so the forage costs could all be allocated to that enterprise, the dairy.

The majority of mixed farms have more than one grazing enterprise, however. The different types of livestock on these farms are not normally confined to separate areas of grassland, but share the total area. Sheep may follow the cows into a pasture to 'clean up', and beef animals may well eat hay from the same field as the dairy heifers. Since it is impossible to allocate the cost of individual fields to particular classes of livestock in such circumstances, an attempt must be made to apportion the total farm forage costs fairly between the grazing enterprises.

The most commonly used means of allocating forage costs is the *grazing livestock unit* (GLU) method. The forage needs of each type of livestock, averaged over the whole of the year, are expressed by reference to the forage needs of a Friesian dairy cow. Thus a Friesian cow is represented by 1.0 GLU, while an animal with one-fifth of a dairy cow's needs is represented by 0.2 GLU (see Table 16.1).

Forage costs may then be allocated to the various grazing enterprises by simple proportion. Take the example of a farm with average annual numbers of 100 dairy cows, 20 dairy young stock over 300 kg, 20 dairy young stock under 300 kg, and 305 lowland ewes and rams. Total forage costs are estimated at £9000, including both winter and summer feeding, and home-grown and purchased fodder. These

**Table 16.1** Grazing livestock units.

| | | |
|---|---|---|
| *Cattle* | | |
| Dairy cows | 1.0 | |
| Beef cows (excl. calf) | 0.8 | } 1.0 |
| Calf under 6 months | 0.2 | |
| Other cattle, under 300 kg | 0.4 | |
| Other cattle, over 300 kg | 0.6 | |
| *Sheep* | | |
| Lowland ewes plus lambs | 0.2 | |
| Hill ewes plus lambs | 0.12 | |
| Ewe replacements 18-month: | | |
| Lowland | 0.14 | |
| Hill | 0.1 | |
| Store lambs, short-keep | 0.02 | |
| Store lambs, long-keep | 0.04 | |
| Other sheep | 0.1 | |

forage costs are apportioned as shown in Figure 16.1. Stocking rates of livestock can be expressed in terms of GLU per hectare (or vice versa), or the GLU used to calculate the stocking rate in terms of head per hectare (or vice versa). Thus if the total forage area represented by the £9000 forage variable costs were 72 ha, the stocking rate would be 2.5 GLU/ha (0.4 ha/GLU). For the sheep flock, this can be expressed as 12.5 adult head/ha (or 0.08 ha/head).

| *Livestock* | *Number of head* | *GLU per head* | *Total GLU* | *% of grand total* | *Share of forage costs (£)* |
|---|---|---|---|---|---|
| Dairy cows | 100 | 1.0 | 100 | 56* | 5000 |
| Heifers, over 300 kg | 20 | 0.6 | 12 | 7 | 600 |
| Heifers, under 300 kg | 20 | 0.4 | 8 | 4 | 400 |
| Ewes plus lambs | 295 | 0.2 | 59 } | 33 | 3000 |
| Rams | 10 | 0.1 | 1 } | | |
| ***Total*** | | | 180 | | 9000 |

* That is (100/180) × 100%.

**Fig. 16.1** Allocation of forage costs using GLU – example.

This method has the enormous practical merit of being simple to use and easy to understand. It will be obvious, however, that the resulting allocation is only approximate. It ignores differences in types of forage, and the fact that better pastures may be reserved for the dairy herd, while the sheep and the heifers are exiled for most of the year to the poorer pastures. The units themselves are based on 'average' relativities, which make no allowance for the breed, condition or productivity of the animals on the farm in question. Moreover, there is disagreement between various authorities as to the GLU attributable to the various classes of livestock.

In most situations, the inaccuracies arising from the use of GLU are relatively small, and do not justify the use of a more complicated method. To a certain extent they can be reduced by adaptation of the units and/or the way in which they are used. As an example of the former, cows in a particularly high-yielding dairy herd could be given a value of more than 1.0. As an example of the latter, an area of forage might be allocated specifically for the use of the dairy herd, and GLU used only to allocate the remainder between the heifers, beef and sheep. Speedy (1980) recommends multiplying the area of poor pasture by 0.5, and of rough grazing by 0.25, to allow for their lower stocking rate potential.

When planning a new unit or contemplating a major change in farm policy, accurate budgeting of forage costs may be absolutely critical. In this case, GLU may be used in the early stages of planning the combination of enterprises. This will then provide the basis for more precise estimates, working from the nutritional needs of the livestock. Once decisions have been made concerning the composition of animal rations, the area, type and allocation of forage crops may be estimated with a reasonable degree of accuracy.

A third method, using *livestock unit grazing days* (LUGD) relates to historical data, and is thus of little direct use in planning the first year's allocation of grazing. It can, however, provide valuable information on which to base forage policy for future years. Records are kept of the number of days' grazing by each group of livestock, and the number and type (and thus forage consumption) of animals in each group. These records can be kept by the field or for the whole farm. The LUGD attributable to each class of livestock is calculated by multiplying the number of animals in each group by the appropriate GLU, and multiplying the product by the number of days spent by that group in grazing. The close recording of the number of grazing days allows the different stages in the life-cycles of livestock to be separated, and more precise GLUs can be used than the annual standards shown in Figure 1.1 (see Table 16.2).

The consumption by each group of home-grown conserved feed should also be recorded to allow accurate allocation. This can be converted to LUGD by multiplying by a factor of 20 per tonne of silage, and 70 per tonne of hay. The total forage costs of the farm may then be apportioned by reference to the LUGD used by each class of livestock, in a similar manner to that used for GLU in Figure 16.1. Separate decisions must be made concerning the allocation of bought-in forage.

This method has the advantages over the simpler GLU method of allowing the division of costs between grazing and conservation, of being adaptable to situations where certain areas of grassland are reserved largely for particular classes of livestock, and of using more precise livestock units which can better allow for variations in size and type of livestock. On the other hand, it is largely restricted to grassland, is historical, can require a high level of recording skill (especially if used on a field basis) and still relies on units calculated for 'average' conditions. It is used frequently where education or research requirements dictate a higher level of accuracy than that provided by the simple GLU method, but is not commonly found on commercial farms.

**Table 16.2** GLU for use in LUGD calculations.

| *Type of stock* | *Specifications* | *Unit values* |
|---|---|---|
| Dairy cows | Holstein, Friesian | 1.0 |
| | Ayrshire, Shorthorn | 0.9 |
| | Guernsey | 0.8 |
| | Jersey | 0.75 |
| Grazing cattle | Under 204 kg (0–11 months) | 0.25 |
| | 204–340 kg (11–20 months) | 0.5 |
| | 340–477 kg (21–30 months) | 0.75 |
| | Over 477 kg (over 30 months | 1.0 |
| Suckler cows | Dry cow | 1.0 |
| | Spring-calving cow with calf | 1.2 |
| | Autumn-calving cow with calf | 1.4 |
| Weaned calves | Weaned calf, spring born | 0.2 |
| | Weaned calf, autumn born | 0.4 |
| Breeding ewes (dry) | 18–36 kg | 0.08 |
| | 36–55 kg | 0.1 |
| | 55–68 kg | 0.13 |
| | 68–77 kg | 0.15 |
| | 77–91 kg | 0.17 |
| | Over 91 kg | 0.2 |
| Weaned lambs | Light lambs (27–36 kg) | 0.08 |
| | Heavy lambs (36–54 kg) | 0.1 |
| Breeding ewes (lactating) | Add 0.01 per ewe to the dry ewe standard for each 25% lambing % | |

Source: Holmes *et al.* (1980).

## 16.4 Estimating potential production

In preparing budgets for livestock enterprises, it is necessary to make an estimate of the likely physical performance of the animals concerned. For meat and wool production, this means estimating the final weight of the product sold. For breeding/suckling livestock it entails estimating numbers of offspring weaned per female parent, and for replacement livestock estimating the number of animals attaining the appropriate age and condition for transfer into the breeding herd/flock (or sale into somebody else's herd or flock). In making such estimates the farmer will be guided by a combination of experience, advice from specialists and neighbours, and published standards.

Special difficulties arise when budgeting for continuous-production livestock, such as those producing milk or eggs. The production pattern of these items is not uniform over time; it is generally characterized by a low level of production at the beginning of the cycle (i.e. after calving in a dairy cow) rising rapidly to a peak production, which is then followed by a slow decline throughout the rest of the cycle. In the dairy cow, this is repeated after each calving, and is known as the *lactation curve*, while in egg production the pattern is repeated after an annual moulting period (although intensively housed poultry are usually replaced before moulting). Figures 16.2 and 16.3 illustrate these production patterns.

If cash flow budgets are to be realistic, and profit budgets are to make the correct allocation of revenue between financial years, these production patterns must be taken into consideration. Since most egg production takes place under closely controlled conditions, predictions of performance can reliably be made by referring to standard curves of the type shown in Figure 16.3, which are usually provided by the breeder of the foundation stock.

The height and shape of a dairy cow's lactation curve will vary according to season of calving, condition of the cow, quality of forage, and other factors. For

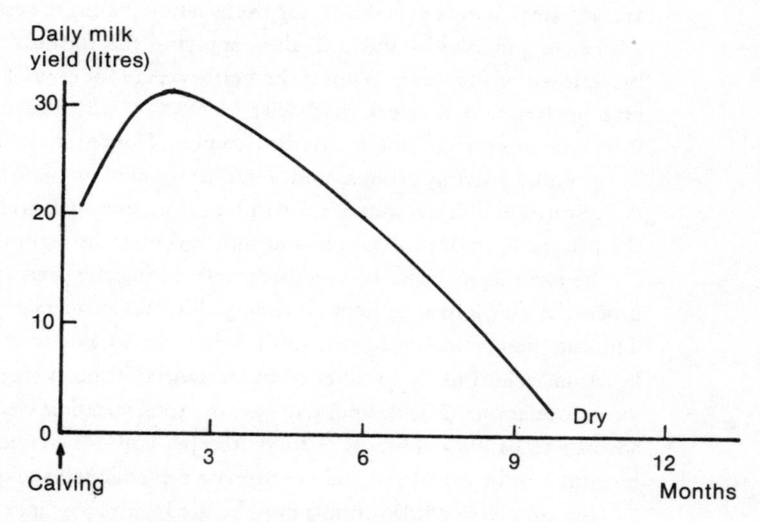


**Fig. 16.2** Typical lactation curve for a dairy cow.

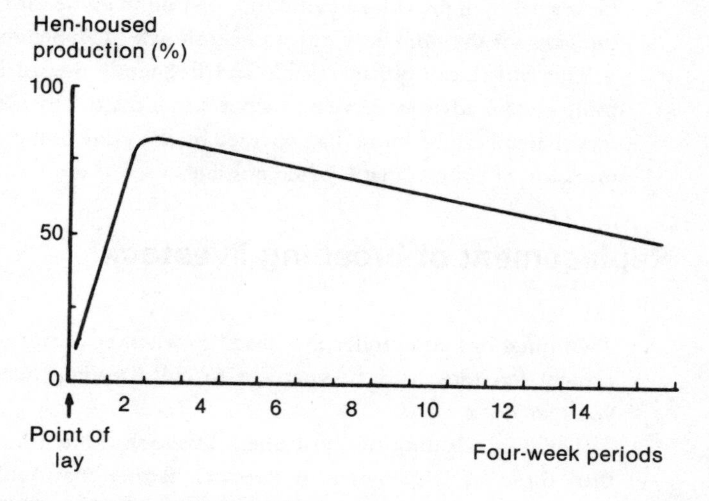


**Fig. 16.3** Typical pattern of egg production.

really precise budgeting, these factors should be taken into account; a valuable guide is Whittemore (1980) on the dairy cow, especially Chapter 5.

A more simplistic approach is to use a standard lactation curve formula. One such formula is described by Russell (1985, p. 153). He identifies three phases:

- A steady rise in milk yield up to a peak at 4 to 6 weeks after calving
- A peak daily yield of 1/200th of expected milk yield for the lactation (1/220 in the case of heifers)
- A steady decline in yield thereafter at the rate of 10% per month for the rest of the 10-month lactation.

This formula may be used in either of two ways. The first way is to estimate the likely average yield of milk per cow during the lactation (basing the estimate on experience, advice, or published standards), then applying the formula to give the monthly breakdown of milk sales. Thus if the herd average for cows is expected to be 6000 litres per lactation, the peak yield will be 6000/200 = 30 litres per day, declining by 3 litres each month until the end of the lactation. This result can be applied to the cows by monthly calving groups, and a similar operation performed for the heifers. Accumulation of the results of the various calculations will give an approximation of the pattern of milk production, and thus revenue, throughout the year.

The formula may also be used in reverse during the year, as a part of the control process. A check may be kept on peak yields, and an average peak yield calculated. This can then be inserted in the formula in order to work out how much the actual lactation yield is likely to differ from the target. If the average peak yield per cow works out at only 28 litres per day, say, the total lactation yield can be estimated in advance to be 28 × (200/1) = 5600 litres per cow. The effect of the difference on revenue can be calculated and appropriate remedial steps taken (see Part 4).

This somewhat crude formula must be used with care, since the peak yield will be affected by month of calving, short-lived feeding or disease problems, and so on. Nevertheless, it provides a useful rule-of-thumb guide for budgeting, and is used as the basis for the 'milkflow' spreadsheet shown in Appendix A.

The farmer can obtain reliable and frequently revised lactation predictions by using certain advisory services, such as those offered by Genus Management. The fees charged can be more than covered by the value of the time saved and the extra precision of control that is made possible.

## 16.5 Replacement of breeding livestock

Two questions arise under this heading: whether or not to rear one's own replacement livestock, and at what age to cull breeding livestock and replace with younger stock.

The home-rearing of replacement livestock is often justified on grounds other than those of simple cost. A livestock farmer may wish to guard against the importing of disease on bought-in animals; he may also be assuring himself of replacements of the required quality at the time he needs them, without having to pay the earth for them. He may wish to raise the standard of his herd or flock by genetic improvement. Where cash flow is particularly tight, home-rearing can ease the process of starting or expanding a breeding herd or flock, at the expense of time.

These are all very sound arguments for home-rearing. But it is important, as always, to test the financial effects of such a decision. The farmer must know how much it will cost him to be swayed by such arguments. It is particularly important for him to remember the principle of opportunity cost; rearing replacements will make demands on scarce resources, such as land and capital, and force the farmer to sacrifice other opportunities for making a return from those resources. Thus a partial budget should be drawn up, as in the Home Farm example in Figure 16.4.

*Decision* Expand grass area of Home Farm by 12 ha at expense of barley in order to rear 20 two-year-old down-calving heifers per year*.

| *Costs of change* | | *Benefits of change* | |
|---|---|---|---|
| *Revenue lost* | £ | *Extra revenue* | £ |
| Barley grain: 4 t/ha @ £95 | 4 560 | 16 heifers transferred into breeding herd at £850 per head | 13 600 |
| Barley straw: 2 t/ha @ £15 | 360 | 4 heifers (in-calf) sold at £750 per head | 3 000 |
| *Sub-total* | 4 920 | *Sub-total* | 16 600 |
| *Extra costs* | | *Costs saved* | |
| Costs of rearing 20 heifers (including (forage and AI) at £460 per head | 9 200 | Variable costs of 12 ha barley @ £130/ha | 1 553 |
| | | Machinery and labour | 300 |
| *Sub-total* | 9 200 | *Sub-total* | 1 853 |
| **Total costs** | 14 120 | **Total benefits** | 18 453 |
| **Extra profit** before interest | 4 333 | | |

* Surplus numbers allow some degree of selection before transfer into the breeding herd.

**Fig. 16.4** Use of a partial budget in deciding whether or not to home-rear replacement breeding livestock.

Even if the partial budget gives a result which is financially unfavourable to home-rearing, the farmer may feel that the non-financial advantages justify the inclusion of home-rearing in the plan. At least he now knows how much that decision is costing him. Partial budgets may also be used to help determine the age at which old breeding stock should be culled and replaced. For a good many animals, replacement is dictated by death or incapacity rather than the farmer, so the most that can be done is to establish a general guideline.

The financial aspects most affected in this decision are the annual cost of buying or rearing livestock, the average annual level of production, and the cost of capital. The cost of replacement animals will increase with decreasing culling age; culling every two years will require twice as many replacements per year as culling every four years. The level of production will tend to decline after the first two or three production cycles. The older an animal is, the more likely it is to be culled or to die in the middle of its production cycle; such events can have a marked effect on the average yields of older age groups.

Figure 16.5 shows the financial effects of a hypothetical replacement rate decision. It assumes that the extra replacement heifers are purchased; were they home-reared their cost would be the variable cost of their production plus the opportunity cost of the extra land used. The example takes account of only one year's costs and benefits; for a more precise assessment, the net present values (see Chapter 18) arising from each of the alternative replacement policies over (say) a ten-year period, should be compared.

| *Decision* Replace Home Farm dairy cows at four years average rather than five. | | | |
|---|---|---|---|
| *Costs of change* | | *Benefits of change* | |
| *Revenue lost* | £ | *Extra revenue* | £ |
| 5th lactation yield from 4 cows at 4600 l/cow (effective*) at 20p per litre | 3680 | 1st lactation yield from 4 heifers at 4800 l/heifer at 20p per litre | 3840 |
| *Sub-total* | 3680 | *Sub-total* | 3840 |
| *Extra costs* | | *Costs saved* | |
| 4 in-calf heifers per year at £850 per heifer | 3400 | | |
| Cost of capital to pay for extra heifers at (say) 15% of £3400 for one year | 510 | | |
| *Sub-total* | 3910 | *Sub-total* | 0 |
| **Total costs** | 7590 | **Total benefits** | 3840 |
| | | **Reduction in profit** (before interest) | –3750 |

* That is, allowing for increased likelihood of mid-lactation mortality and obligatory culling in older cows.

**Fig. 16.5** Use of a partial budget in deciding at what age to replace dairy cows.

## 16.6 Livestock with long production cycles

Certain types of trading livestock, chiefly beef cattle and dairy heifers, take more than one year to reach the point of sale or transfer into another enterprise. In order to provide a regular annual output, a new batch of livestock is usually started each year. This results in an overlap of production cycles, with two, three or even four different age groups on the farm simultaneously.

A farm with an annual requirement of 20 replacement dairy heifers, calving at 2.5 years, will for part of each year be supporting 20 + young calves, 20 + yearlings, and 20 + in-calf heifers. Further complications can develop in beef production if two

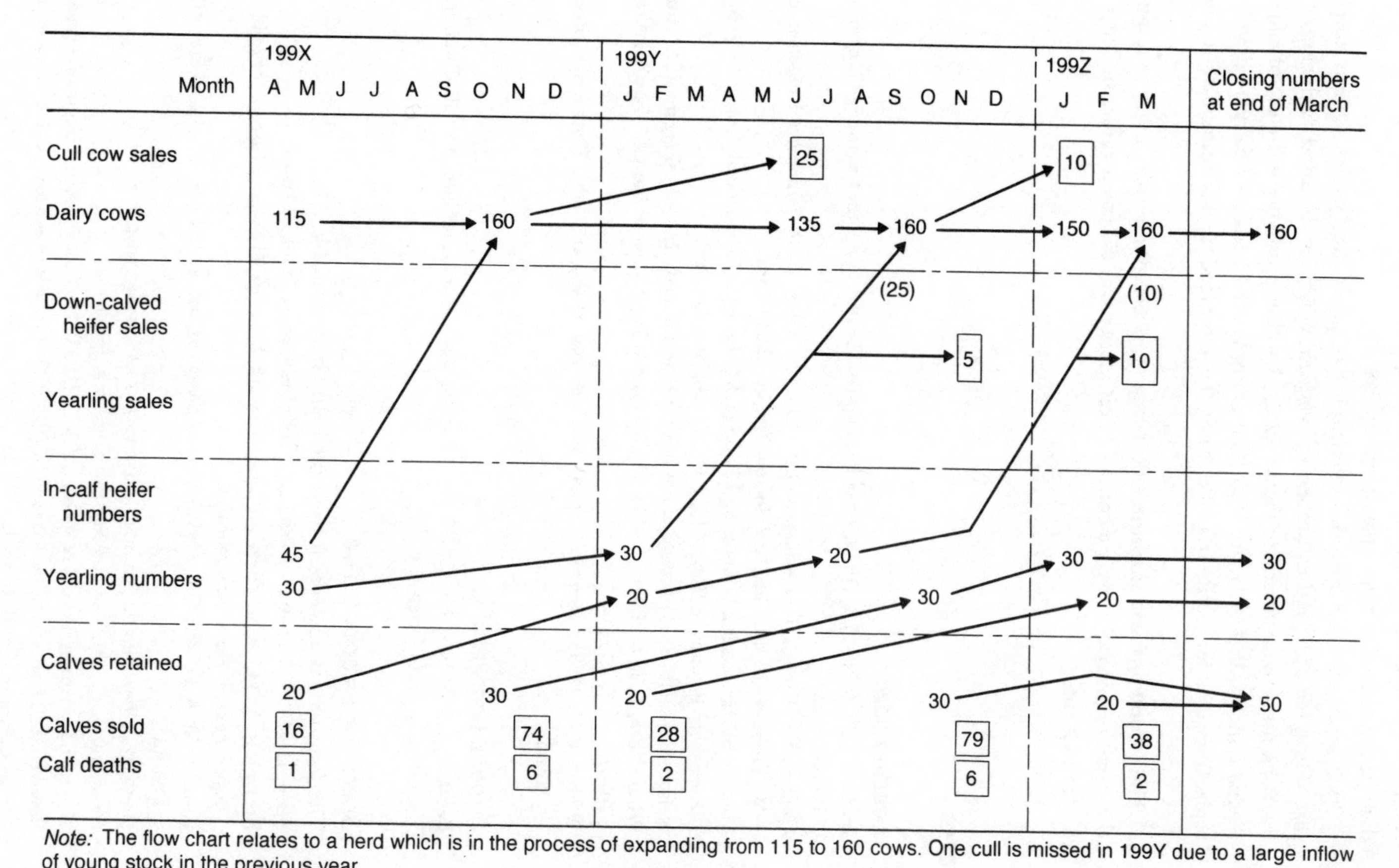


*Note:* The flow chart relates to a herd which is in the process of expanding from 115 to 160 cows. One cull is missed in 199Y due to a large inflow of young stock in the previous year.

**Fig. 16.6** Livestock movement flow chart.

batches, autumn and spring, are started each year, although in this case production conditions will differ sufficiently between the autumn and spring batches to allow them to be treated as two separate enterprises.

Two problems arise from this overlapping of production cycles. One is simply that of keeping track of the numbers of livestock, which can be eased by the use of a livestock movement chart (see Figure 12.2). Another possibility is the use of a flow diagram, such as that in Figure 16.6. The example used is that of a dairy herd with a replacement heifer herd. It could easily be extended to accommodate a beef enterprise.

The second problem is that of calculating gross margins in such a way as to give the detail required for precise control. This matter was discussed earlier in Section 5.1 and Chapter 11.

## References


### *Feeding policy*

Barnard, C.S. & Nix, J.S. (1979) *Farm Planning and Control*, 2nd edn. Cambridge University Press, London.

Holmes, W., Craven, J. & Kilkenny, J.B. (1980) In: *Grass: Its Production and Utilisation* (ed. W. Holmes). British Grassland Society/Blackwells, Oxford.

Ministry of Agriculture, Fisheries and Food (1984) *Energy Allowances and Feeding Systems for Ruminants*. Reference Book 433, 2nd edn. HMSO, London.

Nelson, R.H. (1979) *An Introduction to Feeding Farm Livestock*, 2nd edn. Pergamon, Oxford.

Whittemore, C.T. & Elsley, F.W.H. (1977) *Practical Pig Nutrition*, 2nd edn. Farming Press, Ipswich.

Wilson, P.N. (1995) *Improved Feeding of Cattle and Sheep*, 2nd edn. Blackwell Science, Oxford.

### *General principles*

Barnard, C.S. & Nix, J.S. (1979) *Farm Planning and Control*, 2nd edn. Part 2. Cambridge University Press, London.

### *Specific livestock studies*

The following is a selection from the many texts available.

Allen, D. (1990) *Planned Beef Production*. BSP Professional Books, Oxford.

Brownlow, M.J.C., Carruthers, S.P. and Dorward, P.T. (1995) Financial aspects of finishing pigs on range. *Farm Management*, **9**, 3.

Croston, D. & Pollott, G.E. (1994) *Planned Sheep Production*, 2nd edn. Blackwell Scientific, Oxford.

Drew, B. (1996) New technology – livestock. *Farm Management*, **9**, 5.

Mowlem, A. (1988) *Goat Farming*. Farming Press, Ipswich.

Norman, L., Turner, R.A.E. & Wilson, K.R.S. (1985) *The Farm Business*. Longman, London.

Reinken, G., Hartfiel, W. & Körner, E. (1990) *Deer farming: a practical guide to German techniques*. Farming Press, Ipswich.

Russell, K. (1985) *The Principles of Dairy Farming*, 10th edn. Farming Press, Ipswich.

Soffe, R.J. (ed.) (1995) *The Agricultural Notebook*, 19th edn. Part 3: Animal Production. Blackwell Science, Oxford.
Speedy, A.W. (1980) *Sheep Production*. Longman, London.
Thornton, K. (1988) *Outdoor Pig Production*. Farming Press, Ipswich.
Throup, G. (1994) *Making Profits with Dairy Cows and Quotas*. Farming Press, Ipswich.
Whittemore, C.T. (1980) *Lactation of the Dairy Cow*. Longman, London.
Whittemore, C. (1993) *The Science and Practice of Pig Production*. Longman, Harlow.
Yerex, D. & Spiers, I. (1987) *Modern Deer Management*. Ampersand Publishing Associates Ltd, Carterton, NZ.

# Chapter 17

# Labour and Machinery Planning in Farming

## 17.1 Introduction

Fixed cost control is crucial to the survival of a modern farm business. Of the fixed costs, labour and machinery are the most difficult, and the most important, to plan and control. Unlike such items as rent, mortgage interest, and accountant's fees, they are substantially under the farmer's influence. Since on most farms labour and power represent a major part of total fixed costs, economy in their use is essential.

Planning labour and machinery use is concerned largely with answering two questions. First, what is the optimum combination and level of use of people and machines? Second, what is the cost of implementing the chosen labour/machinery policy? These questions are inextricably linked; one cannot finally budget costs until the policy has been decided, while the choice of policy must reflect the cost of use.

There is thus no easy step-by-step recipe to follow. The simplest approach is probably to start by planning the labour requirements of a given farming system, and using this as a guide to labour and machinery needs. This will result in a basic plan, which can then be budgeted to show the likely costs involved. Alternative policies may then be evaluated by partial budgets showing the net financial effect of deviating from the original plan.

This approach may not be suitable for all situations, and the farmer may be forced into a repetitive trial-and-error process, planning and replanning until a suitable solution is found.

## 17.2 Estimating level of use

### *17.2.1 Simple guesswork*

With experience, a farmer may feel confident enough to look at a proposed farm system and state that it will need a certain number of men, and a certain complement of machinery. In some situations this may be good enough as an initial rule-of-thumb measure. It allows too much room for error, though, to provide reliable estimates for planning and control, except where the labour and machinery force is dictated by the circumstances of the farm (on very small farms, for instance, or all-dairy farms).

### 17.2.2 Standard man-day and standard tractor-hour calculations

An alternative is to estimate the total amount of labour and machinery time required by the farm system in the space of one year, then divide by the annual contribution expected of each man or machine in order to determine the appropriate complement.

To calculate the number of people required, the area of each crop and type of livestock is multiplied by the *standard man-days* (SMD) required per hectare or per head (tables of SMD can be found in various regional Farm Business Survey publications, and in farm management handbooks such as Nix). The SMD for all the enterprises are totalled, and an addition (typically 10%) is made to allow for the contingencies and maintenance work.

The grand total is adjusted for work performed by casual labour and contractors, and is then divided by 300 (the 'average' SMD contribution of a farm worker) to give the number of people required. Appropriate adjustments can be made for workers who are not likely to spend the whole of their working time in physical labour on the farm.

A similar calculation is used in estimating tractor numbers[19], this time using standard tractor-hours (again available in Nix and various Farm Business Survey results). The 'average' number of hours worked per tractor per year is generally taken to be 1000–1500, although there is some sense in using 800 for the first tractor, as suggested by Nix, in order to give greater flexibility.

Figure 17.1 shows the use of SMDs and standard tractor-hours in determining the number of people and tractors required for a mixed farm system. The results of the calculation suggest that the farmer could just manage with six men, plus himself. This leaves him no time for managing the business, which is likely to occupy at least a quarter of his time. The deficiency is relatively small, and could probably be taken up by the judicious use of casual labour and/or contractors.

This type of estimate suffers in precision, since the use of 'average' labour and machinery use per head or hectare, and of 'average' work output per man or tractor, assumes that all farms are alike with respect to methods, soils, topography, building layout, type and size of machinery, and so on. More important, the method makes no allowance for seasonal variation in daylight, fieldwork conditions, and labour demand. Adding further to the complications is the fact that different authorities often quote different 'standard' figures. The actual farm on which Figure 17.1 is based is staffed by four people (including a working farm manager) with weekend relief: total cost was approximately £70 000 in 1996/97; machinery cost was approximately £47 000 including contractors. These drawbacks limit the technique to the role of an initial guide, which should always be followed up by more precise methods.

| Enterprise | Head or ha | SMD per head/ha* | Total SMD | Tractor hours per head/ha* | Total tractor hours |
|---|---|---|---|---|---|
| Winter wheat | 8.0 | 2.00 | 16.0 | 9.00 | 72.0 |
| Winter barley | 25.8 | 2.00 | 51.5 | 9.00 | 231.8 |
| Spring barley | 10.4 | 2.00 | 20.8 | 9.00 | 93.4 |
| Grass – grazing | 48.7 | 0.50 | 24.3 | 2.50 | 121.7 |
| Grass silage (2 cuts) | 30.7 | 4.00 | 122.8 | 24.00 | 736.8 |
| Grass silage (1 cut) | 16.3 | 2.50 | 40.8 | 15.00 | 244.8 |
| Maize (1 cut) – estimate | 11.6 | 3.00 | 34.8 | 18.00 | 208.6 |
| Dairy cows | 108.0 | 5.00 | 540.0 | 6.00 | 648.0 |
| Young stock (1–2 years) | 45.0 | 1.00 | 45.0 | 4.00 | 180.0 |
| Young stock (6–12 months) | 45.0 | 1.00 | 45.0 | 2.25 | 101.3 |
| Calves | 45.0 | 1.25 | 56.3 | 2.25 | 101.3 |
| Breeding sows | 110.0 | 3.00 | 330.0 | 1.75 | 192.5 |
| Boars | 6.0 | 2.00 | 12.0 | 1.00 | 6.0 |
| Other pigs over 1 month | 721.0 | 0.30 | 216.3 | 1.00 | 721.0 |
| Outdoor pigs (estimate) | 60.0 | 3.00 | 180.0 | 1.00 | 60.0 |
| Ewes | 226.0 | 0.50 | 113.0 | 1.25 | 282.5 |
| Rams | 6.0 | 0.50 | 3.0 | 1.25 | 7.5 |
| Sub-total | | | 1851.6 | | 4009.2 |
| plus 10% contingency | | | 185.2 | | 400.9 |
| **Total hours** | | | 2036.8 | | 4410.1 |
| | | | (6 to 7 persons) | | (approx. 4 tractors) |
| | Minimum cost per person: | | £9 135 | Cost per hour: | £12.50 |
| | Labour cost: | | £63 945 | Machy. cost: | £55 125 |

* Based on Nix (1997) *Farm Management Pocketbook*, 27th edn. pp. 93, 122, 123, 192.

**Fig. 17.1** Use of SMD and tractor hours – example.

### 17.2.3 The labour profile

One such precise method is the *labour profile*. The object is to calculate the labour requirement of each enterprise for each month of the year. Examination of the distribution of labour requirements will assist decisions concerning the number of people to employ, the extent to which casual labour and contractors are used, and the number and type of machines required. The clumsy standard man-day is abandoned in favour of the *man-hour*.

Tables published in Nix and similar handbooks give estimates for each enterprise of the man-hours required for each type of operation, and of the month(s) in which each operation is likely to occur. The man-hour requirements are then accumulated month by month, in the form of a table and/or a histogram. This is then compared to the theoretical availability of man-hours. Nix and others give typical monthly availability figures, which can be adjusted in the light of one's own experience to suit the situation of a particular farm. Allowance is made in these figures for poor weather, short days, illness and other contingencies, as well as the possibility of overtime working.

Figure 17.2 shows the application of the labour profile to another mixed farm. This data is shown in different ways in Figures 17.3 and 17.4. The technique may be used in two different ways. The farmer may use the labour profile to estimate how many regular workers he needs to cope with a given system. More usefully, perhaps, he can stipulate that no more than three workers are to be employed (leaving half of his own time for management) in which case the three-worker availability line becomes an absolute boundary, and ways must be found of eliminating or redistributing excess labour at peak times. This is the approach illustrated in Figures 17.2 and 17.3.

Figure 17.3 indicates a significant deficit of labour in March, August and September, with smaller shortfalls in October, November, February and May. Figure 17.4 shows how the various enterprises contribute to that deficit, with winter cereals having a significant effect in the late summer and autumn, and spring cereals and sheep contributing to the problems in March. Using this information, responses may be formulated which might include:

(1) Changes in the farming system (such as switching some winter barley for spring barley, or eliminating the sheep flock)
(2) Use of casual labour and/or contractors (for harvest and autumn cultivations, for instance)
(3) Improving gang organization (such as using larger grain trailers to reduce manpower for transport)
(4) Improving output per worker (by, for instance, putting some evening/weekend jobs, such as bale-carting, on a piece-work basis)
(5) Substitution of machines for labour (such as using a fully mechanized big bale system instead of manhandling small bales).

Each of these possibilities has a cost, of course, but careful use of a labour profile and partial budgets can help in developing a system which balances labour use with other financial considerations. Where machinery is concerned, the labour profile cannot serve as anything more than a rough guide.

### 17.2.4 The gang-work day chart

Apart from the difficulty of determining machinery requirements, the labour profile has the major limitation of ignoring the fact that people and machines are frequently used in teams. One example is that of silage-making, where the team may consist of four tractor/person units; one harvesting, two carting, and one filling the clamp.

For many farms, particularly small farms and those whose workforces are composed of relatively independent, specialist workers (such as stockmen), this is not a significant problem. In such cases, the labour profile or simpler methods are adequate. Where team-working is common, the *gang-work day chart* (GWD chart) should be used. This chart works on a similar principle to that of the labour profile, with one axis measuring the number of people required and the other showing the number of working days available in each month. The time requirements for each operation are measured in hectares per day, rather than per man-hour, and are

| **Hours per hectare/head** | (Average performance; Nix, 1997) | | | | | | | | | | | |
|---|---|---|---|---|---|---|---|---|---|---|---|---|
| *Enterprise* | *ha/hd* | *Oct* | *Nov* | *Dec* | *Jan* | *Feb* | *Mar* | *Apr* | *May* | *Jun* | *Jul* | *Aug* | *Sep* |
| Winter wheat | 51.3 | 2.6 | 0.6 | | | | 0.5 | 1.1 | 0.3 | 0.3 | | 4.3 | 5.2 |
| Winter barley | 32.4 | 2.6 | 0.6 | | | | 0.5 | 1.1 | 0.3 | 0.3 | 3.6 | 5.7 | |
| Spring barley | 27.2 | 0.4 | 0.8 | 0.2 | | | 2.7 | | 1.2 | | | 5.3 | 2.3 |
| Grass – leys | 46.9 | | | | | | 0.6 | 0.6 | 0.6 | 0.6 | 0.6 | 1.9 | 1.4 |
| perm. pasture | 38.2 | | | | | | 0.6 | 0.6 | 0.6 | 0.6 | 0.6 | 0.3 | 0.2 |
| Silage | 40 | | | | | | | | 6.3 | 3.2 | | | |
| Dairy cows | 115 | 2.4 | 2.5 | 2.5 | 2.5 | 2.5 | 2.5 | 2.4 | 2.2 | 2.1 | 2.1 | 2.1 | 2.2 |
| Followers (repl. units) | 15 | 2.9 | 2.9 | 2.9 | 2.9 | 2.9 | 1.2 | 1.2 | 1.2 | 1.2 | 1.2 | 1.2 | 1.2 |
| Ewes | 190 | 0.25 | 0.2 | 0.2 | 0.3 | 0.3 | 1.0 | 0.4 | 0.3 | 0.4 | 0.2 | 0.2 | 0.25 |
| **Total hours** | | | | | | | | | | | | | |
| *Enterprise* | *ha/hd* | *Oct* | *Nov* | *Dec* | *Jan* | *Feb* | *Mar* | *Apr* | *May* | *Jun* | *Jul* | *Aug* | *Sep* |
| Winter wheat | 51.3 | 133.4 | 30.8 | | | | 25.7 | 56.4 | 15.4 | 15.4 | | 220.6 | 266.8 |
| Winter barley | 32.4 | 84.2 | 19.4 | | | | 16.2 | 35.6 | 9.7 | 9.7 | 116.6 | 184.7 | |
| Spring barley | 27.2 | 10.9 | 21.8 | 5.4 | | | 73.4 | | 32.6 | | | 144.2 | 62.6 |
| Grass – leys | 46.9 | | | | | | 28.1 | 28.1 | 28.1 | 28.1 | 28.1 | 89.1 | 65.7 |
| perm. pasture | 38.2 | | | | | | 22.9 | 22.9 | 22.9 | 22.9 | 22.9 | 11.5 | 7.6 |
| Silage | 40 | | | | | | | | 252.0 | 128.0 | | | |
| Dairy cows | 115 | 276.0 | 287.5 | 287.5 | 287.5 | 287.5 | 287.5 | 276.0 | 253.0 | 241.5 | 241.5 | 241.5 | 253.0 |
| Followers (repl. units) | 15 | 43.5 | 43.5 | 43.5 | 43.5 | 43.5 | 18.0 | 18.0 | 18.0 | 18.0 | 18.0 | 18.0 | 18.0 |
| Ewes | 190 | 47.5 | 38.0 | 38.0 | 57.0 | 57.0 | 190.0 | 76.0 | 57.0 | 76.0 | 38.0 | 38.0 | 47.5 |
| **Total hours needed** | | 595.5 | 441.0 | 374.4 | 388.0 | 388.0 | 661.8 | 513.0 | 688.7 | 539.6 | 465.1 | 947.6 | 721.2 |
| Hours available/person/month | | 182 | 131 | 132 | 135 | 122 | 171 | 176 | 216 | 224 | 231 | 216 | 195 |
| **Total available (3 persons)** | | 546.0 | 393.0 | 396.0 | 405.0 | 366.0 | 513.0 | 528.0 | 648.0 | 672.0 | 693.0 | 648.0 | 585.0 |

**Fig. 17.2** Labour profile spreadsheet.

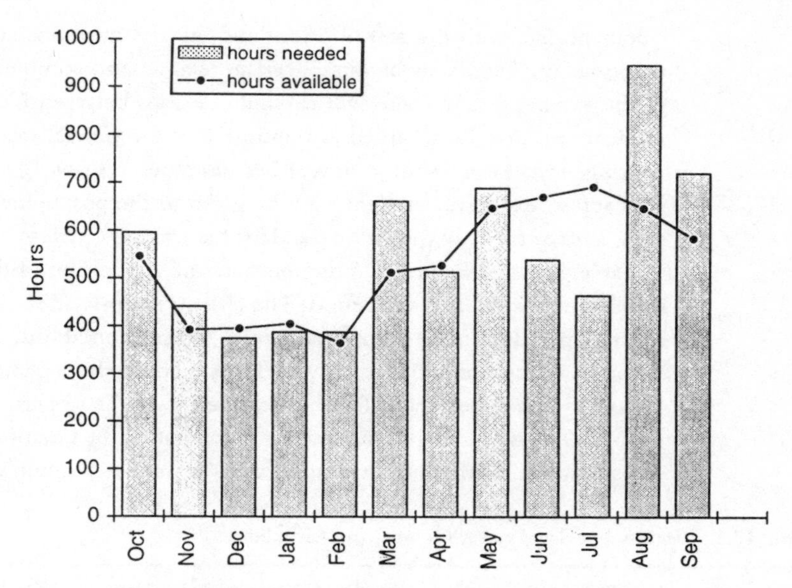


**Fig. 17.3** Labour profile – labour need and availability compared.

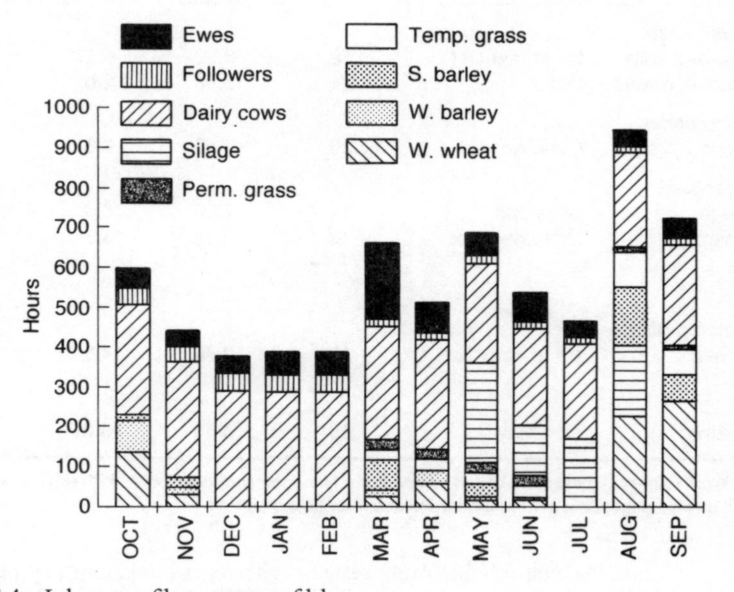


**Fig. 17.4** Labour profile – pattern of labour use.

considered for gangs of people rather than for individuals. The time required for an operation is shown in the GWD chart as a block, with the height determined by the number of people in the gang, and the breadth by the number of days required to complete the operation. It is usually necessary to show only those parts of the year where the labour profile indicates a labour problem.

The blocks representing the various operations are pieced together, rather like a jigsaw puzzle. The resulting picture can be interpreted in a fashion similar to the labour profile, with the aim of smoothing out the most dramatic peaks in labour requirement. There may be some overlaps (e.g. when a combining team carts bales in the evenings), and there will certainly be gaps between blocks. The latter, in moderation, may be regarded as opportunities for general maintenance, and as a contingency allowance for poor weather and other mishaps. If, in a peak season, the gaps appear excessive, thought must be given to the possibilities of changing gang sizes, and/or the way they work, and/or the cropping system.

Table 17.1 and Figure 17.5 demonstrate the application of the technique to an arable farm for the autumn period. The chart (Figure 17.5) shows that it is possible for the operations to be completed, given 'typical' conditions. There is little room for error, though, and the farmer will clearly be doing little management during the period, as he will be needed for physical work. Like the labour profile, the chart can be used as a basis for identifying ways of ameliorating the situation, including the use of contractors, or the introduction of a less ambitious farming system.

**Table 17.1** Gang-work day data for October to December.

| *Enterprise and operations* | *Timing of operations* | *Area (ha)* | *Work rate (ha/GWD)* | *GWD needed* | *Gang size (men)* | *Machinery* |
|---|---|---|---|---|---|---|
| *Winter wheat* | | | | | | |
| Seed-bed cults. | Up to mid-Oct | 50 | 8.00 | 6.25 | 1 | 2 tractors |
| Combine drilling | Oct | 24 | 12.00 | 2.0 | 1 | 2 tractors |
| *Spring barley* | | | | | | |
| Plough | Oct–Feb | 30 | 3.50 | 8.6 | 1 | 1 tractor |
| *Sugar beet* | | | | | | |
| Plough | Sep–Dec | 12 | 2.00 | 6.0 | 1 | 1 tractor |
| Harvest | Up to end Dec | 12 | 0.90 | 13.3 | 3 | 3 tractors<br>harvester<br>2 trailers |
| *Maincrop potatoes* | | | | | | |
| Harvest | Up to early Nov | 10 | 0.70 | 14.3 | 3 | 3 tractors<br>harvester<br>2 trailers |
| Riddling | Oct–May | 10 | 0.40 | 25.0 | 3 | riddle |

*Notes:* Work rates and gang sizes based on Nix, *Farm Management Pocketbook*, 11th edition. Now allowance made for dairy (treated as an independent unit) or for casual labour for potato picking.

As well as being more realistic with regard to labour planning, the gang-work day chart eases decisions concerning the size and type of the machinery complement. Thinking of labour in terms of gangs inevitably involves consideration of the machinery used by the gangs, and the emphasis on type of operation rather than enterprise further clarifies the issues.

### 17.2.5 Other techniques for the planning and control of work

Even with the help of labour profiles and GWD charts, the job of planning and controlling work progress can be one of the most difficult and time-consuming

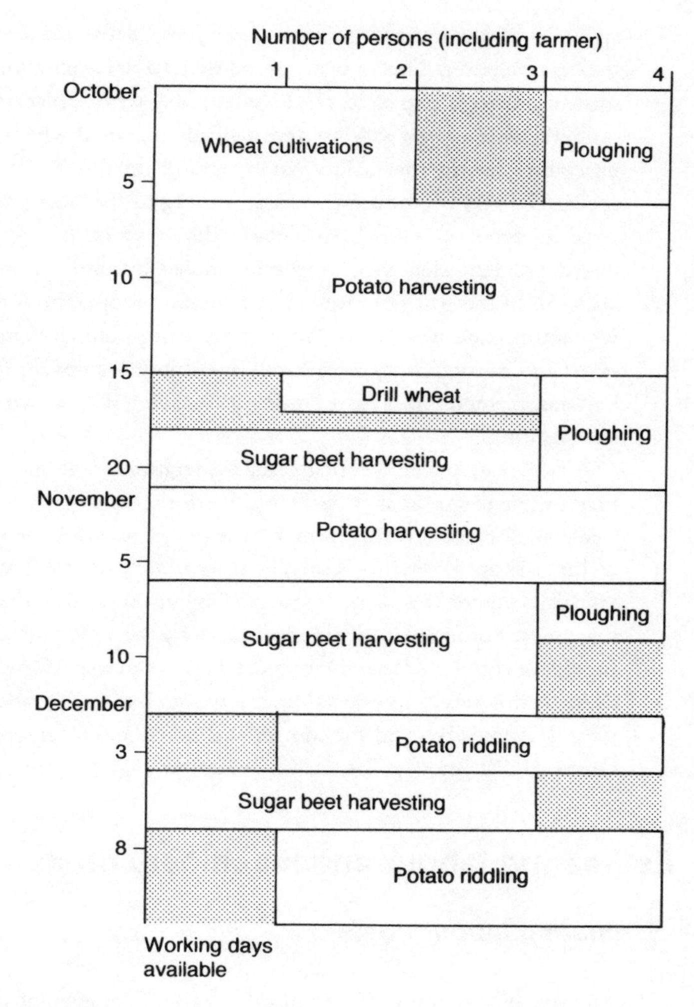


**Fig. 17.5** Gang-work day chart.

aspects of the farmer's job. In the long term it means ensuring that operations are carried out at the optimum point in the production cycle of the crops or livestock concerned. In day-to-day management it entails making sure that jobs are given the appropriate priority, and that each member of the team is in the right place, at the right time, with the right equipment, and knows what is expected of him.

A combination of simple techniques has been suggested by Errington and Hastings (1981) to help the farmer in planning and controlling work progress. For long-term planning, the use of a *year planner* is suggested, on which the farmer can indicate the likely pattern of work in the months ahead. This enables him to monitor his actual progress, and to anticipate the effect of delays on the completion of future operations. It also acts as a valuable aid in the compilation of labour profiles and GWD charts for future years.

With the aid of the year planner, a *job list* may be compiled for each week as it arrives, with the order or priority of the jobs clearly marked, as well as deadlines for their completion. This in turn can be used to draw up a *work schedule* for the farm, showing who is expected to do what, and when. Naturally weather and other unpredictable events will interfere with the planned schedule, but the value of the procedure lies in the facility for weighing up the week's activities quickly and accurately, and reallocating work according to the new priorities.

Some types of work, particularly those which consist of a large number of interdependent elements, can be scheduled by use of *network analysis (critical path analysis)*. In the past the number of agricultural operations to which this technique was appropriate was small, with the greatest potential being in the intensive livestock and horticultural areas. Modern computer software for project management has transformed the ease of using the technique, however, and is well worth investigating.

A farmer may well feel that certain operations could be 'streamlined' in order to speed their progress or to reduce the effort required. A variety of techniques are available for the investigation of the way in which work is performed; these techniques are known collectively as *work study*. Some apply to the work of individuals, some to the work of gangs of men, others to combinations of people and machines. Some are concerned primarily with reducing distance travelled, while others are concerned mainly with the time taken to perform the operation. Many of these techniques can be of value in farming, but to describe them in the detail that they deserve is beyond the aim of this book. Useful references are Fields (1969), Currie (1972) and the International Labour Organization (1979).

## 17.3 Estimating labour and machinery costs

### 17.3.1 Estimating labour costs

The cost of labour on a farm comprises a fixed element and a variable element (these terms are used in their strict sense, and not in the compromise sense of the agricultual gross margin).

| *Fixed* | *Variable* |
|---|---|
| Basic wage | Overtime |
| Benefits in kind[20] | Bonuses |
| Employer's National Insurance contribution | Casual labour[20] |
| Employer's liability insurance | |
| Council Tax (where paid by employer) | |

The fixed element is relatively easy to estimate for a given labour force, since (subject to inflation and the minimum requirements of the law) it is largely determined by the farmer. The basic wage will be the minimum laid down by the current Agricultural Wages Order, plus increments for craftspeople and supervisory grades, and whatever extra the farmer feels is necessary to attract and retain good workers.

Benefits in kind (such as cottages, milk, potatoes, and payment of phone rental) should be priced at the opportunity cost to the farm business (for produce, the normal farm sale price; for cottages, the rent, less expenses, which could be otherwise obtained or, if higher, the return which could be expected if the cottage were sold and the money safely invested). Employer's National Insurance contributions and employer's insurance are dictated by the wages paid. These are both calculated as percentages of the gross earnings, including overtime and bonuses, and thus contain a variable element.

The variable elements of labour costs are more difficult to forecast, since the amounts used of overtime, bonus work and casual labour are likely to be affected by external factors such as weather. Use of labour profiles and GWD charts will help in estimating the hours required, after which the appropriate rates of pay can be applied to find the total cost.

### 17.3.2 Estimating machinery costs

Machinery costs may similarly be classified into fixed and variable elements:

| *Fixed* | *Variable* |
|---|---|
| Depreciation | Fuel and oil |
| Tax and insurance | Spares and repairs |
| Lease payments | Contract hire[20] |
| Cost of capital | |

Depreciation has already been discussed (Section 2.6), and road tax is a set amount, easy to budget. Quotations for insurance are obtainable by phoning insurance agents or brokers, and for leasing terms by contacting the relevant companies. Use of contractors may be estimated with the help of the GWD chart, while phone calls to local contractors will help in pricing. Fuel and oil, and repairs and maintenance, are the most difficult costs to estimate: Nix (1996) gives some help here (Section III, Table 4), but one usually has to rely on a combination of experience and inspired guesswork.

Whether or not the cost of capital is included, and at what rate, depends on the intended use of the costing. If the costs are for inclusion in a whole-business budget, the cost of capital will be calculated on a whole-business basis, by reference to the cash flow budget, and need not be included in the machinery costing. If the costs are intended for comparison between alternative machinery policies (as in the next section), the cost of capital should be calculated by multiplying the average value of the machine (purchase price minus anticipated resale price divided by 2) by the percentage rate of opportunity cost of capital (see Chapter 18).

An alternative to the detailed costing of machinery is to take the total number of tractor hours (as calculated earlier in this chapter) and multiply by the cost per standard tractor hour for the type of farm concerned. The latter may be obtained from regional Farm Business Survey handbooks, but will need updating for use in budgets.

This approach suffers from all the problems of using standard tractor hours, plus those associated with the use of standard costs, which cannot allow for differences between individual holdings in a particular farming type. It can in many situations give a reasonable approximation, however, while needing much less time and effort than the more precise method described above.

## 17.4 Investigating alternatives

### *17.4.1 Introduction*

Once a 'basic' plan has been formulated and costed, various alternatives to that plan may be investigated. This investigation is largely a matter of using partial budgets, although *break-even charts* can be useful. The latter are used to plot the cost of each of the alternatives under consideration against a measure of scale, such as the number of hectares on which a machine is used during the year, or the number of tonnes of a given product processed per annum.

An example of a break-even chart is given in Figure 17.6, comparing the use of a contractor with the purchase of a new 3.6 m cut combine harvester. It is assumed that the use of a contractor will not, in this instance, affect the regular labour cost beyond reducing overtime payments slightly.

The implication of Figure 17.6 is that, for the particular farm and machine under consideration, it would be worth the farmer using contractors if he is planning to harvest less than about 80 ha per year. This is not to say that it would not be worth

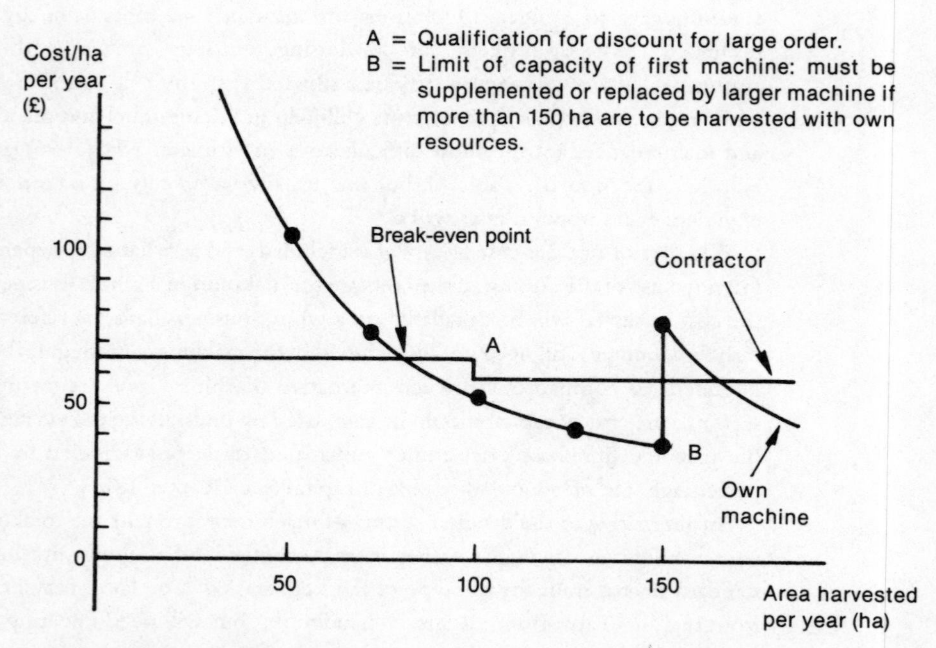


**Fig. 17.6** Break-even chart.

using a smaller and/or second-hand machine for smaller areas, nor does the chart take into account factors such as timeliness.

The following sections list the principal financial and non-financial considerations in a selection of common labour/machinery decisions. The lists are intended to be indicative rather than exhaustive.

### 17.4.2 Increase regular workforce by one man

*Possible benefits*[21]

- Reduction in overtime, casual labour, use of contractors;
- Change in farm system (if labour previously limiting) to give greater profits;
- Opportunity to reduce some machinery costs by use of more labour-intensive methods;
- Improvement in quality of output;
- 'Spare' labour acts as contingency allowance for emergencies;
- Farmer can spend less time on manual work and more on management and/or leisure;
- Maintenance and appearance of farm can be improved.

*Possible costs*[21]

- Extra wages, National Insurance, employer's liability insurance, benefits in kind;
- Need for extra machinery (especially for field work);
- Demands on farmer's organizing ability increased.

### 17.4.3 Replace some regular labour and machinery by use of contractors and/or casual labour

*Possible benefits*

- Reduction in *number* of people and/or machines, saving basic costs of employment and ownership;
- Reduction in *use* of people and/or machines, saving overtime and bonus payments, and machinery running costs;
- Change in farm system (if specialist labour skills and machinery previously limiting) to produce higher profits;
- Improvement in quality of output, e.g. if contractor uses more modern, sophisticated machinery;
- Reduction in capital requirements.

*Possible costs*

- Direct cost of employing contractors and/or casual labour;
- Reduction in quality of output, e.g. contractors in a hurry may be less careful than our drivers;

- Risk of contractor not coming exactly when required, with resulting loss of output;
- Increased worry and demands on organizational ability (outside influences involved; difficult to control);
- Reaction of existing employees to reduction in workforce, and/or loss of overtime and bonus payments.

### 17.4.4 Replace two small machines by one large machine

*Possible benefits*

- Probably only one operator required; save overtime and bonus payments; possibly reduce workforce by one man;
- Lower initial price: one machine of a given capacity usually costs less than two smaller machines with an identical combined capacity;
- Lower running costs (reason same as that given for lower initial price above);
- Change in farm system (if labour previously limiting) to earn higher profits;
- Improvement in timeliness of operations (key tasks performed faster);
- Favourable reaction of employees (especially the potential driver).

*Possible costs*

- Loss of flexibility in use;
- Consequences of breakdown, operator illness, etc. are greater;
- Reaction of employees to reduction in workforce and/or loss of overtime.

### 17.4.5 Buy second-hand rather than new

*Possible benefits*

- Lower purchase price; eases cash flow and may make the difference between having and not having;
- Lower interest costs in consequence.

*Possible costs*

- Higher maintenance and repair costs;
- Reduction in quantity and quality of output, due to age and obsolescence;
- More frequent breakdowns may affect timeliness of operations;
- Lower operator comfort and safety;
- Extra worry generated by unreliability.

### 17.4.6 Replace a machine next year rather than this

*Possible benefits*

- Initial expense delayed; eases cash flow and reduces depreciation expense.

*Possible costs*

- Higher annual repair costs;
- Initial capital allowances delayed;
- Remainder as above for second-hand machine.

### *17.4.7 Lease rather than buy*

*Possible benefits*

- Lower initial payments; eases cash flow and money can be put to good use elsewhere;
- Lower overall costs if profits too low to allow initial capital allowances to be taken;
- Greater security than borrowing on overdraft (though not loan);
- Fixed rental rates (valuable if interest rates rise).

*Possible costs*

- Loss of capital allowances if profits are high enough to allow these to be taken;
- Fixed rental rates (disadvantage if interest rates fall).

## Notes

19. Note that this and the subsequent methods concentrate on the number of tractors, with the assumption that this determines the number and type of the other machines.
20. If the calculation is intended to provide figures for a whole-business profit budget, care should be taken to ensure that these items are not double counted in both fixed and variable costs.
21. The emphasis here is on the work possible – not every one of these aspects will apply to every decision, and some are mutually incompatible.

## References

Barnard, C.S. & Nix, J.S. (1979) *Farm Planning and Control*, 2nd edn. Cambridge University Press, London.

Butterworth, W. & Nix, J.S. (1983) *Farm Mechanisation for Profits*. Granada, St Albans.

Currie, R.M. (revised J.E. Faraday) (1972) *Work Study*, 3rd edn. British Institute of Management and Pitman, London.

Errington, A. & Hastings, M.R. (1981) Work planning in agriculture. *Farm Management*, **4**, 5.

Fields, A. (1969) *Method Study*. Cassell, London.

Gladwell, G. (1985) Labour and machinery analysis and planning. *Farm Management*, **5**, 9.

International Labour Organization (1979) *Introduction to Work Study*, 3rd edn. ILO, Geneva.

Lockyer, K. & Gordon, J. (1991) *Critical Path Analysis*, 5th edn. Pitman, London.

Nix, J.S. (1996) *Farm Management Pocketbook 1997*. Wye College, University of London, Wye, Kent.

Soffe, R.J. (ed.) (1995) Part 4: Farm equipment; Chapter 3: Farm staff management. In: *The Agricultural Notebook*, 19th edn. Blackwell Science, Oxford.

Yule, I.J. (1995) Calculating tractor operating costs. *Farm Management*, **9**, 3.

# Chapter 18
# Capital Planning

## 18.1 Introduction

For the majority of businesses, capital is the most limiting resource in the long term. Given enough capital, other resources such as land, labour or management skills may be bought, although it may take time for the right plot of land to come on the market or the right worker or manager to offer himself for employment. Since capital plays such a key role, it is vital that the manager should be able to calculate its cost, measure the efficiency of its use, and know where to seek additional capital needed for survival and/or expansion of his business.

Discussion of the effects of inflation has been postponed to Chapter 20, in order to simplify the discussion of techniques.

## 18.2 Sources of capital

A prime consideration in any plan which involves the investment of capital is that of where the money is to come from. The most profitable of schemes is as nothing if insufficent cash is available for its implementation. Possible sources of capital may be divided into internal sources and external sources.

Sources of capital internal to the business include personal savings of the owner(s), and profits retained from former years. They also include various forms of capital release: sale of land (and possible lease-back); sale of surplus cottages; hire of machinery, buildings and other assets instead of purchase; change from a system with high working-capital requirements to one with low working-capital requirements; and so on. The advantages of obtaining capital from internal sources are that it does not incur a continuous interest charge, and that the farmer is not indebted to an outsider to the business as a direct result of the use of the capital. On the other hand, it can often incur a heavy opportunity cost (such as sale of land at a heavy discount for lease-back, or lost profitability as a result of a change in system). There is also a limit to its availability.

External sources include gifts, grants, widening of the ownership of the business, and credit. If you have friends or relations who are sufficiently rich and enlightened to make you substantial gifts, all well and good, but in most circumstances it is necessary to look further afield. Grants are available from the Ministry of Agriculture, Fisheries and Food (MAFF) to promote conservation and environmental enhancement, organic farming, and agricultural marketing – but the old days of obtaining capital grants for farm improvements are gone. Details of schemes are available from local offices of MAFF and ADAS, and summaries are found in many farm management handbooks (e.g. Nix, 1996).

For non-agricultural developments, grants or low interest loans may be available from other sources, such as the English Tourist Board, the Rural Development Commission, and the Countryside Commission. Assistance for schemes involving amenity tree-planting, conservation of certain types of natural environment, and similar activities, are often available through bodies such as the Countryside Commission and local authorities. At the time of writing, certain peripheral rural areas of Europe (including several in the UK) qualify for special European grants under the 'Objective 1', 'Objective 5b' and LEADER programmes, aimed at stimulating economic regeneration in those areas: details are available from regional government offices and Business Links.

Bringing other people into the ownership of the business by the formation of a partnership or a limited liability company will also bring money in, but it may not be possible to find someone at the appropriate time who is prepared to risk his money, is trustworthy and with whom one can happily work (see Section 5.2).

The above sources are unlikely to prove sufficient to maintain the pace of development required by most businesses, unless the business is well-established, rich, or lucky. Thus many business owners, particularly the younger ones, are forced to resort to credit. To borrow money exposes the business to extra risk, and means that the owner of the business has to be 'beholden' to someone outside the business. If one is sensible about the level and type of borrowing, however, the risk need not be large. The main (but certainly not only) sources of credit, listed in approximate order of term, are as follows:

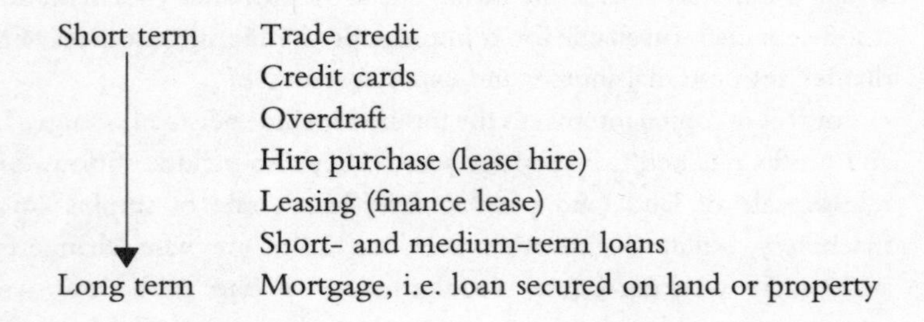


Most businesses make use of *trade credit* to some degree. It is commonplace for businesses to buy and sell on credit, and thus at any one time a business is likely to owe money to a supplier. For a certain period after the invoice date, this credit is free (usually three or four weeks). The cost of extending credit beyond this period can be expensive, however, both in terms of lost goodwill and in terms of cash. A typical credit charge is '2.5% cash discount if payment is made within 28 days of invoice date'. If the bill is settled two weeks late and the discount is lost, the marginal cost of the delay is equivalent to an annual interest rate of 67%.

*Credit cards* work on a similar system to trade credit, and their cost is high unless the balance outstanding is cleared each month. Their cost may be outweighed by their convenience in the purchase of relatively small items.

The largest single source of credit for businesses in the UK is the clearing bank network. The credit supplied by a bank may be in the form of an overdraft and/or loans. An *overdraft* is a negative balance in a bank current account. It fluctuates from day to day as money flows into and out of the account. Interest is calculated on the daily balance outstanding, and is thus payable only on money that is in use. An overdraft is technically repayable on demand and as a result is usually the cheapest source of credit available.

A *loan* is borrowed from the bank on a contractual basis. The bank gives the borrower a sum of money, on the condition that it is repaid in an agreed pattern, and that interest is paid at specified times during the period of the loan (see Section 18.3). A loan is more expensive than an overdraft, since interest rates tend to be higher in reflection of the longer call of the money, and since, unlike an overdraft, interest has to be paid whether or not the money is fully used. It is rather safer than an overdraft, however, due to the existence of a contract and a clearly specified term. A banker will usually prefer to finance major asset purchases by means of loans, using the overdraft to finance seasonal fluctuations in working capital.

Loans are also available from other institutions, often for specific purposes. The Agricultural Mortgage Corporation lends primarily for land purchase, but will also help to finance other capital improvements, diversification, horticulture and woodlands, providing it is granted a first mortgage on agricultural property as security. The Lands Improvement Company also finances land purchase, but on larger properties, and will take other forms of security. Building societies and various merchant banks will provide finance for the purchase of non-agricultural property.

A major source of loans for small businesses is private finance, supplied by family or friends. This usually has the advantage of being relatively cheap, and does not lead to the same worries about being responsible to a commercial organization (an uncle or aunt is often more tolerant over the timing of repayments than a bank manager would be). This is not always the case, however, and there may be other, non-financial costs involved.

There remain various other sources of business credit, including hire purchase (or 'lease purchase') and leasing of machinery and livestock. Generally speaking, these are more expensive than bank finance. From time to time, however, it is possible to find credit bargains in this field, especially when the machinery trade is slow and dealers are anxious to keep their stock moving.

When seeking borrowed money, the manager should first be sure to use a source of credit which is appropriate to the type of investment. The adage, 'never borrow short to invest long' is ancient, but remains true as ever. It is also important for him to remember that he is asking the lender to risk money in the business. The lender may take a lot of persuading, especially if the manager is young and/or inexperienced, and the presentation of the proposal can make all the difference between it being accepted or rejected. Budgets will normally be required, especially monthly cash flow budgets, and honesty and accuracy in these budgets help to boost the confidence of the lender in the proposal.

It is likely that the lender will require some form of *security*, or *collateral*, in case things go wrong. A mortgage for land purchase, for instance, will usually be secured on the title deeds of the land, meaning that if the borrower fails to service the loan, the lender can sell the land in order to recoup his investment. Where the assets are insufficient, or of the wrong type to provide adequate security, it may be possible to secure the credit by guarantee. In other words a third party pledges to pay the amount guaranteed to the lender if the borrower defaults. The role of guarantor is often played by a friend or relative of the borrower. Guarantees may also be obtained from the government through the Loan Guarantee Scheme: details are available from banks.

A good 'track record', combined with a healthy balance sheet and the confidence of the lender, may be sufficient security for overdrafts and short-term loans. By definition this is likely to apply to the well-established manager rather than the young incomer.

Finally, most lenders pride themselves on their judgement of character. Thus personal presentation, in the form of appearance, confidence and enthusiasm, deserves attention.

## 18.3 Estimating the cost of capital

### 18.3.1 Estimating interest costs

Borrowing money from a source outside the business will incur interest charges. The annual interest charge is calculated by applying a percentage rate to the capital borrowed. The exact way in which this is done depends on the form of borrowing, i.e. whether it is in the form of an overdraft, a reducing-balance loan, an annuity loan, or some other variant.

The interest on an *overdraft* is calculated on the daily balance outstanding, and is paid quarterly in arrears. Estimation of likely interest charges is relatively easy as long as a cash flow budget is available (see Chapter 7).

A *reducing-balance loan* is one where the principal, i.e. the capital borrowed, is repaid in equal instalments, and the interest charge is calculated by applying the appropriate percentage interest rate to the amount of capital still unpaid. For example, a three-year loan of £3000 might be paid off in equal instalments of £500 per half year. Assuming that interest is paid six-monthly in arrears at 10% per annum, the payment for the first half-year will be £150 [£3000 × (10% ÷ 2)]; for the second half-year £125 [£2500 × (10% ÷ 2)]; for the third half-year £100 [£2000 × (10% ÷ 2)]; and so on (see Figure 18.1).

Although the reducing-balance type of loan makes interest calculations easy, it has the disadvantage of high interest payments at the beginning of its life (often the time at which an investment project most needs cash). This is so even after allowance has been made for the fact that interest payments can be set against

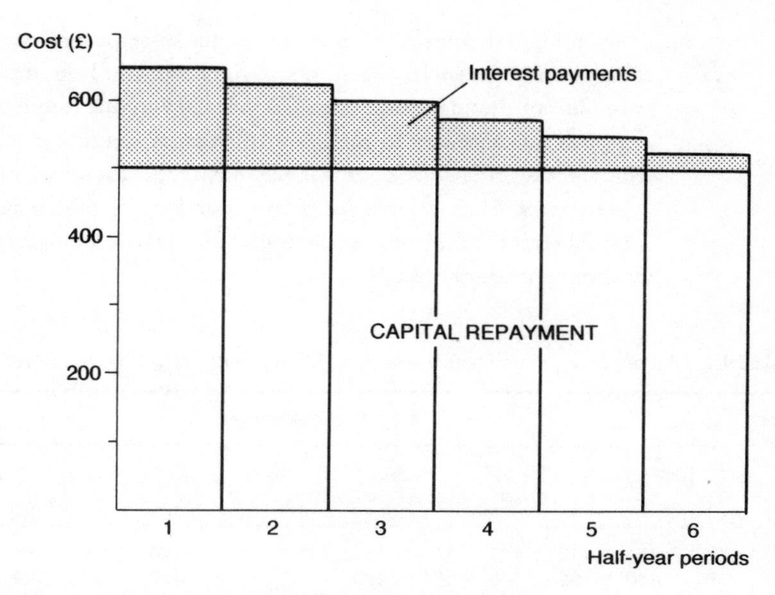


**Fig. 18.1** Reducing-balance loan – illustration (10% interest).

income tax. A commonly-used alternative, which does not suffer from this problem, is the *annuity* form of loan. The aim in using an annuity loan is to arrange payments of capital and interest so that the gross payment for each period is the same. In order to do this and yet keep capital and interest payments in step with each other, a mathematical formula is used which results in smaller payments of capital, and larger payments of interest in the early stages of the loan than in the later stages (see Figure 18.2).

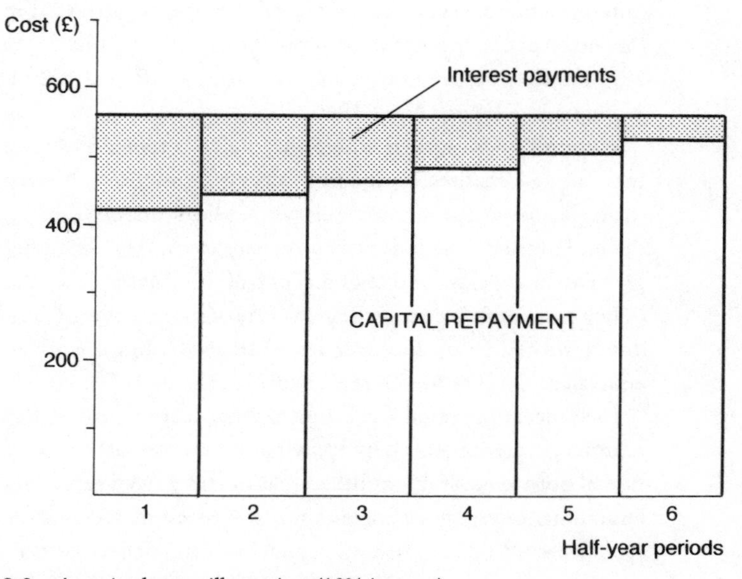


**Fig. 18.2** Annuity loan – illustration (10% interest).

Since interest payments are an allowance expense against income tax, the net effect is one of relatively small net payments at the beginning of the loan, rising as the life of the loan progresses. To calculate annual interest charges for a given loan, it is not necessary to use the mathematical formula, since tables are available which show interest and capital payments due for loans of various terms and interest rates. Such a table for a five-year loan is shown in Table 18.1. More comprehensive tables are to be found in various publications, including farm management handbooks.

**Table 18.1** Amortization table for five-year loan (£ payment per £1000 borrowed).

| Year | Interest rate | | | | | | | | | | | | | |
|---|---|---|---|---|---|---|---|---|---|---|---|---|---|---|
| | 10% | | 11% | | 12% | | 13% | | 14% | | 15% | | 16% | |
| | I | P | I | P | I | P | I | P | I | P | I | P | I | P |
| 1 | 100 | 164 | 110 | 161 | 120 | 157 | 130 | 154 | 140 | 151 | 150 | 149 | 160 | 145 |
| 2 | 84 | 180 | 92 | 178 | 101 | 176 | 110 | 174 | 119 | 172 | 128 | 171 | 137 | 169 |
| 3 | 66 | 198 | 73 | 198 | 80 | 197 | 87 | 197 | 95 | 197 | 102 | 196 | 110 | 196 |
| 4 | 46 | 218 | 51 | 220 | 56 | 222 | 62 | 223 | 67 | 224 | 73 | 225 | 78 | 227 |
| 5 | 24 | 240 | 27 | 243 | 30 | 248 | 33 | 252 | 36 | 256 | 39 | 259 | 42 | 263 |

*Notes:* I = interest payment; P = repayment of capital. Assumes payment in arrears at the end of each year. In practice more frequent payment is likely, giving slightly lower interest costs. Payment half-yearly in arrears gives interest per £1000 in the first year of £96 at 10% and £154 at 16%.

To determine interest payments due in a given year of a loan, the point is found where the year intersects with the appropriate interest rate. The interest due is calculated by multiplying the amount of the loan (in £000s) by the factor in the 'interest' column. The repayment of principal is calculated in the same way, using the factor in the 'principal' column. Thus in year 3 of a 5-year loan of £10 000, at 15% interest, the interest charge is £102 × 10 = £1020 and the repayment of principal is £196 × 10 = 1960.

A third type of loan is where the capital is repaid in full at the end of the term, with interest charged on the full amount throughout the term. It is usual to secure the capital payment with an endowment life assurance policy, designed to mature at the end of the term; hence the term *endowment loan*. The interest may be set against profit as an expense, and tax relief at half the current basic rate is deducted from the policy premiums by the assurance company. The overall cost of such a loan can be lower than either of the types described above, but the initial cost is higher than an equivalent annuity loan – see Figure 18.3.

In budgets prepared for a normalized year, an alternative (cruder) method of estimating interest costs is by applying the appropriate percentage rate to the average capital borrowed. If it is assumed that the borrowed capital is paid off at a more-or-less constant rate, the average capital can be estimated as (initial capital ÷ 2). Initial capital includes the working capital needed to finance the running costs of the proposed system up to the point at which it becomes self-financing. Note that the

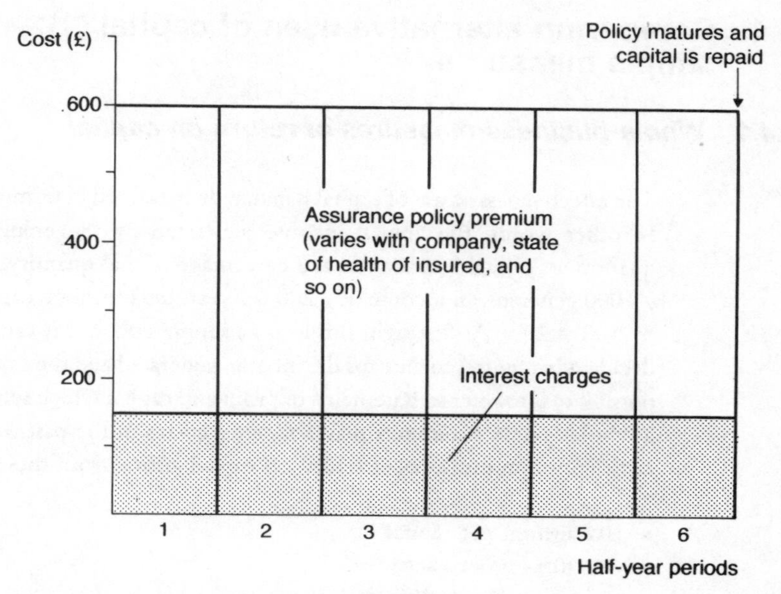


**Fig. 18.3** Endowment loan – illustration (10% interest).

definition of average capital used in this context is slightly different from that used in the context of return on capital later in this chapter.

### *18.3.2 Estimating the opportunity cost of capital*

Any capital invested in the business, whether borrowed or owned, has an opportunity cost; by using it in one particular fashion, the owner is sacrificing the opportunity to use it in other ways. This opportunity cost is represented by the potential return from the best of these other uses, expressed as a percentage of the amount invested. The concept of opportunity cost is particularly useful when making decisions between alternative investments[22], each alternative being assessed by reference to the earning power of all the capital involved. This should provide a stiffer test than a comparison with interest rates on borrowed capital.

Estimating the opportunity cost of capital can be difficult in practice. If the owner is considering only one possible use of the capital within his business, he can take the opportunity cost to be the return that he could obtain by investing the money in, say, gilt-edged securities or a bank deposit account, plus an allowance for the difference in riskiness of the two types of investment.

Where two or more uses of capital within the business are possible, the opportunity cost of any one of those uses is given by comparison with not only the gilt-edged security/deposit account option, but also the potential returns from the other within-business uses. The easiest way of doing this is by comparing marginal rates of return on capital, discussed in the following sections.

## 18.4 Comparing alternative uses of capital (1): simple measures

### 18.4.1 Whole-business measures of return on capital

The effectiveness of use of capital is generally measured in terms of return on capital. In other words, the annual income generated by the employment of a given quantity of capital is expressed as a percentage of that quantity. If an investment of £1000 generates an income of £200 per year, the return on capital is $(200 \div 1000) \times 100 = 20\%$. At first sight this looks a simple task, but it can be far from easy to decide what figures to take for the income generated and the capital employed. One possibility is to refer to statements of profit and capital which relate to the business as a whole. These statements may concern periods in the past or in the future. The most common measures of return on capital arising from these statements are:

- Return on net capital
- Return on net assets
- Return on tenant's capital

The *return on net capital* measures the earning power of the owner's personal stake in the business. It is calculated by using the formula:

$$\frac{\text{Net profit for the year in question}}{\text{Net capital}} \times 100$$

and is expressed as an annual percentage rate. Net profit is taken from the profit and loss account (or budget), and net capital from the closing balance sheet (historic or budgeted). Where opening and closing net capital differ markedly, there is some justification for using the average of the two instead of just the closing figure.

*Return on net assets* measures the return on all the capital employed in the business, after reducing the influence of fluctuations in short-term assets and liabilities. It is calculated by use of the formula:

$$\frac{\text{Net profit for the year before interest on long-term liabilities}}{\text{Net assets}} \times 100$$

where net assets equals total assets minus current liabilities. Profit is used before interest since the interest charge represents part of the return on the assets employed, and to use a profit after interest would be to double-count this part of the return.

*Return on tenant's capital* measures the income generated by those assets which are specific to the businessman's role as a farmer, rather than any role he might have as an owner of land. It is calculated by the formula:

$$\frac{\text{Management and investment income}}{\text{Tenant's capital}} \times 100$$

This is specifically an agricultural measure, and is only meaningful in comparisons between farms (see Chapter 14).

All of the measures described above can be valuable in giving a general impression of the efficiency of capital use in the business, through comparisons between years, between businesses, and with budgets. But consider the case of a owner who wishes to investigate the potential return from a specific use of capital; an investment 'project'. His main concern will be not the effect on the overall return on capital of the whole business, but whether that extra investment will be worthwhile in its own right. He will need to relate the *extra* income generated to the extra *capital employed*; in other words to estimate the marginal return on capital.

### 18.4.2 Simple marginal return on capital

Simple marginal return on capital is also known as 'accounting rate or return' or just 'rate of return'. It is calculated by using this formula:

$$\frac{\text{Average annual extra profit}}{\text{Initial or averagte extra capital invested}} \times 100$$

and is expressed as an annual percentage rate.

The *average annual extra profit* is derived usually from normalized partial budgets. It should be before interest charges, since the latter represent part of the return on capital, and double-counting must be avoided. The profit may be before or after tax. Ideally the after-tax figure should be used, since the various uses of capital under investigation may have different tax implications. In practice, it is usually too difficult to estimate the likely tax effects when using average annual figures.

The *initial capital* is represented by the initial cost of any assets purchased (net of grants and discounts), plus the extra working capital required, less any capital released by the change (such as sale of replaced equipment, or working capital set free). Working capital is the cash required to finance the extra running costs up to the point at which the extra revenue begins to exceed the extra expense. In the example of the expansion of a dairy herd, extra working capital would include the cost of extra feed and other materials consumed during the weeks before the milk cheque reflected the increased production. Estimation of the extra working capital required is made easier by the availability of a full or partial cash flow budget.

*Average capital* may be used as an alternative to initial capital. It reflects the decline in value of wasting assets over the life of the project and can be estimated by using the formula:

$$NIC(ND) + \frac{[NIC(D) - WDV]}{2} + WDV$$

where *NIC(ND)* is the net initial cost of non-depreciating assets (including working capital), *NIC(D)* is the net initial cost of depreciating assets and *WDV* the anticipated written-down value of depreciating assets at the end of the project life. This assumes a straight-line decline in value of depreciating assets.

Return on initial capital is the more stringent of the two tests, since average capital will never be greater than initial capital. When compared with the more precise discounted cash flow techniques, return on initial capital tends to give a pessimistic result, and return on average capital an optimistic one. For this reason alone it is usually best to use initial capital.

Once a simple marginal rate of return has been calculated, it may be compared with the cost of capital in order to determine whether or not the investment is likely to be worthwhile. The first such comparison must always be with the cost of interest payable on borrowed money, since if the project cannot even cover the cash costs of using the capital, it would be financially foolish to consider it further. The other comparison must be with the opportunity cost of the capital employed, i.e. with the marginal return which could be obtained from other investments of the capital concerned.

Despite its relative ease of calculation, simple marginal return has several limitations. The most important of these is illustrated by the example shown in Figure 18.4. The two projects each give a simple marginal return on capital of 25%. This implies that the decision-maker will be indifferent between the two projects, and yet in practice he may have a strong preference for one or the other. He may, for instance, prefer project 1 because he only has £500 to spend, or project 2 because it has a longer life, and will go on producing 25% for three years more than project 1. He is certainly likely to be influenced by the fact that by investing in project 1 he is likely to recover his initial investment more quickly than by investing in project 2.

| | *Project 1 (£)* | *Project 2 (£)* |
|---|---|---|
| Initial investment | 500 | 1000 |
| Extra profit | | |
| Year 1 | 250 | 250 |
| 2 | 150 | 250 |
| 3 | 125 | 250 |
| 4 | 70 | 250 |
| 5 | 30 | 250 |
| 6 | | 250 |
| 7 | | 250 |
| 8 | | 250 |
| Average annual extra profit | 125 | 250 |
| Marginal return on capital | 25% | 25% |

**Fig. 18.4** Comparison of two projects by simple marginal return on initial capital.

Of these three problems, the first two are common to all percentage rates of return and are, in any case, reasonably easy to overcome when interpreting the results of the calculations. The third, the inability to discriminate between projects on the basis of the timing of their income flow, is a major fault of this particular method. It is in the nature of humans to prefer the receipt of £1 now to the receipt of £1 in the future, even in the absence of inflation. A method of investment appraisal which ignores this 'time preference' characteristic is of dubious value, except as an initial rule-of-thumb screening device.

### 18.4.3 Payback

One method of judging investment projects in a way that allows for time preference is to calculate their *payback period*. The payback period of a project is simply the time that it takes to recoup the initial investment from the net cash flow generated by that investment. As an example, consider the project shown in Figure 18.5. The payback time is three years, since by the end of the third year enough cash has been generated to recover the initial investment with £100 to spare.

| | £ |
|---|---|
| *Initial investment* | 1000 |
| Net cash flows | |
| Year 1 | 600 |
| 2 | 300 |
| 3 | 200 |
| 4 | 200 |
| 5 | 100 |

**Fig. 18.5** Payback – example.

The initial investment is calculated net of grants, discounts, and any capital released by superseded projects. Annual cash flow should include interest payments (but not repayments of borrowed capital) and any fixed or working capital that is likely to be set free at the end of the project's life. It should ideally also include an allowance for changes in tax liability arising from the project. In practice it is rarely worth making anything more than a rough guess at the tax implications, given the crudity of the technique.

Although it is not expressed as a percentage rate of return, payback may still be used to compare the income generated by an investment with the cash and opportunity costs of the capital employed. Comparison with the cost of borrowed money is effected by the inclusion of interest charges in the annual net cash flows; if the project does not pay back, it cannot support the interest charges. Comparison with other opportunities is via comparison of the payback period with those of other potential uses of the capital. The use with the lowest payback is, by implication, the most worthwhile.

Payback takes account of time preference, and is quick and simple to calculate. It is in common use as a rule-of-thumb measure for the initial screening of investment possibilities, and is especially useful when liquidity is a problem. Its simplicity makes it a rather crude measure, however. In particular, it ignores cash flows after the payback period. In the example in Figure 18.6, project 1 comes out as 'best' by the payback method, even though project 2 is likely to produce a steady stream of cash for the next 20 years.

| | Project 1 (£) | Project 2 (£) |
|---|---|---|
| Initial investment | 1000 | 1000 |
| Net cash flows | | |
| Year 1 | 600 | 200 |
| 2 | 300 | 200 |
| 3 | 200 | 200 |
| 4 | 200 | 200 |
| 5 | 100 | 200 |
| . | | . |
| . | | . |
| . | | . |
| 20 | | 200 |
| Payback (years) | 3 | 5 |

**Fig. 18.6** Payback comparison of projects with widely differing lives.

The only way of allowing for time preference in comparisons between investment projects, while at the same time incorporating a reasonable degree of precision, is to use discounted cash flow techniques.

## 18.5 Comparing alternative uses of capital (2): discounted cash flow techniques

### *18.5.1 The discounting principle*

Section 18.3 discussed a device which implicitly takes time preference into account: interest. An interest charge reflects the time preference of the lender, since he will not part with his money unless the fee he is offered is sufficient to compensate him for his sacrifice of present consumption. Similarly it reflects the time preference of the borrower, since he will be prepared to borrow the money only if the benefit he expects to gain from its use is greater than the interest charge.

The formula for calculating compound interest states that a sum of money $X$ placed in an interest-bearing investment will, if left untouched, grow to a value of $X(1 + r)^n$ over $n$ years at a compound interest rate of $r$ (expressed as a decimal). Thus a sum of £100, if invested at 15% interest, will grow to £100 $(1 + 0.15)^5$, or approximately £200, by the end of five years. Reversing the calculation, a sum of £200 which is likely to be received in five years' time is equivalent to £200 $\div$ $(1 + 0.15)^5$, or approximately £100 in the present.

It is possible to think of the time preference of an owner in terms of a percentage compound rate. Assume for the moment that the time preference rate of a given owner is 15% (we will worry about how this is determined at a later stage). If he expects to receive £200 in five years' time, we can refer to the formula above and deduce that he is likely to be indifferent between £100 now and £200 in five years' time. Put another way, the *present value* of £200 in five years' time is £100. This principle forms the basis for discounted cash flow (DCF) techniques such as net present value (NPV) and internal rate of return (IRR).

### *18.5.2 The net present value method*

The idea of expressing future receipts in terms of 'present values' may be applied to an investment project such as project 1 in Figure 18.7. Given a certain time preference rate, each year's net cash flow can be reduced to a present value by multiplying it by $1 \div (1 + r)^n$. This process is known as 'discounting'. Luckily tables such as Table 18.2 are available, giving the value of $1 \div (1 + r)^n$ for a

**Table 18.2** Table of discount factors.

| *Years* | *Percentage* | | | | | | | | | | |
|---|---|---|---|---|---|---|---|---|---|---|---|
| | 1 | 4 | 5 | 6 | 7 | 8 | 9 | 10 | 11 | 12 | 13 |
| 1 | 0.990 | 0.961 | 0.952 | 0.943 | 0.934 | 0.925 | 0.917 | 0.909 | 0.900 | 0.892 | 0.884 |
| 2 | 0.980 | 0.924 | 0.907 | 0.889 | 0.873 | 0.857 | 0.841 | 0.826 | 0.811 | 0.797 | 0.783 |
| 3 | 0.970 | 0.888 | 0.863 | 0.839 | 0.816 | 0.793 | 0.772 | 0.751 | 0.731 | 0.711 | 0.693 |
| 4 | 0.960 | 0.854 | 0.822 | 0.792 | 0.762 | 0.735 | 0.708 | 0.683 | 0.658 | 0.635 | 0.613 |
| 5 | 0.951 | 0.821 | 0.783 | 0.747 | 0.712 | 0.680 | 0.649 | 0.620 | 0.593 | 0.567 | 0.542 |
| 6 | 0.942 | 0.790 | 0.746 | 0.704 | 0.666 | 0.630 | 0.596 | 0.564 | 0.534 | 0.506 | 0.480 |
| 7 | 0.932 | 0.759 | 0.710 | 0.665 | 0.622 | 0.583 | 0.547 | 0.513 | 0.481 | 0.452 | 0.425 |
| 8 | 0.923 | 0.730 | 0.676 | 0.627 | 0.582 | 0.540 | 0.501 | 0.466 | 0.433 | 0.403 | 0.376 |
| 9 | 0.914 | 0.702 | 0.644 | 0.591 | 0.543 | 0.500 | 0.460 | 0.424 | 0.390 | 0.360 | 0.332 |
| 10 | 0.905 | 0.675 | 0.613 | 0.558 | 0.508 | 0.463 | 0.422 | 0.385 | 0.352 | 0.321 | 0.294 |
| 11 | 0.896 | 0.649 | 0.584 | 0.526 | 0.475 | 0.428 | 0.387 | 0.350 | 0.317 | 0.287 | 0.260 |
| 12 | 0.887 | 0.624 | 0.556 | 0.496 | 0.444 | 0.397 | 0.355 | 0.318 | 0.285 | 0.256 | 0.230 |
| 13 | 0.873 | 0.600 | 0.530 | 0.468 | 0.414 | 0.367 | 0.326 | 0.289 | 0.257 | 0.229 | 0.204 |
| 14 | 0.869 | 0.577 | 0.505 | 0.442 | 0.387 | 0.340 | 0.299 | 0.263 | 0.231 | 0.204 | 0.180 |
| 15 | 0.861 | 0.555 | 0.481 | 0.417 | 0.362 | 0.315 | 0.274 | 0.239 | 0.209 | 0.182 | 0.159 |

| *Years* | *Percentage* | | | | | | | | | | |
|---|---|---|---|---|---|---|---|---|---|---|---|
| | 14 | 15 | 16 | 17 | 18 | 19 | 20 | 25 | 30 | 35 | 40 |
| 1 | 0.887 | 0.869 | 0.862 | 0.854 | 0.847 | 0.840 | 0.833 | 0.800 | 0.769 | 0.740 | 0.714 |
| 2 | 0.769 | 0.756 | 0.743 | 0.730 | 0.718 | 0.706 | 0.694 | 0.640 | 0.591 | 0.548 | 0.510 |
| 3 | 0.674 | 0.657 | 0.640 | 0.624 | 0.608 | 0.593 | 0.578 | 0.512 | 0.455 | 0.406 | 0.364 |
| 4 | 0.592 | 0.571 | 0.552 | 0.533 | 0.515 | 0.498 | 0.482 | 0.409 | 0.350 | 0.301 | 0.260 |
| 5 | 0.519 | 0.497 | 0.476 | 0.456 | 0.437 | 0.419 | 0.401 | 0.327 | 0.269 | 0.223 | 0.185 |
| 6 | 0.455 | 0.432 | 0.410 | 0.389 | 0.370 | 0.352 | 0.334 | 0.262 | 0.207 | 0.165 | 0.132 |
| 7 | 0.399 | 0.375 | 0.353 | 0.333 | 0.313 | 0.295 | 0.279 | 0.209 | 0.159 | 0.122 | 0.094 |
| 8 | 0.350 | 0.326 | 0.305 | 0.284 | 0.266 | 0.248 | 0.232 | 0.167 | 0.122 | 0.090 | 0.067 |
| 9 | 0.307 | 0.284 | 0.262 | 0.243 | 0.225 | 0.208 | 0.193 | 0.134 | 0.094 | 0.067 | 0.048 |
| 10 | 0.269 | 0.247 | 0.226 | 0.208 | 0.191 | 0.175 | 0.161 | 0.107 | 0.72 | 0.049 | 0.034 |
| 11 | 0.236 | 0.214 | 0.195 | 0.177 | 0.161 | 0.147 | 0.134 | 0.085 | 0.055 | 0.036 | 0.024 |
| 12 | 0.207 | 0.186 | 0.168 | 0.151 | 0.137 | 0.124 | 0.112 | 0.068 | 0.042 | 0.027 | 0.017 |
| 13 | 0.182 | 0.162 | 0.145 | 0.129 | 0.116 | 0.104 | 0.093 | 0.054 | 0.033 | 0.020 | 0.012 |
| 14 | 0.159 | 0.141 | 0.125 | 0.111 | 0.098 | 0.087 | 0.077 | 0.043 | 0.025 | 0.014 | 0.008 |
| 15 | 0.140 | 0.122 | 0.107 | 0.094 | 0.083 | 0.073 | 0.064 | 0.035 | 0.019 | 0.011 | 0.006 |

wide range of values of *r* and *n*[(23)]. The value of $1 \div (1 + r)^n$ at a discount rate of 15% is thus 0.497.

The present values of all the annual net cash flows may then be summed to give the total present value. An example of this calculation is shown in Figure 18.7, using project 1 from Figure 18.6. The next stage of the calculation is to deduct the capital cost of the investment from the total present value of the net cash flows. If the resulting NPV is positive, the rate of return on the capital invested is greater than the discount rate used in the calculation. The reverse applies if the NPV is negative. In the case of project 1, the NPV is positive, so the investment is worthwhile even after the time preference effect has been taken into account.

Project 1: Initial investment £1000

'Hurdle' discount rate (%): 15

| *Year* | *Annual net cash flow* | *Factors for 'hurdle' discount rate* | *Present values at 'hurdle' rate* |
|---|---|---|---|
| 0 | −1000 | 1.000 | −1000 |
| 1 | 600 | 0.870 | 522 |
| 2 | 300 | 0.756 | 227 |
| 3 | 200 | 0.658 | 132 |
| 4 | 200 | 0.572 | 114 |
| 5 | 100 | 0.497 | 50 |
| | | | Net present value: 45 |

**Fig. 18.7** Calculation of net present value – example.

Although this calculation does not result in a percentage rate of return, it can still be used in comparisons with the cost of capital. For comparisons outside the business the discount rate can be set to reflect interest rates. Thus if all the money for the project was being borrowed at a rate of 10%, the discount rate would need to be at least 10%, and possibly as much as 15% to give a safety margin in view of the risk involved. This can be thought of as the 'hurdle' rate which any project has to exceed to be considered. For comparisons *within* the business, all projects under consideration should be ranked in order of magnitude of their NPVs. The 'best' investment is indicated by the project within the highest NPV.

We have in the NPV technique a method of investment appraisal which enables implicit comparison of return on captial with the cost of that capital, while allowing for the pattern of cash flow over time. It reflects growth in wealth brought to the business by the investment, and thus consistently ranks projects in order of the size of that growth. The technique has various limitations, however. One of these is that the size of the NPV is affected by the size of the investment. Take two projects, one of which gives an NPV of £100 and the other an NPV of £150. According to the procedure described above, the second of the two should be chosen. If it is then revealed that the first involves an investment of £500 and the second an investment of £1000, however, commonsense suggests that the first gives the higher rate of return.

This drawback can be overcome to some extent by use of the *profitability index* (PI). The PI is calculated by dividing the total present value of the cash flows by the total present value of the investment required; the higher the PI the better. Thus in the example above, the PI of the first investment is £600 ÷ £500 = 1.20, while that of the second is £1150 ÷ £1000 = 1.15. This measure helps to overcome one of the problems of the NPV method, in relating the NPV to the size of the initial investment. It can be particularly valuable in ranking projects when capital is rationed (see Mott, 1993, Chapter 5), but being a relatively crude device is best used in conjuction with, rather than instead of, NPV.

A more central problem is that the concept of an NPV is awkward for a layman to understand. Most managers have a good idea of what is meant by 'return on capital', but few will have a grasp of the implications of an NPV. This is not specifically a reflection on the education of the business community; it is just that NPV is not a convenient yardstick. Ideally, we need an investment appraisal technique which will incorporate the discounting principle and yet give a percentage rate of return on capital. Such a technique is the internal rate of return method.

### 18.5.3 The internal rate of return method

In Figure 18.7 the NPV of project 1 was calculated to be +£45. The fact that the NPV is positive implies that the actual rate of return is greater than the rate of discount (15%) used in the calculation. If a discount rate of 20% is used, the resulting NPV is negative – see Figure 18.8. From this we can deduce that the actual rate of return lies somewhere between 15% and 20%. This actual rate of return is known as the IRR or *discounted yield*. To determine the IRR with more precision, this trial and error process is continued until a discount rate is found which results in an NPV of zero. In this case, a discount rate of 18% gives an NPV of –£8; this is near enough in most circumstances for us to state that the IRR is about 18%.

Where tables of sufficient detail are not available, or where greater precision is required, it is necessary to use *interpolation*. This is a simple proportions sum which assumes that the relationship between discount rate and NPV is linear. The process

Project 1: Initial investment £1000

| Year | Annual net cash flow | Factors for 20% discount rate | Present values at 20% discount rate | Factors for 18% discount rate | Present values at 18% discount rate |
|---|---|---|---|---|---|
| 0 | –1000 | 1.000 | –1000 | 1.000 | –1000 |
| 1 | 600 | 0.833 | 500 | 0.847 | 508 |
| 2 | 300 | 0.694 | 208 | 0.718 | 215 |
| 3 | 200 | 0.579 | 116 | 0.609 | 122 |
| 4 | 200 | 0.482 | 96 | 0.516 | 103 |
| 5 | 100 | 0.402 | 40 | 0.437 | 44 |
| | | Net present values: | –40 | | –8 |

**Fig. 18.8** Calculation of internal rate of return – example.

is illustrated by Figure 18.9. We know that, for project 1, the IRR lies somewhere between 15% and 18%. This point is represented by $x$ in the diagram, and corresponds, by definition, to an NPV of 0. By simple proportion, the ratio of $a$ to $b$ is the same as that of $c$ to $d$. Distance $a$ is thus given by $(c \div d) \times b$, i.e:

$$(45 \div 53) \times 3\% = 2.5\%$$

and the IRR is given by adding this result to the lower of the two discounted rates used:

$$15\% + 2.5\% = 17.5\%$$

The formal expression of this calculation is:

$$\frac{\text{NPV at 'hurdle' rate}}{\text{NPV at hurdle rate} - \text{NPV at second rate}}$$

*times* difference between discount rates (%)
*plus* lower of the two discount rates
*equals* internal rate of return

Once the IRR has been calculated, it can be compared with the cost of capital in the manner described in relation to the simple marginal rate of return earlier in this chapter.

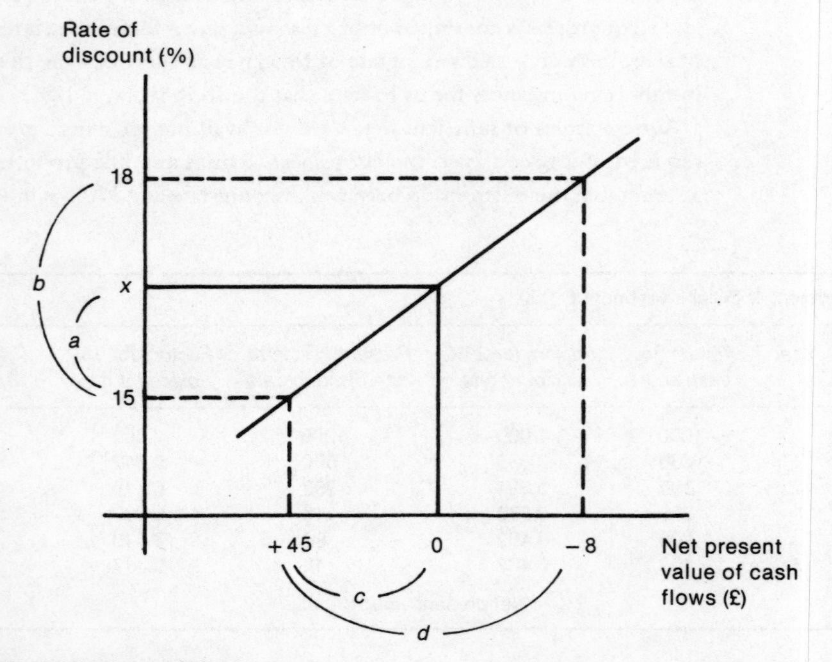


**Fig. 18.9** Interpolation.

Like NPV, the IRR technique has its limitations. In particular, when the cash flow of a project changes from positive to negative during the project life, IRR can give more than one solution (see Lumby,1991, Chapter 5). In most cases the effect will not be significant, but it nevertheless casts doubts on the reliability of the method. Furthermore, the rate of return does not reflect the size of a project, a characteristic which becomes especially limiting when capital is rationed (see Mott, 1993, Chapter 5). However, it does provide a way of expressing the outcome of DCF calculations which is easy to understand and to communicate, and allows easy allowance for risk (see Chapter 19).

In general, NPV (ideally supported by the profitability index) is preferable on grounds of consistency, with IRR having values of convenience and ease of communication.

### 18.5.4 Discounting techniques – the pros and cons

As a whole, discounting techniques are time-consuming and are based on a concept which is initially difficult to grasp. Because it is possible to manipulate cash flow data in a precise manner, there is a huge danger of being carried away with the technique, forgetting that the results must be considered in the light of the owner's objectives, and that the data itself, being forecast for many years ahead, is far from precise. The risk that events may not turn out as forecast should be incorporated in the calculations, using the techniques described in Chapter 20. Nor should the ramifications of the project for the *monthly* cash flow be ignored; it is easy to be lured into overtrading by an apparently profitable, but cash-hungry project.

Despite these drawbacks, the fact remains that the only reliable way of assessing the worth of an investment, taking account of time preference, is by the use of discounted cash flow techniques. The methods force the owner to consider investment projects thoroughly, looking at annual net cash flow rather than vague measures of annual profitability, and making allowance for the implications of such aspects of taxation and increases in asset values.

To describe fully every aspect of the use of NPV and IRR would require far more space than is available here. The reader who wishes to use the techniques on 'real life' examples is recommended to make use of the references listed below.

## Notes

22. In whole-business budgets it is more appropriate to use the actual interest paid, if any.
23. This example is from the Scottish Agricultural College's *Farm Management Handbook* – others can be found in Nix and ABC.

# References


### *Sources of capital*

Agro Business Consultants (latest edition) *The Agricultural Budgeting and Costing Book*. Agro Business Consultants, Melton Mowbray (published half-yearly).

Bright, G. (1987) The current economics of machinery leasing. *Farm Management*, **6**, 6.

Dyer, L.S. (1977) *A Practical Approach to Bank Lending*. Institute of Bankers, London.

Mitchell, W. (1986) Essential information for bank borrowing. *Farm Management*, **5**, 12.

Nix, J.S. (latest edition) *Farm Management Pocketbook*. Wye College, University of London, Wye, Kent (published annually).

Poole, A.H. (1986) Farm indebtedness: stimulus or millstone. *Farm Management*, **6**, 2.

Soffe, R.J. (1987) How much does an overdraft really cost? *Farm Management*, **6**, 6.

### *Examples of the use of DCF techniques*

Bateman, I. (1988) The financial potential for farm-based forestry in the UK following recent government policy developments. *Farm Management*, **6**, 11.

Hayton, R. (1990) Business strategy and the apple business. *Farm Management*, **7**, 6.

Lingard, J. (1980) Investing in irrigation. *Farm Management*, **4**, 11.

Meikle, S.M. (1991) A management approach for orchard evaluation. *Farm Management*, **7**, 10.

Savill, P.S., Crockford, K.J. & Spilsbury, M.J. (1988) Economic returns from farm woodlands. *Farm Management*, **6**, 9.

Sells, J.E. (1991) A financial assessment of a long-term project: coppicing. *Farm Management*, **7**, 10.

### *Investment appraisal*

Attrill, P. & McLaney, E. (1994) *Management Accounting: an active learning approach*. Unit 3. Blackwell Business, Oxford.

Barnard, C.S. & Nix, J.S. (1979) *Farm Planning and Control*, 2nd edn. pp. 50–79. Cambridge University Press, London.

Kerr, H.W.T. Farm Management Notes (Nottingham University). *An Introduction to the Technique of Discounting Cash Flows to Assess Investment Projects* (No. 36, 1968). *An Example of Discounting Cash Flows to Assess the After-tax Return on an Agricultural Investment* (No. 38, 1968). *Handling Working Capital in Discounted Cash Flow Calculations* (No. 41, 1970).

Lumby, S. (1991) *Investment Appraisal and Financing Decisions*, 5th edn. Chapman & Hall, London.

Mott, G. (1993) *Investment Appraisal*, 2nd edn. Pitman Publishing, London.

# Chapter 19
# Allowing for Risk and Uncertainty

## 19.1 Introduction

Any attempt to plan ahead is subject to risk and/or uncertainty[24]; in other words there is a strong possibility that the predictions will not be realized. The budgets for Home Farm used earlier in this book, for instance, rely heavily on assumptions relating to yields, prices, numbers of livestock, etc. The budgets indicate that if these assumptions are justified, the prospects for the business in the short term are good. But what if feed costs per cow turn out to be 20% higher than expected and wheat yields are 10% down? Naturally some account of such possibilities has to be taken in the planning process.

Before examining ways of allowing for risk and uncertainty in the planning process, it is worth taking a brief look at the causes of risk, and ways in which the business might be adapted in order to minimize its impact. The environment of a business has several components. One way of classifying these components is to group them into physical, economic, political/social, and human aspects. The risk attached to the running of a business derives from variation in all these categories.

The physical environment is the first one thinks of in this respect when considering a rural business; weather, disease and pests can have enormous impact on the outcome of even the best-laid business plans. Almost as great are the influences of economic, political and social factors, such as changes in consumption patterns, introduction of new government policy, and resistance to a new tourist facility on the grounds of noise and other nuisance. The aspect more often taken for granted is the human factor, yet this can contribute at least as much to the risk level in a business as any of the others. A business is owned and staffed by people, any one of whom may unexpectedly fall ill, have an accident, have his judgement affected by personal worries, or just make an uncharacteristic error.

The impact of risk on the business may be reduced in various ways. Some risks, such as fire, crop failure or personal injury, may easily be covered by normal commercial insurance. Other risks can be spread by diversification; in other words, running a variety of enterprises so that a poor year for one product does not cripple the whole business. The manager can avoid enterprises which are known to be particularly risky, and can safeguard prices by buying futures or selling produce forward on contract. He can adopt a policy of maintaining reserves of various forms, such as a larger labour force, machinery complement, or bank balance than he would normally need, or a higher equity percentage than strictly necessary. A common feature of agriculture is the farmer who is never happy unless he has one or two years' supply of hay in the barn, ready for the bad year to end all bad years.

The common factor in all of these strategies is that each has a price, in terms of either lost potential or of increased costs. The price limits the freedom of the owner of a business to reduce the adverse effects of risk upon his business. A certain level of risk is therefore something a businessman has to live with. This does not mean that he should be fatalistic about it; he should be able to adapt his financial management so as to cope with most eventualities.

## 19.2 Allowing for risk and uncertainty in planning

### *19.2.1 The role of careful budgeting and control*

The first line of defence against risk when planning should be a meticulous attention to detail in the budgeting process. This applies in particular to forecasting; the more informed one's guesses are about the future, the less risk is attached to those guesses. It is always wise to be slightly conservative in forecasting, to counteract the inevitable tendency towards over-optimism in budgeting. Things *will* go wrong; costs *will* be higher than expected; yields *will* be lower than hoped.

Just as important as informed forecasting is the operation of a system of budgetary control, so that as soon as it begins to look as if the original assumptions of the plan are likely to be proved wrong, alternative courses of action may be formulated (see Part 4 of this book). These devices help to minimize the degree to which risk affects the outcome of a particular plan, but they give the manager no impression of the risk that may be inherent in that plan, and no means of comparing alternative plans with respect to the risk levels involved. Techniques which do tackle these problems are the break-even budget, flexible budgeting and sensitivity analysis.

### *19.2.2 The break-even budget*

A *break-even budget* is used to determine the amount of variation that can be tolerated in one of the key elements of a profit budget before the anticipated margin is eliminated, i.e. the break-even point is reached. Although it is possible to use this technique in conjunction with a whole-business profit budget, it is most suitable for use with a partial budget.

The first step in the process is to express the anticipated margin (i.e. the increase or reduction in profit in the case of a partial budget; net profit or loss in the case of a whole-business budget) in terms of the variable under consideration. This variable could be a price, a unit quantity, or a number of head or hectares. Deducting this expression of the anticipated margin from the original level of the variable gives the break-even level of that variable.

An example will help to clarify the process. In the normalized partial budget used to investigate the introduction of a caravan site to Home Farm (Figure 9.2) it was assumed that the average occupancy rate would be 60%. The resulting increase in profit after interest (Figure 9.3) was expected to be £3568. A 1% change in occupancy rate is worth £240 (£14 400 ÷ 60) and so £3568 represents £3568 ÷ £240 = 14.9%. This is the amount by which the occupancy rate would have to fall before the change resulted in a decrease, rather than an increase in profits. The break-even occupancy rate is thus 60% – 14.9%, or 45.1%.

The same technique can be used to investigate the variability in prices, as well as quantities. The nightly charge originally assumed for the caravan site was £4. Each £1 change in the charge is worth £3600 (£14 400 ÷ 4). The change needed to merely break even is thus a fall of £3569 ÷ £3600 = £0.99 per night. The break-even charge would thus be £4 – £0.99 = £3.01 per night.

The break-even point gives the manager a reference against which to measure risk. Adam Haynes can, for instance, look at the break-even occupancy rate of 45% and ask himself how likely it is that actual yields will drop that far. He may decide that such an event is unlikely, and so be reassured that the risk arising from variation in occupancy rates is small. Similarly, he may look at the break-even nightly charge and ask himself how likely it is that the price will have to drop that far. The way in which he reacts will depend to some extent on his character, since some people are more nervous of taking risks than others.

The break-even budget has a valuable role as a quick rule-of-thumb measure of the risk involved in a given plan. It also has the advantage that its implications can be readily understood. It has a major flaw, however, in that it can only be used to investigate variability in one element of a budget at a time; all the other possible varieties are assumed to stay constant. In the Home Farm example, for instance, we can look at the effect of changes in the occupancy rate or the nightly charge, but not the effect of both changing at the same time.

### 19.2.3 Flexible budgeting

*Flexible budgeting* takes all variables into account by the preparation of more than one budget for a particular plan. Budgets such as those for Home Farm in Part 2 are based upon the manager's expectation of the most likely occurrences. Flexible budgeting would involve preparation of at least one other version, based upon his feelings of what the worst possible outcome would be (the 'downside risk').

This again gives the manager a set of reference points, enabling him to see the possible range of financial results arising from his plans. It can be used with partial and whole-business budgets, for profit, cash flow and capital. Its particular value is that it forces the manager to consider explicitly the effects of a bad year or years, and thus brings home the financial consequences of risk. It may discourage him from pursuing some of his wilder schemes, or at least encourage him to prepare contingency plans in case the worst happens.

The biggest limitation of this approach is the amount of work involved, although it need not be used on every budget, and the use of partial budgets around a 'central' budget can reduce the effort (see Chapter 9). Use of computerized budgeting techniques, particularly spreadsheets, virtually eliminates this limitation, and makes this a powerful planning tool (see Appendix A).

### 19.2.4 Sensitivity analysis

*Sensitivity analysis* measures the responsiveness, or 'sensitivity', of a margin to changes in one or more of the elements comprising that margin. It can be characterized as a question: 'What changes in *x* will result from a given change in the level of *y*?', where *x* is a measure of profit, cash flow or capital and *y* is one of the components of that measure, such as fertilizer price, labour use, crop yield, interest rate, and so on. The technique is illustrated in Figure 19.1, using the normalized partial budget for the Home Farm caravan site (Figure 9.2). The most significant elements of the budget are identified, and estimates made of their likely levels under the best and worst assumptions. The figures in the right-hand column show which of the elements cause the most variation in the balance of the budget[25].

| *Variable* | | *Level* | *Effect on extra profit after interest* | *Range (£)* |
|---|---|---|---|---|
| Occupancy rate (%) | B | 70 | +2400 | |
| | M | 60 | — | 9600 |
| | W | 40 | –4800 | |
| Site charges (£ per night) | B | 4.5 | +1800 | |
| | M | 4 | — | 5400 |
| | W | 3 | –3600 | |
| Running costs (£) | B | 3000 | +300 | |
| | M | 3300 | — | 1500 |
| | W | 4500 | –1200 | |
| Fixed investment costs (£) | B | 25000 | +1250 | |
| | M | 30000 | — | 2500 |
| | W | 35000 | –1250 | |
| Barley output (£) | B | 360 | +50 | |
| | M | 410 | — | 100 |
| | W | 460 | –50 | |

B = Best likely, M = Most likely, W = Worst likely.

**Fig. 19.1** Sensitivity analysis.

The analysis will give an impression of the riskiness of the plan but, perhaps more important, will also show the elements to which the balance is most sensitive. In the example, these are occupancy rate and nightly charge. This is useful in two ways. First, the manager can see clearly which items need closest monitoring during the course of the plan, to ensure that the hoped-for results materialize. Second, the most variable items can be combined into two-way and three-way analyses in order to show the likely effect of simultaneous variations in more than one variable.

The two-way analysis in Figure 19.2 shows the effect of simultaneous variation in occupancy rate and nightly charge, using the same data as for Figure 19.1. The manager can immediately see the potential outcome of his worst fears, and his best hopes, being realized.

| Occupancy rates (%) | 40 | 50 | 60 | 70 |
|---|---|---|---|---|
| Site charges (per night) | | Extra profit after interest (£) | | |
| £4.50 | –32 | 2668 | 5368 | 8068 |
| £4.00 | –1232 | 1168 | 3568 | 5968 |
| £3.00 | –3632 | –1832 | –32 | 1768 |

**Fig. 19.2** Two-way variance table – example.

It is possible to take the process further, and build up tables showing three-way or even four-way analysis. Such tables tend to be cumbersome and difficult to interpret, and thus much of the benefit is lost. Use of a *tree diagram* can overcome some of the problems by showing the analysis in an easily assimilable, pictorial form. An example of a tree diagram is shown in Figure 19.3. This shows a three-way sensitivity analysis, incorporating variation in initial investment.

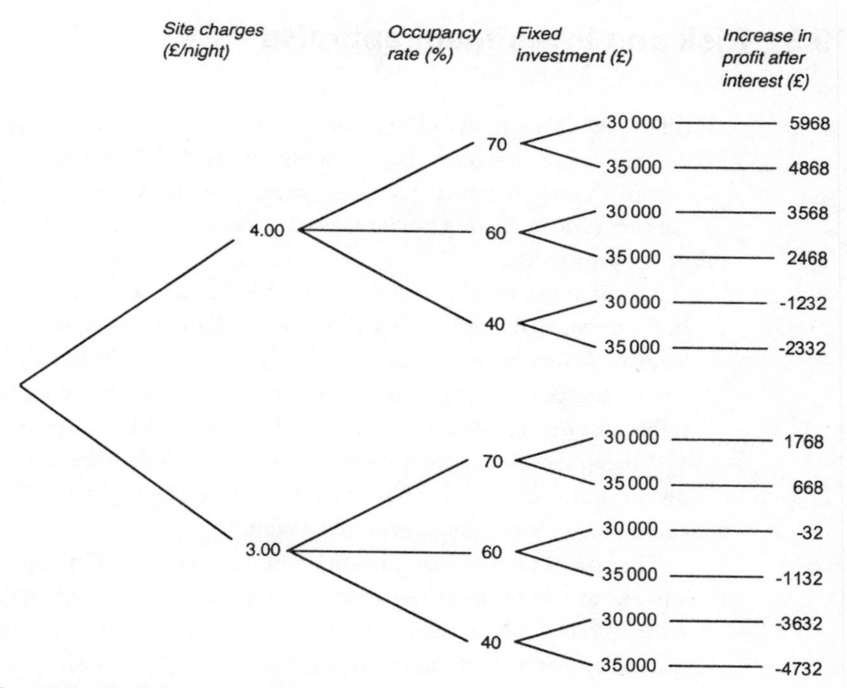


**Fig. 19.3** Tree diagram – example.

Like all management techniques, sensitivity analysis has its limitations. It gives an impression of the effects of risk, without attempting to quantify those effects. It is limited to the investigation of simultaneous variation in three or four assumptions only, if the results are to be comprehensible. It does not include the effect of alternative decisions which could be made by the manager of a business, but examines the outcome of only one decision at a time. Problems can arise in a two-way or three-way analysis in cases where the variables are independent.

On the other hand, it is a technique which allows the manager to examine the likely effects of his worst, best, and most likely assumptions on the outcome of a budget. It is a highly flexible device, and can be used in relation to any financial measure, including gross margins, profit, cash flow, bank balance, net capital, and rate of return on capital.

Techniques are available which either extend the capability of sensitivity analysis or provide a more sophisticated alternative. These techniques include probability analysis, decision trees, game theory, and various types of computer-based simulation models. Further information can be found in the references at the end of the chapter.

The development of cheap computerized budgeting techniques – particularly spreadsheets – has radically transformed the ease and effectiveness of use of sensitivity analysis during the last decade. Once a budget is compiled on the computer, key figures can be changed at the press of a button, and outcomes calculated within seconds. Variance tables and tree diagrams have their place even here in allowing the clear presentation of the results to the decision-maker.

## 19.3 Risk and investment appraisal

All the above methods can be incorporated into the investment appraisal procedures described in the previous chapter. Sensitivity analysis, for instance, can be applied to net cash flows, marginal tax rates, initial investment, cost of capital, and so on. Conservative budgeting can be applied to financial forecasts, as well as to the length of the project life.

An additional means of allowing for risk is to add a margin to the cost of capital. If, for instance, the only comparison for an investment project was with the return from a safe investment (in, say, a building society) at 10% per annum, several extra percentage points should be added to this rate before its use as a comparison. Otherwise there is danger of a highly risky project with an expected rate of return of 11% being accepted, even though the safe 10% is likely to be more acceptable to the owner in the long term. Thus, the comparison rate might be 15%, or even 20% to compensate for the differences in riskiness.

The size of the risk margin varies with the riskiness of the project, and with the attitude to risk of the investor. It is subjectively determined, and is thus prone to error. While more precise methods of allowing for risk are normally to be preferred (particularly if a computer is being used for calculations), this is nevertheless a convenient and neat means of incorporating an allowance for risk into investment appraisal.

## Notes

24. Risk is the effect of variations in the business and its environment whose occurrence can be predicted with some degree of probability. Uncertainty is the effect of variations whose likelihood of occurrence is entirely unknown. In practice it is difficult (and usually

pointless) to distinguish between the two, and the terms are treated here as interchangeable.

25. Figures 19.1 and 19.2 allow for changes in interest arising from changes in fixed capital but, in the interests of clarity, not working capital.

# References

## *Risk and uncertainty in planning*

Barnard, C.S. & Nix, J.S. (1979) *Farm Planning and Control*, 2nd edn. Cambridge University Press, London.

Hardaker, B., Anderson, J.R. & Huirne, R.B.M. (1997) *Coping with Risk in Agriculture*. CAB International, Wallingford.

McCrimmon, K.R. & Wehrung, D.A. (1986) *Taking Risks*. Collier MacMillan, London.

Moore, P.G. & Thomas, H. (1976) *The Anatomy of Decisions*. Penguin, Harmondsworth.

Sadgrove, K. (1996) *The Complete Guide to Business Risk*. Gower, London.

## *Risk and investment appraisal*

Attrill, P. & McLaney, E. (1994) *Management Accounting: an active learning approach*. Blackwell Business, Oxford.

Bull, R.J. (1990) *Accounting in Business*, 6th edn. London, Butterworth Heinemann, London.

Hull, J.C. (1980) *The Evaluation of Risk in Business Investment*. Pergamon, Oxford.

Lumby, S. (1994) *Investment Appraisal and Financing Decisions*, 5th edn. Chapman & Hall, London.

# Chapter 20

# Allowing for Inflation

## 20.1 Introduction

The term 'inflation' refers to a rise in the level of prices in *general*. It may alternatively be considered as a fall in the purchasing power of the currency of an economy. The usual measure of inflation in the United Kingdom is the rise in the Retail Price Index, which shows the change in cost of a 'typical' household's purchases. Ideally, this general rise in prices should be distinguished from *specific* price movements; changes in relative prices of individual types of goods and services in response to changes in supply and demand, government policy, and so on. In practice, it is often difficult to make this distinction.

The aim of this chapter is to outline the main effects of inflation on the financial management of a business, and to suggest ways in which allowances may be made for these effects.

## 20.2 Inflation and financial accounts

Financial accounts are those which are required by law from every business. They are prepared, usually by a professional accountant, in accordance with various conventions. One of these is the 'historic cost' convention; in general, assets are valued on the basis of their cost when originally purchased. In times of inflation, the use of this convention inevitably creates distortions in the 'message' imparted by the accounts. Among these distortions are that the values of physical assets are understated; the decline in purchasing power of financial assets (mainly cash and debtors) and liabilities is ignored; and depreciation allowances, being calculated on the basis of understated asset values, are insufficient to finance the replacement of wasting assets.

When inflation is high, these distortions can affect the usefulness of the financial accounts in decision-making. The principal danger is that 'paper' increases in profit, caused by inflation, encourages corresponding increases in cash withdrawals, whether for taxation, personal drawings or new investments. The consequence is likely to be a decline in liquidity (and possibly an increase in gearing) since the inflationary increase in profit is needed to pay for increases in input prices over the next year. If it has been spent elsewhere, cash will have to be obtained from another source in order to maintain production. This 'other source' is likely to be some form of borrowing. Where inflation is sufficiently high to cause significant problems of this sort, the most sensible approach for the typical business is to rely on management accounts for decision-making, and use financial accounts solely for taxation purposes.

In the late 1970s and early 1980s, when the inflation rate in the UK was always in the teens, and at times over 20%, much consideration was given to establishing a form of accounting which would correct the resulting distortions in financial accounts, giving a more 'accurate' picture of financial performance (see the References at the end of the chapter). The convention adopted, *current cost accounting* (CCA), was imposed on public companies for a three-year trial period from 1981. The CCA seeks to eliminate the effects of inflation-induced increases in stock values, to adjust depreciation calculations so that the annual charge is based on the cost of replacing assets, and to make corresponding adjustments to the value of physical assets in the balance sheet. It also attempts to take account of the reduction in the value of liabilities and of financial assets.

For various reasons, including falling inflation rates, difficulty of comprehension by shareholders, and distortions created by CCA in trying to correct those of inflation, companies were allowed to revert to reliance on the historic cost convention at the end of the trial period. The CCA has quietly fallen out of use in the UK, but it or its main rival, the *current purchasing power* (CPP) method may still be found in other countries with high inflation rates.

## 20.3 Inflation and management accounts

### *20.3.1 Alternative approaches*

The difficulties caused by high inflation for the use of management accounts are the same as those noted in respect of financial accounts; profit becomes an unreliable guide to the availability of cash, and the accuracy of both profit and capital statements as measures of performance is reduced.

The first problem is overcome easily by recognizing the inherent limitations of profit as an indicator of disposable income at the best of times, let alone in an inflationary period. The obvious alternative is to rely on scrupulous attention to cash flow budgeting. Ideally a five-year rolling cash flow budget should be used with, for instance, the first year budgeted on a monthly basis, the second quarterly, and the remainder in outline only.

Being more detailed and, in many respects, more accurate than financial accounts, management accounts allow the effects of inflation to be more easily seen and, therefore, to be distinguished from the effects of changes in the performance of the business. In using management accounts in times of inflation, two alternative courses of action are open – allow for inflation only in the *interpretation* of the accounts, or adjust for inflation in their *calculation*.

For convenience of analysis, the latter may be preferable, with CCA being the best documented of the various methods (see References ). On the other hand, the majority of managers find the preparation of management accounts in historic form a sufficient burden, without adding the complexity of inflation accounting. Even if a computer is available to perform the calculations, there is a risk of error in the interpretation of the adjusted accounts, due to insufficient understanding on the manager's part of the operations that have been performed on the data.

For these reasons this section concentrates on the first of the two alternatives, which involves preparation of accounts in historic cost form, i.e. as described throughout this book, and allowing for the effects of inflation within the interpretation process. Thus depreciation is based on the historic cost of the asset concerned, and the values of stocks reflect their market value at the time of their valuation.

### 20.3.2 The balance sheet

In interpretation of opening and closing balance sheets, a 'commonsense' revaluation of physical assets will allow ratio analysis to proceed in the usual way. This revaluation could rely entirely on the manager's own knowledge of current market values, but reference to journals, or the making of a few phone calls, e.g. to a local estate agent for an indication of land values, may be necessary. Where a machinery complement is large and varied, it will probably be sufficient to determine the approximate percentage difference between written-down value (WDV) and sale value of a few representative machines, and apply this percentage to the WDV of the complement as a whole.

This may seem a crude technique, but it is similar to the procedure used by a bank manager when assessing the credit-worthiness of a farming client. All the clearing banks use forms called 'farmers' balance sheets' or 'confidential statements' on which the manager enters the current market value of assets and liabilities. These values are determined in discussion with the farmer.

If this method is good enough for the critical gaze of a bank manager, it should be sufficient for the business owner's needs. It is an *impression* of stability that is sought, after all – measures such as the current, solvency and gearing ratios are, in most rural businesses, relatively insensitive to small errors in valuation.

To avoid confusion, these adjustments should normally be made after the balance sheet has been reconciled with the profit and loss account. If assets are revalued *before* reconciliation, a *holding gain* equivalent to the amount by which the asset has been upvalued should be credited to the capital account. This enables reconciliation of the revalued account, and at the same time allows the capital account to indicate the source of that part of the increase in net capital.

In flow of funds analysis, inflationary increases in asset values should be balanced by an extra 'use' category labelled 'holding gain'. As well as ensuring that the calculation works out, this provides an indication of the impact of inflation on the business. If desired, it can be subdivided to show separately the gains due to revaluation of stocks, machinery, and property.

### 20.3.3 *Enterprise accounts*

Analysis of full cost or gross margin accounts relies on three types of comparison; with previous years' accounts for the same business; with published standard or survey data; and with budgets for the current year (see Chapter 14). Of these, the first is the worst affected by inflation, since discrepancies in costs and returns between years reflect not only differences in management, climate, specific price levels, etc., but also the general upward movement of prices. The most sophisticated of inflation accounting methods cannot avoid this problem. Some value may be derived from a comparison of physical performance between the years, but financial comparisons are of dubious worth.

Comparisons with survey or standard data are less afflicted in this respect, since, in theory, the comparison figures should be exactly contemporary with the original account. In practice this is difficult, if not impossible, to achieve, but any resulting errors are likely to be relatively small. It is important that both account and comparison are calculated on the same basis; some FBS regions make adjustments for inflation in their publications, especially with respect to machinery depreciation and breeding livestock. Where this is the case, similar adjustments must be made in the account to be analysed if the comparison is to have any meaning.

The third form of comparison, that with the original budgets, is at any time the most useful of the three. It is the least affected by the presence of inflation, since general price rises may be built into the budget in the first place. It may turn out that the actual inflation rate is different from that expected, but such differences will usually be small, and their impact can be minimized by frequent monitoring and rebudgeting where necessary. As long as the account is calculated on the same basis as the budget, no adjustments are necessary.

The methods outlined above lack the precision of inflation accounting methods such as CCA, but will be sufficient for all normal purposes. Since the onus for allowing for inflation is thrown on to the manager's judgement, he must be sure to incorporate adequate detail in enterprise accounts, and to study their make-up with particular care. It is not enough, for instance, for him to register that the dairy gross margin is £600; he must be aware that £100 of that gross margin arose from an inflation-induced increase in breeding stock values. Moreover, distortion of the financial values renders physical data all the more important, and every effort is needed to ensure that details of quantities and qualities of inputs and outputs are recorded with precision.

The use of variance analysis with account comparisons (Chapter 14) enables isolation of the influence of price changes, as opposed to physical variations. The manager may then use his judgement as to how much of the price difference is due to specific price changes rather than inflation.

## 20.4 Inflation and planning

The comments made with respect to management accounting also apply to the preparation of budgets. In most circumstances the sensible course is to avoid specific adjustments for inflation, letting closing stock values reflect their forecast net realizable value at the time of valuation. Planning techniques other than budgeting, such as gross margin planning and linear programming, tend to rely on normalized data, and are thus largely unaffected by problems of increases in stock valuation, understated depreciation charges, etc.

A problem which is peculiar to planning techniques is the increased difficulty of forecasting prices. Having to predict specific price changes is bad enough, but in times of inflation it is necessary to forecast general price rises as well. An associated difficulty is the uncertainty of timing of price increases. The prices of many goods and services move in an irregular fashion; suppliers will hold prices constant for several months and then increase them by an amount that reflects both past and expected levels of inflation. With the majority of sales and purchases, the effects are relatively insignificant, and tend to balance each other out. With large purchases, such as machinery, the accuracy with which the timing of price rises is predicted can be critical.

The consequence of both effects is that precision in planning becomes even more difficult to achieve. Cash reserves are particularly susceptible to this loss of precision. Take the example of a farmer who, in a given year, buys store cattle at £350 per head, and sells them six months later for £450 per head. If, the next year, he has to pay £420 per head, and receives £540 at sale, he will probably feel that the higher sale price is adequate compensation for the higher purchase price. To finance this operation, however, he will need another £70 cash per head over six months (and this is ignoring increases in other costs, such as feed). If his budgets are based on last year's prices, he may find himself very short of cash before the six months are up. It is even possible that he will have to sell some of the cattle before they are finished, at little or no profit, in order to raise cash for essentials such as rent or living expenses.

To avoid such eventualities, even more energy than usual should be devoted to forecasting prices for the year ahead. In addition, flexible budgeting and sensitivity analysis can be used to test the effect of different inflation rates and timings of price rises. Budgetary control measures should be reviewed, with the possibility of increasing the frequency of budget/actual checks. In this way, the effects of any unforeseen fluctuations in the inflation rate may be spotted at the earliest possible stage, and remedial action taken.

## 20.5 Inflation and investment appraisal

The presence of inflation can distort the message given by the calculation of return on capital of investment projects. As an example, take the owner who calculates that a certain safe investment is likely to give him a return on capital of 15% per annum. In an inflation-free economy, such a return would indicate a very worthwhile project. If, on the other hand, this calculation relates to a period with an inflation rate of 20% per annum, the project looks less attractive. The returns from the project are not sufficient to keep pace with inflation; in fact the project will lose money in 'real' terms.

The *real rate of return* is an 'inflation-free' rate; it can be estimated by deducting the annual inflation rate from the nominal rate of return that we have used up to now[26]. The real rate of return corresponding to a nominal rate of 15% and an inflation rate of 20% is thus about –5%.

The estimate of the annual inflation rate is based usually on expectations of changes in the retail price index, although this may be an imperfect guide to the decline in purchasing power of money owned by someone whose spending pattern differs markedly from the 'typical' pattern assumed by the index.

When an investment project is tested by the techniques described in Chapter 18, its annual percentage rate of return is compared to the cost of the capital invested. It matters little whether this comparison is in real or nominal terms, as long as both parts of the comparison are expressed in the same currency. The most common error in investment appraisal is to calculate the rate of return in real terms, i.e. building no inflation into cash flow or profit estimates, testing against a cost of capital expressed in nominal terms. The result is the rejection of many projects which would in fact have been well worthwhile.

Thus if no inflation is built into the basic calculation, the result should be compared with the cost of capital in real terms. In many ways this is the simplest solution to the inflation problem, as long as it can be assumed that the values of all inputs and outputs will rise in price at the same constant rate over the project life, and that the rate is identical to the investor's 'personal' inflation rate. If these conditions do not apply (for instance, if costs are likely to rise faster than prices of outputs – a common story in agriculture), the simplicity is totally lost in attempts to adjust cash flows for differential price movements in real terms. In such cases it is more appropriate to use nominal values throughout, building anticipated price increases into cash flows, and discounting by the nominal (i.e. inflation-boosted) cost of capital. This has the further advantage of making the result easier to understand: if the appraisal is carried out on behalf of, rather than by the owner, the use of a real rate of return complicates a communication process which is already fraught with problems, especially if discounted cash flow techniques are being used.

# Notes

26. More precisely, real rate of return is $1 - [(1 - r) \div (1 - I)]$, where $r$ is the monetary rate of return and $I$ the inflation rate, both expressed as fractions.

# References

## General

Bull, R.J. (1990) *Accounting in Business*, 6th edn. Butterworth Heinemann, London.
Kirkman, P.R.A. (1974) *Accounting under Inflationary Conditions*. Allen & Unwin, London.
Scapens, R.W. (1977) *Accounting in an Inflationary Environment*. Macmillan, London.

### Agricultural

Bright, G. (1996) An exploration of profit measurement. *Farm Management*, **9**, 8, pp. 383–391.

Lewis, R.W. & Jones, W.D. (1980) Current cost accounting and farm businesses. *Journal of Agricultural Economics*, **31**, 1.

### Investment appraisal

Lumby, S. (1994) *Investment Appraisal and Financing Decisions*. 5th edn. Chapman & Hall, London.

Mott, G. (1993) *Investment Appraisal*, 2nd edn. Pitman Publishing, London.

# Appendix A
# Personal Computers in Management

## A.1 Introduction

The 15 years since the publication of the first edition of this book have seen enormous advances in business computing, and what was then regarded as a luxury justified only on the largest farms is now well within the reach of all but the smallest. Microcomputers (now often referred to as *personal computers* or PCs) have increased dramatically in computing power while becoming cheaper in real terms, and it is now possible to buy a computer system suitable for most small businesses in virtually any high street electrical store. Software (the programs that make the computer equipment (hardware) do its stuff) has in turn become much more sophisticated and, best of all, much more 'user-friendly', so that an intelligent novice can tap the power of a computer without being daunted by programming languages, codes, and so on.

While in the early days of microcomputers each make of machine had its own operating system, making transfer of software between machines difficult (if not impossible), there are now just two main types of operating system: the IBM-compatible system (now mostly using Windows and/or MS-DOS) and the Apple Macintosh system. The latter is an extremely easy-to-use system, very good for producing high quality print and graphics as well as performing all the basic computer functions. It has been slower to catch on in business circles than the IBM standard, however, with the result that Windows/MS-DOS machines tend to dominate this market. Moreover the large volume and more open marketing approach of IBM have ensured that costs are generally lower for IBM-type machines and software. All the major UK farm business software is designed for use with Windows/MS-DOS computers.

## A.2 Computers and farming

Any person involved in financial management to the extent implied by this book should seriously consider enlisting the aid of a computer in one form or another. Not to do so is to risk spending valuable time in performing mechanical calculations that could be better used for thinking, or worse still, being ill-informed for want of time to do those calculations. The competitive position of the non-computer-using business is being rapidly eroded in the increasingly competitive 1990s.

Computers can be used in various ways by a farm business. A computer bureau service can be used (offered by an accountant or specialist consultant): the farmer sends information on request to the bureau, where it is processed by the bureau's computer and the results sent back to the farmer. In this way management accounts for enterprises and complete businesses can be provided with little effort on the farmer's part, without the cost of ownership of a computer and without the need for anyone in the farm business having to come to terms with computer operation.

On the other hand, the fees are likely to be costly, and the business loses the flexibility of being able to process information when needed, and in the specific form it requires. An alternative is to purchase a complete business management 'package', including computer, software and a certain amount of introductory training and support. Several firms offer specialized farm accounting and management packages in the UK, typically based round a financial recording process (either cash analysis or double-entry). Once the income and expenditure has been entered for the period in question, reports can quickly be produced to show financial accounts (profit and loss accounts, balance sheet, flow of funds), gross margins, cash flows and detailed analysis of overhead costs. It is usually possible to produce budgets for the coming year, and it may be possible to link this directly with the accounts to enable the computer to perform comparisons of actual and budgeted performance, including variance analysis. Additional facilities may include specialist physical recording and management programs, such as records of a pig fattening herd, records of dairy herds, breeding and productive efficiency and arable crop recording with field mapping. The computer will also be able to cope with the standard office tasks such as word-processing, described below.

Modern business packages are easy to use and once set up can save an enormous amount of repetitive work while producing a precision of management reporting that would otherwise be impossible. One deterrent for the smaller farm is likely to be the cost of the package, though. At the time of writing, a very basic specialist farm system is likely to cost at least £3000. Not much compared to a tractor but still, even after tax relief (it is a business asset), enough to make the owner of a family farm think twice. Costs can be substantially reduced by using standard business (i.e. non-agricultural) software, for accounts and payroll calculations, for instance, though such software may need to be adapted to the peculiarities of a farm business.

A related problem is that the likely financial benefit can be difficult to assess and farmers are understandably reluctant to take the advantages on trust. Another, often stronger deterrent is the unjustified mystique that has built up around computers: many potential users are frightened of approaching a computer in case its 'cleverness' will show up their 'ignorance'. Help is available to counter both problems. The excellent article by Morley (1989), though dated, gives very clear guidance to a prospective user: ADAS has specialists who advise and publish useful booklets; and ATB LandBase, local agricultural colleges and other organizations run courses in most areas on using computers on the farm.

## A.3 General business applications

One way of reducing the cost, and thus the risk, of buying a computer is to start with a cheap, standard office machine combined with off-the-shelf software applications. These will typically include a *word processor*, a program that allows the computer to act as a superior sort of typewriter, able to store the typing in its memory so that it can be recalled for alteration and reuse at a later date. Automatic spelling checkers, and the facility to correct typing on the computer screen before printing, make it possible for even the most ham-fisted of typists to produce a professional result.

More relevant to this book, perhaps, is the *spreadsheet* software, allowing calculations to be set out on the computer screen in such a way that the figures are recalculated automatically when any element of the calculation is changed. In a cash flow budget, for instance, changing the interest rate or one of the revenue items would result in the whole of the budget being recalculated in an instant. A graph can be produced of the final bank balance (see Figure 7.7, for instance) which will be redrawn automatically every time an alteration is made in the budget.

Every table in this book was produced using such a spreadsheet program. The management accounting system of one of the largest farming companies in southwest England was built up from small beginnings using a standard spreadsheet package. The modern spreadsheet is a most powerful and flexible management tool. Moreover, because it works in a very obvious manner, it is extremely easy for even a novice to use. This appendix contains a number of sample spreadsheets intended to provide a starting-point for the farm-based user (see Figures A.1–A.11).

Another standard type of software application is the *database*, available at many levels of sophistication but basically a way of storing records in a form that enables them to be easily retrieved and the information sorted. For instance, if a database is used to store information about the cows in a dairy herd, the program can then be subsequently used to identify all the cows giving less than 5000 litres of milk per year, all those put to a certain bull, all those with a history of mastitis, and so on. Databases can be used for field records, accounting records and machinery records, and they can also be used in combination with a word processor to produce a whole series of identical letters, personally addressed to, say, potential customers for the new organic beef enterprise.

Programs are available which combine word processor, spreadsheet and database in one *integrated software* package. Popular examples at the time of writing include *Microsoft MSWorks*™ and *ClarisWorks*™. Such packages allow easy transfer of information between the component parts, and allow the acquisition of all three functions (and a few others besides) for a relatively low cost. This is an ideal place for a low budget user to start, allowing progression on to more specialized applications as he grows in confidence, and possibly eventually making the transition to a full-blown agricultural management package.

An application that has dramatically increased in popularity in the 1990s is the facility for communicating with other computers. By connecting a device called a 'modem' to a PC it is possible to transmit data from that computer across telephone lines to other computers. By paying an annual fee to an agency or service provider, the user of the computer can connect it to a network of other computers, and send information to one or all of those computers. If the service provider is within the 'local call' area of the user, all calls to the network are at local rates, the service provider carrying the onward costs within their monthly provision charge.

The network may be a closed one, ensuring that information is shared only between specified machines. It may, however, be connected to a vast array of interlinked networks across the world – the *Internet.* This allows the computer user to send messages to other users, wherever they are, for the cost of a local telephone call: electronic mail or 'e-mail'.

A further development has been that of devices to make use of the Internet easier, such as the 'World-Wide Web', or WWW, which provides a user-friendly method of linking the user's PC with other computers dedicated as 'servers': repositories of information which can be interrogated using special computer applications designed to make the job as easy as possible. Servers can be maintained by individuals, or public bodies responsible for provision of information, or commercial firms wanting to advertise their wares

A major potential problem with this vast, unstructured repository of information is locating specific and relevant information quickly and easily. The WWW provides search facilities within the system which allow searching by words or phrases (two examples are Netscape Navigator and Altavista). Pages are often linked by 'highlighted' words (picked out in colour on the screen): selecting a word will divert the user to other sources of associated information.

The WWW has various other features to help the user, including 'bookmarks' which make it easy to return to a particularly useful 'page'. Another possibility is the use of an electronic 'bulletin board', allowing messages to be exchanged in a very active way between those with access to the board – rather like sticking notices on a pinboard, some of which may be pure information, but others being responses to earlier statements. This creates a form of electronic debate or conference without the contributors needing to be physically present in the same room. 'Video-conferencing' goes one step further than this, with small, cheap video cameras (£100) connected to the computers allowing the participants to see and hear each other. As yet the quality of some telephone lines makes this difficult, but developments in land-line and satellite communications may soon solve that problem.

Harrison and Williams (1996) have given a comprehensive overview of the Internet in an agricultural context, and there is a vast array of books on using the Internet in general terms. Like most such innovations, agriculture has been slower to adopt it than other industries (Warren *et al.* 1996), but is now making up for lost time, with all the major companies, universities, and government departments having their 'Web sites'. Acquiring MAFF statistics, for instance, has never been easier.

On the negative side, using the Internet can be frustratingly slow, depending partly on the amount of use of the Internet (throughout the world) at a particular time of day, and also on the quality of telephone cabling and exchanges in your area.

Also some of the most useful (and therefore commercially valuable) information is now being provided only to closed subscriber groups (such as the NFU/ADAS/CLA Rural Business Network in the UK, available only through the closed group 'Farming Online').

Nevertheless, this is probably the most significant and exciting development in management technology for small businesses in the last decade, and promises to help considerably in overcoming some of the problems arising from geographical isolation of the rural business and its manager.

## References

Damms, I.M. & Stone, M.A.H. (1995) The role of video and new communication technologies in agricultural communication and training. *Farm Management*, **9**, 3.

Harrison, S.R. & Course, C.P. (1985) Wordprocessing in the farm office. *Farm Management*, **5**, 9.

Harrison, S.R. & Williams, N.T. (1996) The Internet: a business necessity or time-consuming curiosity. *Farm Management*, **9**, 4.

Houseman, I. (1990) Communications in agriculture: the role of information technology. *Farm Management*, **7**, 7.

Morley, A. (1989) Practicalities of farm computing. *Farm Management*, **7**, 1.

Temple, M. (1986) Spreadsheet programs for farm management advisory work. *Farm Management*, **5**, 12.

Warren, M.F., Soffe, R.J., Stone, M.A., Mackwood, B. and Walbank, M. (1996) The uptake of new communications technologies in farm management: a case study from the south west of England. *Farm Management*, **9**, 7.

Williams, N.T. (1986) Using database management systems in farm management. *Farm Management*, **6**, 4.

| | A | B | C | D | E | F | G | H | I | J |
|---|---|---|---|---|---|---|---|---|---|---|
| 1 | ENTERPRISE dairy | | | | ha or head? 80 | | YEAR 199X-9Y | | | |
| 2 | | | | | | | VERSION 2 | | | |
| 3 | products>> | milk | | calves | | culls | | | | |
| 4 | | qty. | £ | qty. | £ | qty. | £ | qty. | £ | |
| 5 | cash received | 400500 | 76095 | 74 | 7400 | 20 | 11000 | | | |
| 6 | closing debtors | 11000 | 1980 | 7 | 700 | | | | | |
| 7 | -opening debtors | –12000 | –2280 | –5 | –500 | | | | | |
| 8 | sales | =SUM(B5:B7) | =SUM(C5:C7) | =SUM(D5:D7) | =SUM(E5:E7) | =SUM(F5:F7) | =SUM(G5:G7) | =SUM(H5:H7) | =SUM(I5:I7) | |
| 9 | transfer out | | | | | | | | | |
| 10 | benefits inkind | 700 | 133 | | | | | | | |
| 11 | closing valuation | | | 10 | 800 | | | | | |
| 12 | -opening valuation | | | –10 | –800 | | | | | |
| 13 | OUTPUT | =SUM(B8:B12) | =SUM(C8:C12) | =SUM(D8:D12) | =SUM(E8:E12) | =SUM(F8:F12) | =SUM(G8:G12) | =SUM(H8:H12) | =SUM(I8:I12) | |
| 14 | | | | | | | | | | |
| 15 | inputs>> | purch. livestock | | conc. feed | | straw | | vet/med | | |
| 16 | | qty. | £ | qty. | £ | qty. | £ | qty. | £ | |
| 17 | cash paid | 25 | 17500 | 86 | 12900 | | | 1 | 2500 | |
| 18 | closing creditors | | | 20 | 3000 | | | | | |
| 19 | '-opening creditors' | | | –25 | –4000 | | | | | |
| 20 | purchases | =SUM(B17:B19) | =SUM(C17:C19) | =SUM(D17:D19) | =SUM(E17:E19) | =SUM(F17:F19) | =SUM(G17:G19) | =SUM(H17:H19) | =SUM(I17:I19) | |
| 21 | transfers in | | | 42 | 3780 | 26 | 390 | | | |
| 22 | '-benefits in kind' | | | | | | | | | |
| 23 | opening valuation | | | 20 | 2800 | | | | | |
| 24 | -closing valuation | | | –20 | –2800 | | | | | |
| 25 | VARIABLE COST | =SUM(B20:B24) | =SUM(C20:C24) | =SUM(D20:D24) | =SUM(E20:E24) | =SUM(F20:F24) | =SUM(G20:G24) | =SUM(H20:H24) | =SUM(I20:I24) | |
| 26 | | | | | | | | | | |
| 27 | MISCELLANEOUS VALUATION CHANGES | | | | breeding livestock | | | crops in gd. | | total |
| 28 | | | | | | no. | £ | | £ | £ |
| 29 | closing valuation | | | | | 82 | 53300 | | | =G29+I29 |
| 30 | -opening valuation | | | | | –77 | –50050 | | | =G30+I30 |
| 31 | valuation change | | | | =SUM(E29:E30) | =SUM(F29:F30) | =SUM(G29:G30) | =SUM(H29:H30) | =SUM(I29:I30) | =SUM(J29:J30) |
| 32 | | | | | | | | | | |

**Fig. A.1** Spreadsheet formulas for gross margin worksheet (Figure 6.11).

| | A | B | C | D | E | F | G | H | I | J |
|---|---|---|---|---|---|---|---|---|---|---|
| 33 | =B1 | | | GROSS MARGIN | | | | | | |
| 34 | head | TOTAL | | | Qty/hd | | £/UNIT QTY | | £ | £/hd |
| 35 | =F1 | QUANTITY | | | | | | | TOTAL | |
| 36 | =B3 | | =B13 | | =C36/(A35) | | =IF(C36=0,0,I36/C36) | | =C13 | =I36/A35 |
| 37 | =D3 | | =D13 | | =C37/(A35) | | =IF(C37=0,0,I37/C37) | | =E13 | =I37/A35 |
| 38 | =F3 | | =F13 | | =C38/(A35) | | =IF(C38=0,0,I38/C38) | | =G13 | =I38/A35 |
| 39 | =H3 | | =H13 | | =C39/(A35) | | =IF(C39=0,0,I39/C39) | | | =I39/A35 |
| 40 | misc.valn. changes | xxxxxxxxx | xxxxxxxxx | xxxxxxxxx | xxxxxxxxx | xxxxxxxxx | xxxxxxxxx | | =G31+I31 | =I40/A35 |
| 41 | OUTPUT | | | | | | | | =SUM(I36:I40) | =SUM(J36:J40) |
| 42 | | | | | | | | | | |
| 43 | VARIABLE | TOTAL | | | Qty/hd | | £/UNIT QTY | | £ | £/hd |
| 44 | COST | QUANTITY | | | | | | | TOTAL | |
| 45 | =B15 | | =B25 | | =C45/A35 | | =IF(C45=0,0,I45/C45) | | =C25 | =I45/A35 |
| 46 | =D15 | | =D25 | | =C46/A35 | | =IF(C46=0,0,I46/C46) | | =E25 | =I46/A35 |
| 47 | =F15 | | =F25 | | =C47/A35 | | =IF(C47=0,0,[illegible] | | [illegible] | [illegible]35 |
| 48 | =H15 | | =H25 | | =C48/A35 | | =IF(C48=0,0,[illegible] | | [illegible] | [illegible]35 |
| 49 | =J15 | | =J25 | | =C49/A35 | | =IF(C49=0,0,[illegible] | | [illegible] | [illegible]35 |
| 50 | =K15 | | =K25 | | =C50/A35 | | =IF(C50=0,0,[illegible] | | [illegible] | [illegible]35 |
| 51 | =L15 | | =L25 | | =C51/A35 | | =IF(C51=0,0,I51/C51) | | =#REF! | =I51/A35 |
| 52 | | | | | | | | | =#REF! | =I52/A35 |
| 53 | VARIABLE COST | | | | | | | | =SUM(I45:I52) | =SUM(J45:J52) |
| 54 | | | GROSS MARGIN | | | | | | =I41-I53 | =J41-J53 |

This formula means 'if C38 is 0, then enter 0 here- otherwise enter I38/C38'. It stops the program trying to divide by 0 (a logical impossibility)

**Fig. A.1 (cont'd)**

| | A | B | C | D | E |
|---|---|---|---|---|---|
| 1 | CASH FLOW BUDGET | | | YEAR | 199X/9Y |
| 2 | | | | | |
| 3 | % interest rate 15 | | TOTAL | Oct | Nov |
| 4 | | | | | |
| 5 | *RECEIPTS* | | | | |
| 6 | Milk | 1 | =SUM(D6:O6) | 2280 | 3800 |
| 7 | Calves | 2 | =SUM(D7:O7) | 500 | 3500 |
| 8 | Cull cows | 3 | =SUM(D8:O8) | | |
| 9 | | 4 | =SUM(D9:O9) | | |
| 10 | Sheep | 5 | =SUM(D10:O10) | | |
| 11 | | 6 | =SUM(D11:O11) | | |
| 12 | Pigs | 7 | =SUM(D12:O12) | 6300 | 6300 |
| 13 | | 8 | =SUM(D13:O13) | | |
| 14 | Poultry | 9 | =SUM(D14:O14) | | |
| 15 | | 10 | =SUM(D15:O15) | | |
| 16 | Wheat | 11 | =SUM(D16:O16) | | |
| 17 | Wheat straw | 12 | =SUM(D17:O17) | | |
| 18 | Oats | 13 | =SUM(D18:O18) | | |
| 19 | Apples | 14 | =SUM(D19:O19) | | |
| 20 | Potatoes | 15 | =SUM(D20:O20) | | |
| 21 | | 16 | =SUM(D21:O21) | | |
| 22 | | 17 | =SUM(D22:O22) | | |
| 23 | Sundries | 18 | =SUM(D23:O23) | | |
| 24 | VAT charged | 19 | =SUM(D24:O24) | | |
| 25 | VAT refunds | 20 | =SUM(D25:O25) | | |
| 26 | Capital - grants | 21 | =SUM(D26:O26) | | |
| 27 | machinery sales | 22 | =SUM(D27:O27) | | 4000 |
| 28 | | 23 | =SUM(D28:O28) | | |
| 29 | Personal receipts | 24 | =SUM(D29:O29) | | |
| 30 | TOTAL RECEIPTS | | =SUM(D30:O30) | =SUM(D6:D29) | =SUM(E6:E29) |
| 31 | | | | | |
| 32 | | | | =D3 | =E3 |
| 33 | *PAYMENTS* | | | | |
| 34 | Feed: cows | 25 | =SUM(D34:O34) | 3750 | |
| 35 | young stock | 26 | =SUM(D35:O35) | | |
| 36 | pigs | 27 | =SUM(D36:O36) | 2800 | 14000 |
| 37 | | 28 | =SUM(D37:O37) | | |
| 38 | Vet and Med. cows | 29 | =SUM(D38:O38) | 212 | 208 |
| 39 | Vet and Med. pigs | 30 | =SUM(D39:O39) | 100 | 100 |
| 40 | L/stock purch: cows | 31 | =SUM(D40:O40) | | |
| 41 | pigs | 32 | =SUM(D41:O41) | 2500 | 2500 |
| 53 | | | | | |
| 54 | | | | | |
| 55 | Overdraft interest | 46 | =SUM(D55:O55) | | |
| 56 | Loan interest | 47 | =SUM | | |
| 57 | Hire purchase | 48 | =SUM | | |
| 58 | Leasing charges | 49 | =SUM | | 525 |
| 59 | Overhead sundries | 50 | =SUM | | 1725 |
| 60 | | 51 | =SUM | | |
| 61 | Capital - orchard | 52 | =SUM | | 7000 |
| 62 | machinery | 53 | =SUM | | |
| 63 | capital repayment | 54 | =SUM(D63:O63) | 960 | 980 |
| 64 | Personal - drawings | 56 | =SUM(D64:O64) | | |
| 65 | tax | 57 | =SUM(D65:O65) | | |
| 66 | VAT paid | 58 | =SUM(D66:O66) | | |
| 67 | TOTAL PAYMENTS | | =SUM(D67:O67) | =SUM(D34:D66) | =SUM(E34:E66) |
| 68 | N. CASH FLOW | 59 | =C30-C67 | =D30-D67 | =E30-E67 |
| 69 | OPENING BALANCE | 60 | -21000 | =C69 | =D70 |
| 70 | CLOSING BALANCE | 61 | =C68+C69 | =D68+D69 | =E68+E69 |
| 71 | monthly interest | | =IF(C69<0,-C69*B3/1200,0) | =IF(D70<0,-D70*B3/1200,0) | =IF(E70<0,-E70*B3/1200,0) |
| 72 | | | | | |
| 73 | | | | | |
| 74 | | | | | |

This means: 'If the number in cell E70 is negative, the interest charge is -E70 multiplied by the interest rate (i.e. cell B3 divided by 100 to convert it to a decimal, and divided again by 12 to bring it to a monthly rate). If it is not negative, the interest charge is 0'

**Fig. A.2** Spreadsheet formulas for cash flow budget (Figure 7.4).

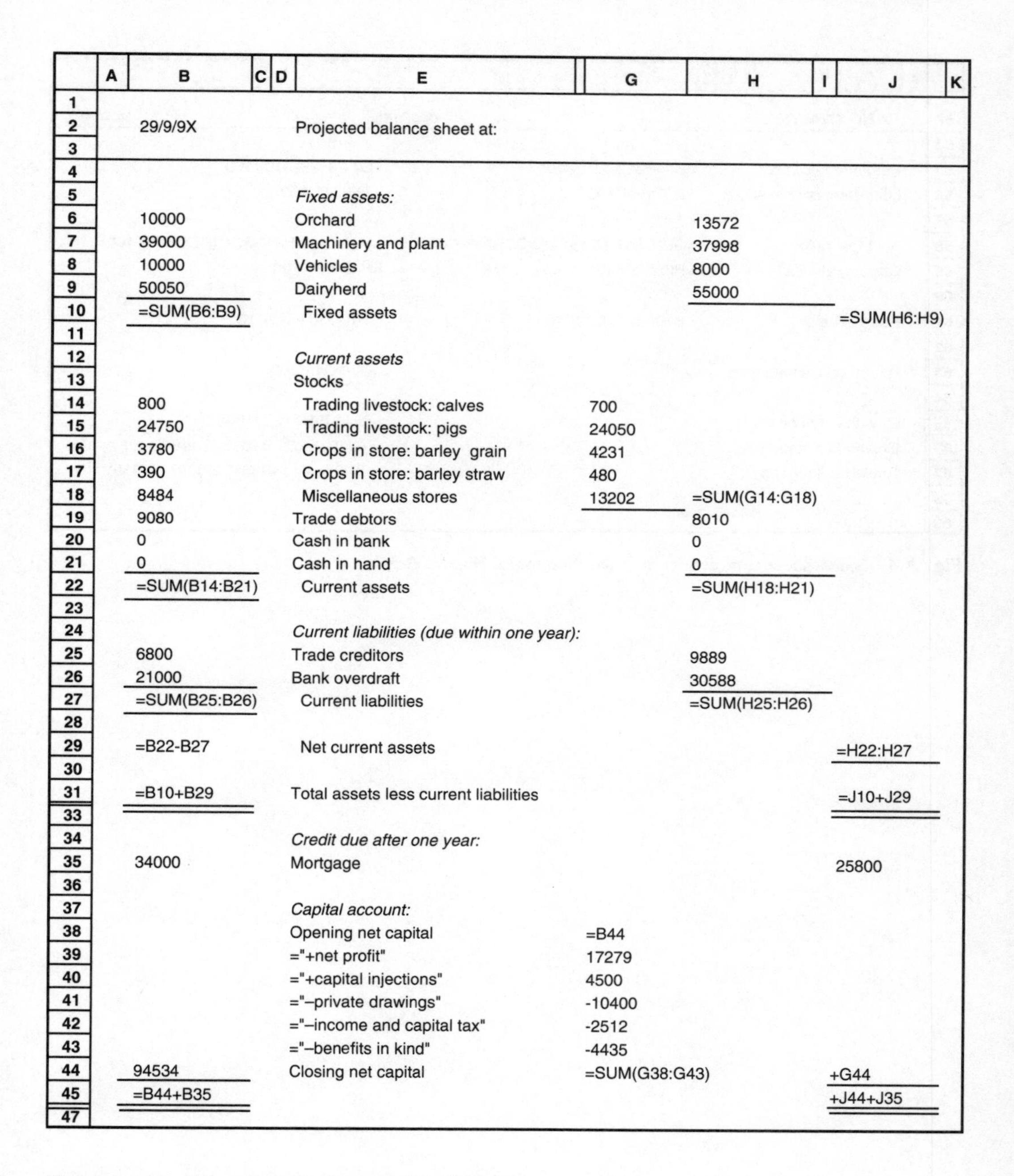

| | A | B | C | D | E | G | H | I | J | K |
|---|---|---|---|---|---|---|---|---|---|---|
| 1 | | | | | | | | | | |
| 2 | | 29/9/9X | | | Projected balance sheet at: | | | | | |
| 3 | | | | | | | | | | |
| 4 | | | | | | | | | | |
| 5 | | | | | *Fixed assets:* | | | | | |
| 6 | | 10000 | | | Orchard | | 13572 | | | |
| 7 | | 39000 | | | Machinery and plant | | 37998 | | | |
| 8 | | 10000 | | | Vehicles | | 8000 | | | |
| 9 | | 50050 | | | Dairyherd | | 55000 | | | |
| 10 | | =SUM(B6:B9) | | | Fixed assets | | | | =SUM(H6:H9) | |
| 11 | | | | | | | | | | |
| 12 | | | | | *Current assets* | | | | | |
| 13 | | | | | Stocks | | | | | |
| 14 | | 800 | | | Trading livestock: calves | 700 | | | | |
| 15 | | 24750 | | | Trading livestock: pigs | 24050 | | | | |
| 16 | | 3780 | | | Crops in store: barley grain | 4231 | | | | |
| 17 | | 390 | | | Crops in store: barley straw | 480 | | | | |
| 18 | | 8484 | | | Miscellaneous stores | 13202 | =SUM(G14:G18) | | | |
| 19 | | 9080 | | | Trade debtors | | 8010 | | | |
| 20 | | 0 | | | Cash in bank | | 0 | | | |
| 21 | | 0 | | | Cash in hand | | 0 | | | |
| 22 | | =SUM(B14:B21) | | | Current assets | | =SUM(H18:H21) | | | |
| 23 | | | | | | | | | | |
| 24 | | | | | *Current liabilities (due within one year):* | | | | | |
| 25 | | 6800 | | | Trade creditors | | 9889 | | | |
| 26 | | 21000 | | | Bank overdraft | | 30588 | | | |
| 27 | | =SUM(B25:B26) | | | Current liabilities | | =SUM(H25:H26) | | | |
| 28 | | | | | | | | | | |
| 29 | | =B22-B27 | | | Net current assets | | | | =H22:H27 | |
| 30 | | | | | | | | | | |
| 31 | | =B10+B29 | | | Total assets less current liabilities | | | | =J10+J29 | |
| 33 | | | | | | | | | | |
| 34 | | | | | *Credit due after one year:* | | | | | |
| 35 | | 34000 | | | Mortgage | | | | 25800 | |
| 36 | | | | | | | | | | |
| 37 | | | | | *Capital account:* | | | | | |
| 38 | | | | | Opening net capital | =B44 | | | | |
| 39 | | | | | ="+net profit" | 17279 | | | | |
| 40 | | | | | ="+capital injections" | 4500 | | | | |
| 41 | | | | | ="–private drawings" | -10400 | | | | |
| 42 | | | | | ="–income and capital tax" | -2512 | | | | |
| 43 | | | | | ="–benefits in kind" | -4435 | | | | |
| 44 | | 94534 | | | Closing net capital | =SUM(G38:G43) | | | +G44 | |
| 45 | | =B44+B35 | | | | | | | +J44+J35 | |
| 47 | | | | | | | | | | |

**Fig. A.3** Spreadsheet formulas for vertical balance sheet.

| | A | B | C | D | E | F | G |
|---|---|---|---|---|---|---|---|
| 53 | | *RATIO ANALYSIS* | | | | *opening* | *closing* |
| 54 | | | | | | | |
| 55 | | Percentage owned | | | | =B44/(B22+B10)*100 | =G44/(H22+J10)*100 |
| 56 | | Long-term debt to equity | | | | =B35/B44*100 | =J35/J44*100 |
| 57 | | | | | | | |
| 58 | | Acid-test ratio | | | | =SUM(B19:B21)/SUM(B25:B26)*100 | =SUM(H19:H21)/SUM(H25:H26)*100 |
| 59 | | Current ratio (%) | | | | =B22/B27*100 | =H22/H27*100 |
| 60 | | | | | | | |
| 61 | | Fixed assets | | | | =B10/(B10+B22)*100 | =J10/(J10+H22)*100 |
| 62 | | | | | | | |
| 63 | | Return on Owner equity | | | | | =G39/G44*100 |
| 64 | | | | | | | |
| 65 | | Growth in net capital | | | | | =(G44-B44)/B44*100 |
| 66 | | Growth in assets | | | | | =(((J10+H22)-(B10+B22))/(B22+B10))*100 |
| 67 | | Growth in liabilities | | | | | =(((H27+J35)-(B35+B27))/(B27+B35))*100 |
| 68 | | | | | | | |
| 69 | | | | | | | |

**Fig. A.4** Spreadsheet formulas for balance sheet ratios (Figure A.3).

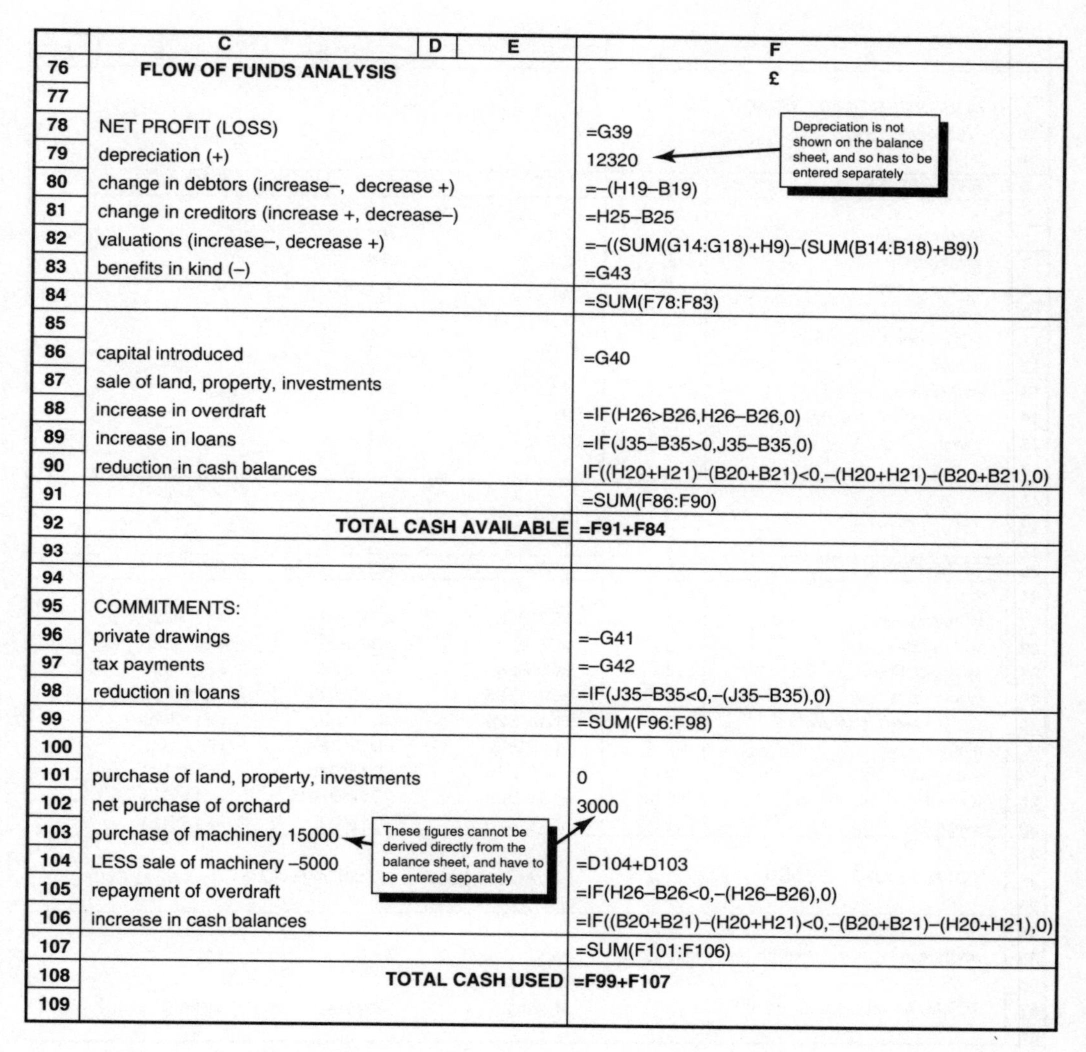

| | C | D | E | F |
|---|---|---|---|---|
| 76 | **FLOW OF FUNDS ANALYSIS** | | | £ |
| 77 | | | | |
| 78 | NET PROFIT (LOSS) | | | =G39 |
| 79 | depreciation (+) | | | 12320 |
| 80 | change in debtors (increase–, decrease +) | | | =–(H19–B19) |
| 81 | change in creditors (increase +, decrease–) | | | =H25–B25 |
| 82 | valuations (increase–, decrease +) | | | =–((SUM(G14:G18)+H9)–(SUM(B14:B18)+B9)) |
| 83 | benefits in kind (–) | | | =G43 |
| 84 | | | | =SUM(F78:F83) |
| 85 | | | | |
| 86 | capital introduced | | | =G40 |
| 87 | sale of land, property, investments | | | |
| 88 | increase in overdraft | | | =IF(H26>B26,H26–B26,0) |
| 89 | increase in loans | | | =IF(J35–B35>0,J35–B35,0) |
| 90 | reduction in cash balances | | | IF((H20+H21)–(B20+B21)<0,–(H20+H21)–(B20+B21),0) |
| 91 | | | | =SUM(F86:F90) |
| 92 | **TOTAL CASH AVAILABLE** | | | **=F91+F84** |
| 93 | | | | |
| 94 | | | | |
| 95 | COMMITMENTS: | | | |
| 96 | private drawings | | | =–G41 |
| 97 | tax payments | | | =–G42 |
| 98 | reduction in loans | | | =IF(J35–B35<0,–(J35–B35),0) |
| 99 | | | | =SUM(F96:F98) |
| 100 | | | | |
| 101 | purchase of land, property, investments | | | 0 |
| 102 | net purchase of orchard | | | 3000 |
| 103 | purchase of machinery | 15000 | | |
| 104 | LESS sale of machinery | –5000 | | =D104+D103 |
| 105 | repayment of overdraft | | | =IF(H26–B26<0,–(H26–B26),0) |
| 106 | increase in cash balances | | | =IF((B20+B21)–(H20+H21)<0,–(B20+B21)–(H20+H21),0) |
| 107 | | | | =SUM(F101:F106) |
| 108 | **TOTAL CASH USED** | | | **=F99+F107** |
| 109 | | | | |



**Fig. A.5** Spreadsheet formulas for flow of funds analysis (Figure 14.10).

| | A | B | C | D | E | F |
|---|---|---|---|---|---|---|
| 1 | | | | | | |
| 2 | *HOURS PER HECTARE/HEAD* | | | | | |
| 3 | *(from Nix, 1990)* | | | | | |
| 4 | | | | | | |
| 5 | *ENTERPRISE* | *ha/hd* | *OCT* | *NOV* | *DEC* | |
| 6 | | | | | | |
| 7 | *winter wheat* | *51.3* | *3.6* | *0.6* | | |
| 8 | *witer barley* | *32.4* | *5.1* | *0.6* | | |
| 9 | *spring barley* | *27.2* | *0.7* | *1.1* | *0.7* | |
| 10 | *grass - leys* | *46.9* | | | | |
| 11 | *- perm. pasture* | *38.2* | | | | |
| 12 | *silage* | *40* | | | | |
| 13 | *dairy cows* | *115* | *2.5* | *2.5* | *2.5* | |
| 14 | *followers (repl. units)* | *15* | *3.7* | *3.7* | *3.7* | |
| 15 | *ewes* | *190* | *0.2* | *0.2* | *0.2* | |
| 16 | | | | | | |
| 17 | | | | | | |
| 18 | *TOTAL HOURS* | | | | | |
| 19 | | | | | | |
| 20 | ENTERPRISE | ha/hd | OCT | NOV | DEC | |
| 21 | | | | | | |
| 22 | winter wheat | =B7 | =C7*B22 | =D7*B22 | =E7*B22 | |
| 23 | witer barley | =B8 | =C8*B23 | =D8*B23 | =E8*B23 | |
| 24 | spring barley | =B9 | =C9*B24 | =D9*B24 | =E9*B24 | |
| 25 | grass - leys | =B10 | =C10*B25 | =D10*B25 | =E10*B25 | |
| 26 | - perm. pasture | =B11 | =C11*B26 | =D11*B26 | =E11*B26 | |
| 27 | silage | =B12 | =C12*B27 | =D12*B27 | =E12*B27 | |
| 28 | dairy cows | =B13 | =C13*B28 | =D13*B28 | =E13*B28 | |
| 29 | followers (repl. units) | =B14 | =C14*B29 | =D14*B29 | =E14*B29 | |
| 30 | ewes | =B15 | =C15*B30 | =D15*B30 | =E15*B30 | |
| 31 | | | | | | |
| 32 | TOTAL HOURS NEEDED | | =SUM(C22:C30) | =SUM(D22:D30) | =SUM(E22:E30) | |
| 33 | | | | | | |
| 34 | | | | | | |
| 35 | hours available/man/month (from Nix, 1990) | | 182 | 131 | 132 | |
| 36 | | | | | | |
| 37 | TOTAL AVAILABLE (3 men) | | =C35*3 | =D35*3 | +E35*3 | |
| 38 | | | | | | |

**Fig. A.6** Spreadsheet formulas for labour profile (Figure 17.2).

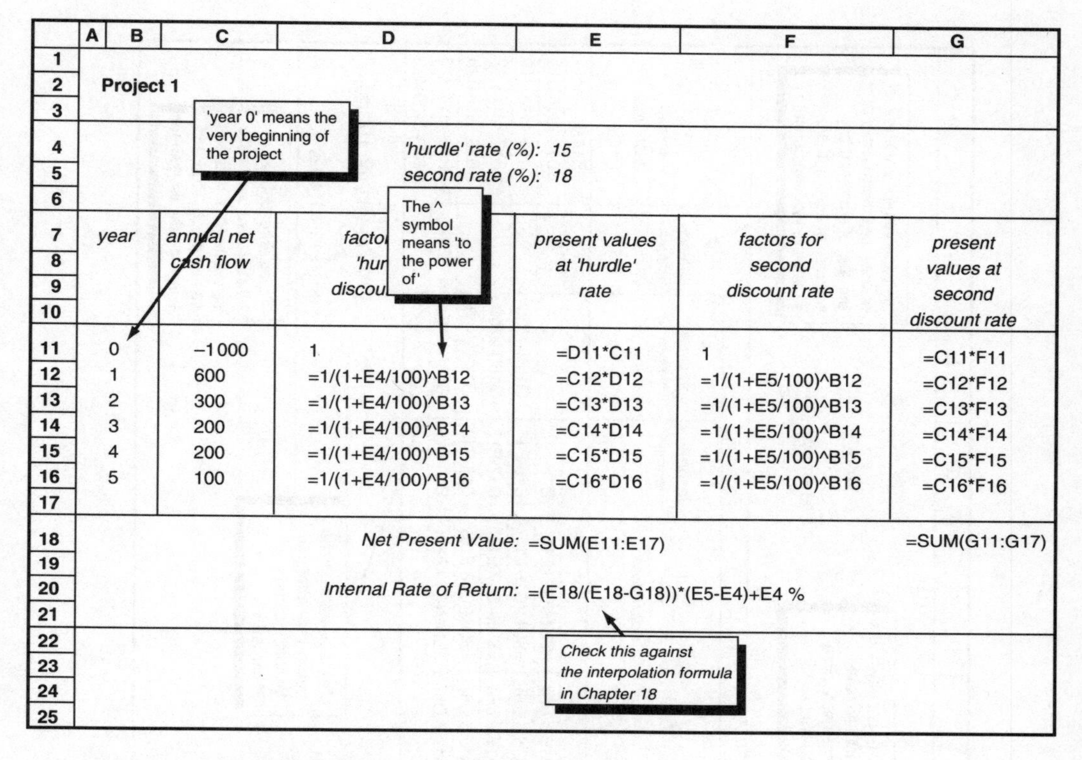


**Fig. A.7** Spreadsheet formulas for NPV and IRR calculation (Figure 18.10).

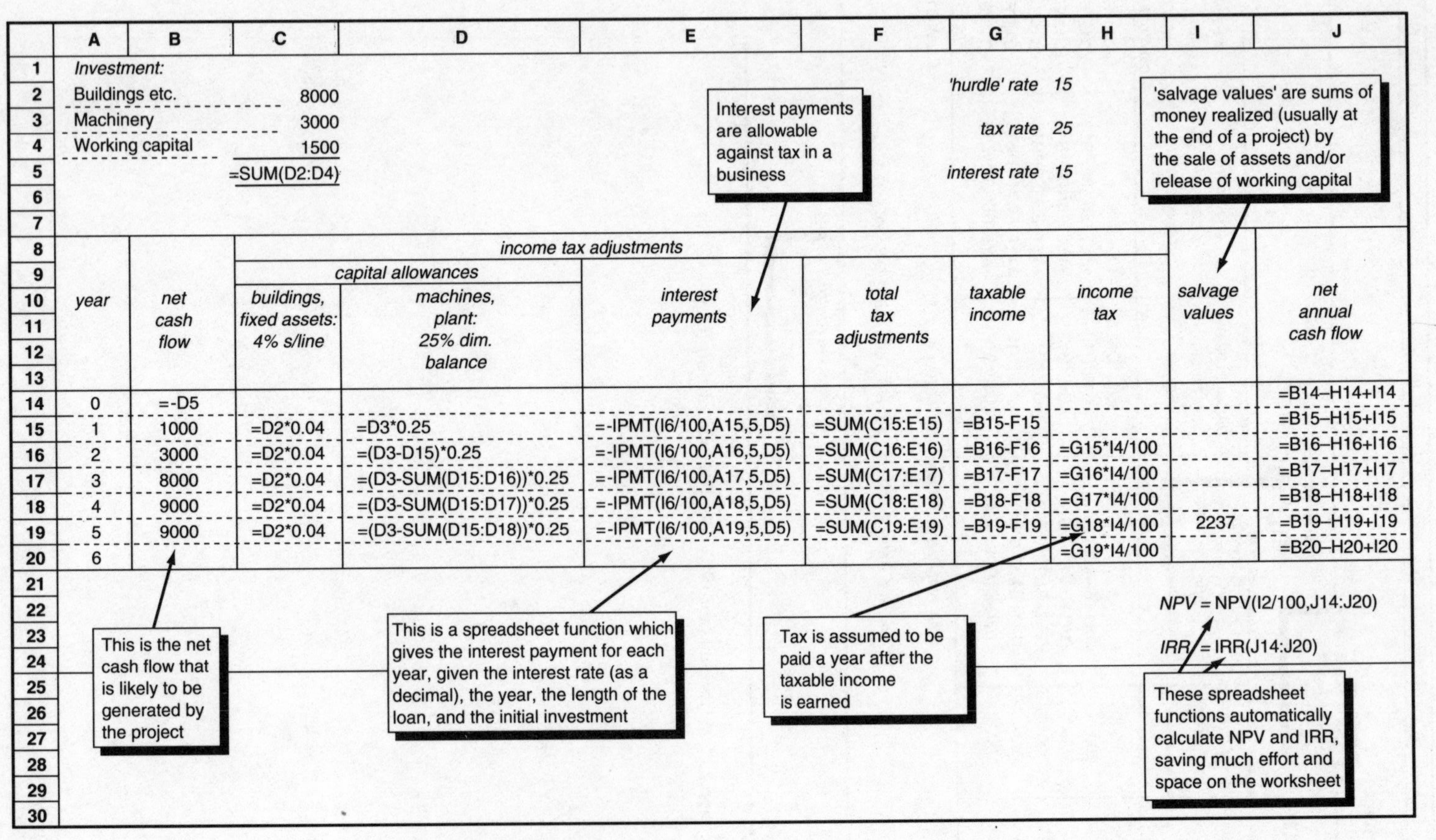


**Fig. A.8** Spreadsheet for more advanced DCF calculation (see Figure A.9).

*Initial investment:*

| | |
|---|---|
| Building and fixtures | 8000 |
| Machinery | 3000 |
| Extra working capital | 1500 |
| | 12500 |

'Hurdle ' discount rate (%): 15

Marginal tax rate (%): 25

Loan interest rate (%): 15

| | | *Income tax adjustments* | | | | | | | |
|---|---|---|---|---|---|---|---|---|---|
| | | *Capital allowances* | | | | | | | |
| *Year* | *Net cash flow* | *Buildings, fixed assets: 4% s/line* | *Machines, plant: 25% dim. balance* | *Interest payments* | *Total tax adjust-ments* | *Taxable income* | *Income tax* | *Salvage values* | *Net annual cash flow* |
| **0** | −12500 | | | | | | | | −12500 |
| **1** | 1000 | 320 | 750 | 1875 | 2945 | −1945 | | | 1000 |
| **2** | 3000 | 320 | 563 | 1597 | 2479 | 521 | −486 | | 3486 |
| **3** | 8000 | 320 | 422 | 1277 | 2019 | 5981 | 130 | | 7870 |
| **4** | 9000 | 320 | 316 | 909 | 1546 | 7454 | 1495 | | 7505 |
| **5** | 9000 | 320 | 237 | 486 | 1044 | 7956 | 1864 | 2237 | 9373 |
| **6** | | | | | | | 1989 | | −1989 |

Net present value at hurdle rate: 3714

Internal rate of return: 26%

**Fig. A.9** More advanced DCF calculation (based on Figure A.8).

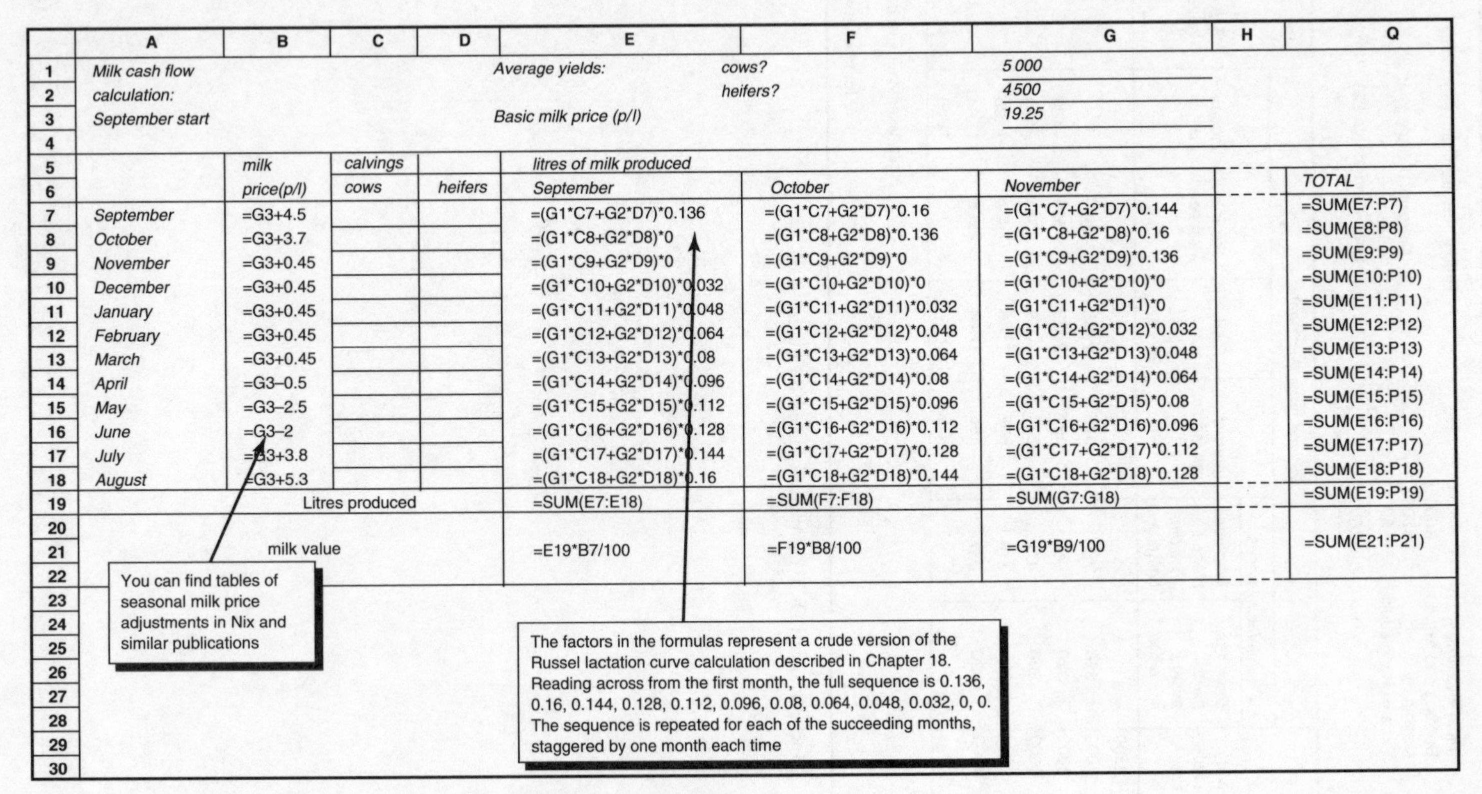

| | A | B | C | D | E | F | G | H | Q |
|---|---|---|---|---|---|---|---|---|---|
| 1 | *Milk cash flow* | | | | *Average yields:* | *cows?* | *5 000* | | |
| 2 | *calculation:* | | | | | *heifers?* | *4500* | | |
| 3 | *September start* | | | | *Basic milk price (p/l)* | | *19.25* | | |
| 4 | | | | | | | | | |
| 5 | | *milk* | *calvings* | | *litres of milk produced* | | | | |
| 6 | | *price(p/l)* | *cows* | *heifers* | *September* | *October* | *November* | | *TOTAL* |
| 7 | *September* | =G3+4.5 | | | =(G1*C7+G2*D7)*0.136 | =(G1*C7+G2*D7)*0.16 | =(G1*C7+G2*D7)*0.144 | | =SUM(E7:P7) |
| 8 | *October* | =G3+3.7 | | | =(G1*C8+G2*D8)*0 | =(G1*C8+G2*D8)*0.136 | =(G1*C8+G2*D8)*0.16 | | =SUM(E8:P8) |
| 9 | *November* | =G3+0.45 | | | =(G1*C9+G2*D9)*0 | =(G1*C9+G2*D9)*0 | =(G1*C9+G2*D9)*0.136 | | =SUM(E9:P9) |
| 10 | *December* | =G3+0.45 | | | =(G1*C10+G2*D10)*0.032 | =(G1*C10+G2*D10)*0 | =(G1*C10+G2*D10)*0 | | =SUM(E10:P10) |
| 11 | *January* | =G3+0.45 | | | =(G1*C11+G2*D11)*0.048 | =(G1*C11+G2*D11)*0.032 | =(G1*C11+G2*D11)*0 | | =SUM(E11:P11) |
| 12 | *February* | =G3+0.45 | | | =(G1*C12+G2*D12)*0.064 | =(G1*C12+G2*D12)*0.048 | =(G1*C12+G2*D12)*0.032 | | =SUM(E12:P12) |
| 13 | *March* | =G3+0.45 | | | =(G1*C13+G2*D13)*0.08 | =(G1*C13+G2*D13)*0.064 | =(G1*C13+G2*D13)*0.048 | | =SUM(E13:P13) |
| 14 | *April* | =G3–0.5 | | | =(G1*C14+G2*D14)*0.096 | =(G1*C14+G2*D14)*0.08 | =(G1*C14+G2*D14)*0.064 | | =SUM(E14:P14) |
| 15 | *May* | =G3–2.5 | | | =(G1*C15+G2*D15)*0.112 | =(G1*C15+G2*D15)*0.096 | =(G1*C15+G2*D15)*0.08 | | =SUM(E15:P15) |
| 16 | *June* | =G3–2 | | | =(G1*C16+G2*D16)*0.128 | =(G1*C16+G2*D16)*0.112 | =(G1*C16+G2*D16)*0.096 | | =SUM(E16:P16) |
| 17 | *July* | =G3+3.8 | | | =(G1*C17+G2*D17)*0.144 | =(G1*C17+G2*D17)*0.128 | =(G1*C17+G2*D17)*0.112 | | =SUM(E17:P17) |
| 18 | *August* | =G3+5.3 | | | =(G1*C18+G2*D18)*0.16 | =(G1*C18+G2*D18)*0.144 | =(G1*C18+G2*D18)*0.128 | | =SUM(E18:P18) |
| 19 | | | Litres produced | | =SUM(E7:E18) | =SUM(F7:F18) | =SUM(G7:G18) | | =SUM(E19:P19) |
| 20 | | | | | | | | | |
| 21 | | milk value | | | =E19*B7/100 | =F19*B8/100 | =G19*B9/100 | | =SUM(E21:P21) |
| 22 | | | | | | | | | |
| 23 | | | | | | | | | |
| 24 | | | | | | | | | |
| 25 | | | | | | | | | |
| 26 | | | | | | | | | |
| 27 | | | | | | | | | |
| 28 | | | | | | | | | |
| 29 | | | | | | | | | |
| 30 | | | | | | | | | |



**Fig. A.10** Spreadsheet formulas for milk production calculator (see Figure A.11).

Milk cash flow calculation:
September start

Average yields: cows? 5000
heifers? 4500
Basic milk price (p/l): 19

| | *Milk* | *Calvings* | | *Litres of milk produced* | | | | | | | | | | | | |
|---|---|---|---|---|---|---|---|---|---|---|---|---|---|---|---|---|
| | *price (p/l)* | *Cows* | *Heifers* | *Sep.* | *Oct.* | *Nov.* | *Dec.* | *Jan.* | *Feb.* | *Mar.* | *Apr.* | *May* | *Jun.* | *July* | *Aug.* | ***Total*** |
| September | 23.75 | 26 | 8 | 22576 | 26560 | 23904 | 21248 | 18592 | 15936 | 13280 | 10624 | 7968 | 5312 | 0 | 0 | 166000 |
| October | 22.95 | 34 | 4 | 0 | 25568 | 30080 | 27072 | 24064 | 21056 | 18048 | 15040 | 12032 | 9024 | 6016 | 0 | 188000 |
| November | 19.7 | 22 | | 0 | 0 | 14960 | 17600 | 15840 | 14080 | 12320 | 10560 | 8800 | 7040 | 5280 | 3520 | 110000 |
| December | 19.7 | 15 | | 2400 | 0 | 0 | 10200 | 12000 | 10800 | 9600 | 8400 | 7200 | 6000 | 4800 | 3600 | 75000 |
| January | 19.7 | 8 | 2 | 2352 | 1568 | 0 | 0 | 6664 | 7840 | 7056 | 6272 | 5488 | 4704 | 3920 | 3136 | 49000 |
| February | 19.7 | 4 | | 1280 | 960 | 640 | 0 | 0 | 2720 | 3200 | 2880 | 2560 | 2240 | 1920 | 1600 | 20000 |
| March | 19.7 | 7 | | 2800 | 2240 | 1680 | 1120 | 0 | 0 | 4760 | 5600 | 5040 | 4480 | 3920 | 3360 | 35000 |
| April | 18.75 | 2 | | 960 | 800 | 640 | 480 | 320 | 0 | 0 | 1360 | 1600 | 1440 | 1280 | 1120 | 10000 |
| May | 16.75 | 0 | | 0 | 0 | 0 | 0 | 0 | 0 | 0 | 0 | 0 | 0 | 0 | 0 | 0 |
| June | 17.25 | 1 | | 640 | 560 | 480 | 400 | 320 | 240 | 160 | 0 | 0 | 680 | 800 | 720 | 5000 |
| July | 23.05 | 5 | 16 | 13968 | 12416 | 10864 | 9312 | 7760 | 6208 | 4656 | 3104 | 0 | 0 | 13192 | 15520 | 97000 |
| August | 24.55 | 8 | 12 | 15040 | 13536 | 12032 | 10528 | 9024 | 7520 | 6016 | 4512 | 3008 | 0 | 0 | 12784 | 94000 |
| Total milk produced (litres) | | | | 62016 | 84208 | 95280 | 97960 | 94584 | 86400 | 79096 | 68352 | 53696 | 40920 | 41128 | 45360 | 849000 |
| Value of milk produced | | | | 14729 | 19326 | 18770 | 19298 | 18633 | 17021 | 15582 | 13465 | 10578 | 8061 | 8102 | 8936 | 172501 |

**Fig. A.11** Milk production calculator (based on Figure A.10).

# Index


accounting equation, 17
accruals, 12–13, 144
accrued revenue, 13
acid-test ratio, 39
activity ratios, 32–3
aims, 203
amortization table, 244
annual net cash flow, 99
annuity loan, 243
assets
  appreciation of, 26
  classification, 18
  current, 18
  examples, 18
  fixed, 18
  historic cost, 28
  life, 21
  liquid, 18
  wasting, 20

background records, 168
balance sheet, 18, 20, 156
  analysis and interpretation, 33–6
  budgeted, 106–10
  closing, 9, 18
  farmers', 266
  gearing, 39–40
  horizontal, 18–19
  interfirm comparison, 37
  opening, 9, 18
  ratio analysis, 37
  vertical (narrative) form, 20
bank statement, 130
barn record, 164
benefits in kind, 13, 69
book value, 22
break-even budget, 75
break-even chart, 234
break-even point, 48
break-even pricing, 50
breeding livestock, 11, 55
  replacement, 218
budget, *see* profit budget, cash flow budget, budgeted balance sheet, partial budget
budgetary control, 176–81, 190
budgeting, 63–120

capital, 16–26
  account, 19–20, 57, 156
  average (loan), 117
  average (return on capital), 247
  cost, 242
  employed, 20
  initial, 117, 247
  monitoring, 197–8
  opportunity cost, 245
  payback, 249
  planning, 239–56
  recording, 156–9
  return on, *see* return on capital
  sources, 239
cash analysis
  book (CAB), 125, 177
  choice of headings, 125
  contra-transactions, 126–7
  discounts, 129
  entry of transactions, 127
  income tax records, 142
  petty cash, 131
  reconciliation with bank statement, 130
  Value Added Tax, 136
cash balance, 9
  opening and closing, 17
cash flow, 9
cash flow budget, 89–105
  assumption sheet, 97
  actual columns, 17
  headings, 96
  partial, 118
  recording, 123–43
  reconciliation with profit budget, 99
  rolling, 90
  timing, 96
  Value Added Tax, 90
cash flow profile, 104
cash flow recording
  cash analysis, 125
  double entry method, 124

monitoring, 175–82
profile, 100
reconciliation with profit, 147
companies, 17, 58
comparative analysis, *see* profit
complete budget, 77, *see also* profit budget
computer, 171, 271–5
accounts, 272
book-keeping, 125
bulletin board, 274
bureau, 272
business applications, 273
database, 273
email, 274
farming, 271
internet, 274
payroll, 272
spreadsheet, 273
video-conferencing, 274
world-wide web, 274
word-processor, 273
contractors, 235
contribution, 46
control, 6, *see also* budgetary control
short-term, 166
stock, 167
co-operatives, 60
cost
behaviour, 45
centre, 42
direct, 43
of sales, 15
relevant, 111–20
cost-profit-volume analysis, 47
costs, *see* fixed costs, variable costs, opportunity costs, feed costs, labour and machinery
credit, 98, 239–45, *see also* loan, finance lease, hire purchase, leasing, overdraft, security, trade credit
credit cards, 240
creditors, 12, 98
adjustments for, 144
turnover, 33
critical path analysis, 232
crop year, 16, 148
current account, 57
current assets, 18, 39
current liabilities, 18, 39
current ratio, 39

debtors, 12, 69, 98
adjustments for, 144
turnover, 33
decision making, 6
deferred revenue, 13
delaying payments, 101
depreciation, 13, 20–26
diminishing balance method, 23–5
fractional, 26
in partial budget, 115
pool calculations, 24–6, 76
profit or loss on sale, 26
reducing balance method, 23–5
straight line, 22–3
digestible energy (DE), 212
direct costs, 43
discounted cash flow techniques (DCF), 250–55
pros and cons, 255
discounted yield, 253
discounts, 129
diversification, 257
dividends, 59

efficiency ratios, 32–3
enterprises, 42, 53, 66
enterprise output, 69, 146
enterprise record, 164
equity, 31, 37
expense, 10

Farm Business Survey (FBS), 186
farm management accounting, 53–6
feed cost, 213
feeding policy selection, 212
field record, 164
filing, 169
finance lease, 29
financial accounts
adjusting for realism, 28
interpretation, 28–42
financial period, 16
financial year, 16, 148
fixed assets, *see* assets
fixed costs, 45, 66, 146
agricultural, 54
analysis, 187
flow of funds analysis, 34, 266
forage costs allocation, 213
forecasting, 65
full budget, 67
full cost absorption, *see* full costing
full costing, 44, 147
futures, 257

gang work day chart (GWD), 227
gearing ratios, 37
grazing livestock enterprise, 75
grazing livestock unit (GLU), 213

gross margin, 32, 43, 146
  accounting, 43
  agricultural, 53
  analysis, 188
  assumption worksheet, 74
  livestock enterprise, 53
  worksheets, 70, 72
gross profit, 15, 32

herd basis, 158
hire purchase, 241
historic cost, 28
holding gain, 106, 266
Home Farm details, 1

income and expenditure account, 60
Income Tax records, 142
inflation, 264–9
  accounts, 264
  balance sheet, 266
  current cost accounting, 265
  current purchasing power accounting, 265
  management accounts, 265–7
  investment appraisal, 268
indirect cost, 66
interest cover, 39
interest: compound, 250
interfarm comparisons, 30, *see also* profit, capital
interfirm comparisons, *see* profit, capital
internal rate of return (IRR), 253
interpolation, 253
investment appraisal, 246
invoice, 12, 161

labour and machinery
  estimating costs, 232–8
  estimating level of use, 224
  gang work day chart, 227
  investigating alternatives, 234–7
  job list, 231
  labour profile, 226
  network analysis, 232
  planning, 224–38
  standard man day and tractor hour calculation, 225
  work schedule, 232
  work study, 232
labour profile, 226
lactation curve, 217
leasing, 29, 237, 241
lease purchase, 241
liabilities, 17
  classification, 18
  current, 18
  deferred, 18
  examples, 18
  long-term, 18
limited liability companies, 58
linear programming (LP), 209, 212
liquid assets, 39
liquidity ratio, 39
livestock
  estimating potential production, 216
  feeding policy, 211
  long production cycles, 220
  stocking rate, 214
livestock movement record, 164–5
livestock movement flow chart, 221
livestock unit grazing days (LUGD), 215
loan, 241
  annuity, 243
  endowment, 244
  reducing balance, 242
long-term debt to capital employed, 39
long-term debt to equity, 37
long-term liabilities, 18, 37
loss, *see* profit, profit and loss account, profit budget, profit or loss on sale
lumpy costs, 47

machinery, *see* labour and machinery
machinery register, 166
management, 7
management accounts, 29, 42–52
  farms', 53–6
Management and Investment Income (MII), 186–7
management by objectives, 210
margin of safety, 48
margin over concentrates, 190
marginal costing, 45
mark-up, 50
metabolizable energy (ME), 212
milk quota, 28
mission, 203–4
mortgage, *see* loans

National Insurance, 166
net assets, 20
net capital, 17, 37
net cash flow, 89
net farm income (NFI), 186
net margin, 32, 43
net present value (NPV), 251
net profit, 10, 15, 32, *see also* profit
net worth, 17
network analysis, 232
normalized budget, 75

objectives, 6–8, 203
  setting, 204–7
office, 169, 171
operating profit, 15, 32
opportunity cost
  capital, 245
  partial budget, 112
optimal plan selecting, 208
order, 161
outline budgets, 75
output, 54, 66
  unallocated, *see* gross margin
overdraft, bank, 9
  cost, 242
  interest, 92, 98, 242
  renegotiating, 101
overhead cost, 66
owner equity, 17, 31

paperwork organization, 168
partial budget, 113–20, 219, 234
  cash flow, 118
  principle, 113–20
  profit, 114
  satellite, 113
  use, 113–20
partnership, 17, 56–60
PAYE records, 166
payback period, 249
percentage equity, 37
percentage owned, 37
personal transactions, 14
petty cash, 131
  imprest method, 136
physical records, 162–7
planning, 6
'pool' depreciation, 76
prepayments, 12–13, 144
present value, 250
pricing, 50–51
  break-even, 50
  cost-based, 50
  decisions, 46, 51
  direct cost, 51
  loss-leader, 52
  market-orientated, 51
  mark-up, 51
  penetration, 52
  profit-orientated, 50
  skimming, 52
prime costs, 43
production year, 16, 148
profit
  budgetary comparisons, 190–91
  centre, 42, 66
  comparison with previous years, 185
  gross, 32
  inter-farm comparisons, 186–91
  monitoring, 183–99
  net, 10
  operating, 15
  recording, 144–56
  reconciliation with cash flow, 99, 147
  statements, 14–16
  variance analysis, 191–4, 267
profit and loss account, 15, 144, *see also* profit
  common-size comparison, 30
  conventional form, 144
  crop year adjustment, 149
  full costing form, 147
  gross margin form, 146
  historical analysis, 29–30
  interfirm comparison, 30
  ratio analysis, 31
  reserve, 59
profit and loss appropriation account, 55
profit and loss budget, *see* profit budget
profit budget, 65
  compiling, 66
  whole-business, 65
profit or loss on sale, 26
profit statements, 14–16
profitability index (PI), 253
profitability ratios, 31
project planning, 210

ratio analysis, 198
real rate of return, 269
reconciliation
  livestock, 164–5
  profit and capital, 19–20
  profit and cash flow, 99, 147
records, *see* cash flow recording, profit and loss account, balance sheet, physical records, transaction records, background records
reducing balance loan, 242
relevant costs and benefits, 51, 111
relevant range, 47
reserves, 58
return on capital
  accounting rate, 247
  average capital, 247
  discounted cash flow techniques, 250–55
  discounted yield, 253
  inflation, 268
  initial capital, 247
  internal rate of return, 253

interpolation, 253
marginal, 247
net present value, 251
net assets, 246
net capital, 246
present value, 251
profitability index, 253
real rate of return, 269
tenant's, 187, 246
revenue, 10, 66
non-cash, 99
risk, 257–63
breakeven budget, 258
causes, 257
flexible budgeting, 259
gearing, 38
investment appraisal, 262
planning, 258–62
reducing, 257
sensitivity analysis, 260
tree diagram, 261
variance table, 261
rolling plan, 71–5

satellite budget, 115
security, 241
sensitivity analysis, 75, 260
shares, 58
shareholders, 58
short list and budget method, 209
short-term stability, 39
sole proprietor, 17
solvency ratio, 39
source and allocation of funds, *see* flow of funds analysis
spreadsheets: examples, 276–87
standard man day (SMD), 225
standard tractor hour, 225
stepped costs, 47
stocking rate calculation, 214
stocks, 11
AVCO, 152
change in valuation, 11–12
classification, 11
control, 167
economic order quantity, 168
farm, 152–3
FIFO, 152
just-in-time systems, 168
LIFO, 152
methods of valuation, 150
net realizable value, 151
reorder level, 168
valuation, 69
stock turnover period, 32
SWOT analysis, 207

tax
assessment, 28
exclusion from profit, 14
tenant's capital, return on, 187
times covered, 39
total assets, 37
trade credit, 240
trade creditors, 12, *see also* creditors
trade debtors, 12, *see also* debtors
trading account, 14, *see also* profit and loss account
trading and profit and loss account, 14, *see also* profit and loss account
trading livestock
production cycle, 55
valuation, 153
transaction records, 160
transfers between enterprises, 69
tree diagram, 261
turnover, 30

uncertainty, *see* risk

Value Added Tax (VAT), 90, 136
exempt transactions, 137
modified cash analysis, 139–42
standard rate, 137
zero rate, 137
variable costs, 45, 66, 69, 146
agricultural, 53
variance analysis, 191–4
voluntary/non-profit organizations, 60

wages book, 166
wasting assets, 13
work schedule, 232
work study, 232
written-down value (WDV), 22, *see also* depreciation

year planner, 231